THE UNFIT

A History of a Bad Idea

THE UNFIT
A History of a Bad Idea

Elof Axel Carlson
State University of New York, Stony Brook

COLD SPRING HARBOR LABORATORY PRESS
Cold Spring Harbor, New York

THE UNFIT
A History of a Bad Idea

Developmental Editor	Judy Cuddihy
Project Coordinator	Joan Ebert
Production Editor	Patricia Barker
Interior Designer	Denise Weiss
Cover Designer	Ed Atkeson, Berg Design
Desktop Editor	Susan Schaefer

Front Cover Art: Judy Cuddihy and Ed Atkeson
Section Title Pages: Reprinted from "Exhibits Book: Second International Exhibition of Eugenics" (Courtesy of CSHL Archives).

Library of Congress Cataloging-in-Publication Data

Carlson, Elof Axel.
    The unfit : a history of a bad idea / Elof Axel Carlson.
       p. cm.
    Includes bibliographical references and index.
    ISBN 0-87969-587-0 (alk. paper)
    1. Eugenics--History. I. Title.
HQ751 .C27 2001
363.9'2'09--dc21

                    2001032347

10 9 8 7 6 5 4 3 2

All Cold Spring Harbor Laboratory Press publications may be ordered directly from Cold Spring Harbor Laboratory Press, 500 Sunnyside Boulevard, Woodbury, New York 11797-2924. Phone: 1-800-843-4388 in Continental U.S. and Canada. All other locations: (516) 422-4100. FAX: (516) 422-4097. E-mail: cshpress@cshl.org. For a complete catalog of all Cold Spring Harbor Laboratory Press publications, visit our World Wide Web Site http://www.cshl.org/

Dedicated to H.J. Muller
(1890–1967)

---

*Whose work and ideals led him to support, denounce,
and reflect deeply on efforts to protect or improve
humanity's genes*

Contents

Preface

I STARTED *THE UNFIT: A HISTORY OF A BAD IDEA* about ten years ago. I did not intend to write a book on degeneracy theory and its relation to eugenics at that time, nor did I know how far back in history the idea of unfit people had existed. Fortunately, the notion of writing a book emerged in the process of preparing a formal lecture while I was on sabbatical leave at Indiana University. The leisure time to pursue that notion eventually led to this book. In writing this book, I was fortunate to have the time provided as a Fellow of the Institute for Advanced Study at Indiana University to use their research library and the Kinsey Library, whose librarians were very helpful obtaining scarce items on interlibrary loans. Portions of this book were written at Indiana University, at Tougaloo College, on the *SS Universe*, in my study in Setauket, and in the Honors College lounge at Stony Brook University. I especially benefited from discussion of the work in progress with faculty in the NSF-Chautauqua Short Course on the Unfit that I have taught for three years during spring break. Of particular help for their comments on some or all of the manuscript were A. Peter Gary, Rabbi Howard Diamond, Michael Kramer, Owen Debowy, Paul Bingham, Frederick Brown, and Ruth Cowan. I am also grateful to the many comments in class from my students in the Honors College at Stony Brook who had read an earlier draft of this work. Others who were helpful include Jack Scovil for his enthusiasm for this book, Leon Sokoloff for references on anti-Semitism, Larry Slobodkin for his legalistic interpretations of Jewish traditions, and David Smith for his many insights into ethical issues related to eugenics. In the Honors College, both Arthur Bozza and Adam Weinberger were particularly helpful. I appreciate the skill in tracking references by the able staff of the Stony Brook University library.

I am grateful to James D. Watson for reading the manuscript and recommending the publication of this book by Cold Spring Harbor Laboratory Press. This book benefited from the many devoted hours of hard work by Judy Cuddihy, Developmental Editor, and Joan Ebert, Project Coordinator. I also acknowledge the advice of John Inglis, Director of the Press; Patricia Barker, Production Editor; Susan Schaefer, Desktop Editor; and Clare Bunce, who guided me through the Cold Spring Harbor Laboratory Library Archives. It was a pleasure to see the book emerge through their talents and enthusiasm. I was helped immensely by discussions with the advisory members of the Eugenics Archive Project: David Micklos, Director of the DNA Learning Center; Jan Witkowski, Director of the Banbury Center; and those members of the Advisory Board who discussed aspects of the book with me, including Paul Lombardo, Steven Selden, Martin Levitt, Garland Allen, Philip Reilly, and Henry Friedlander.

I much appreciated the Friday night dinners with my wife Nedra, and her many helpful comments as she read portions of the manuscript and discussed the issues of reproduction biology and patient reactions from her prespective as a cytogenetic technician and as an in vitro fertilization embryologist. Michael Kramer was also of immense help in converting my penciled flow diagrams and pedigrees into computer graphics, and he was able to retrieve and produce a comprehensive and uniform computer manuscript from a sometimes-corrupted set of documents on old disks in Word Star, Word Perfect, and Word.

E.A.C.
April 2001

Chronology of the Biological Concept of Unfit People

THIS CHRONOLOGY PROVIDES A TIME LINE for the history of the idea of unfit people from the publication of *Onania* in 1710 to the revelation of the Nazi death camps in 1945. Not all people mentioned were malicious. Many had no notion their ideas would be used to justify vicious programs. Still others were self-deceived and did not think through the implications of their biases. As one can see from following the time line, there is no chain of causality. For a variety of different reasons, people were classified as unfit, and different, often contradictory, responses were made to claims that these degenerate or unfortunate groups existed and that something should be done for or to them. This is a selection of some of the major (and minor) players in this story to give a sense of what thinking was like among educated classes in two and a half centuries of biological theories of human inferiority.

1710: Publication of the anonymous *Onania, or the Heinous Sin of Self-Pollution and All Its Frightful Consequences in Both Sexes* leads to the idea of onanism as a cause of degeneracy in the self-abuser and progeny.

1758: Publication of Samuel Tissot's *Onania, or a Treatise upon the Disorders Produced by Masturbation* shifts masturbation to a medical problem and makes masturbatory degeneracy a theme of medical school teaching until the end of the 19th century.

1798: Thomas Robert Malthus publishes *An Essay on the Principles of Population*, blaming the poor for their misfortunes.

1837: The Elizabethan Poor Laws are largely abandoned. Charles Dickens writes *Oliver Twist* to describe the consequences.

1850: Herbert Spencer publishes *Social Statics*, the founding document that led to what was later called Social Darwinism.

1853: Joseph Arthur, Comte de Gobineau, publishes *The Inequality of Human Races*, launching scientific racism, which considers races biologically inferior or superior, with Teutons (Nordic or Aryan) as the prized race.

1857: Benedict Morel's *Dégénérance* is published. It argues that degeneracy caused by unfavorable environments leading to progressively worsening heredity is self-extinguishing within five generations.

1859: Charles Darwin publishes *The Origin of Species*.

1866: Gregor Mendel's paper on patterns of inheritance in pea plants is published.

1867: Richard Dugdale extends Elisha Harris's study of a criminal family and publishes *The Jukes*. Michigan marriage act is passed, making it a crime for idiots, the insane, uncured syphilitics, and people with uncured cases of gonorrhea to marry or live together.

1869: Francis Galton founds the eugenics movement (not yet by that name) with publication of *Hereditary Genius* and stresses what would be later called positive eugenics.

1872–1892: Emile Zola publishes the Rougon-Macquart series of 20 novels exploring hereditary pathology in two families.

1879: Wilhelm Marr publishes *The Victory of Jewry over Germany* and establishes modern anti-Semitism.

1880s: Oscar McCulloch studies the *Tribe of Ishmael*, publicizing it as a socially degenerate collection of families. His views are popularized by essays of David Starr Jordan, then active in Indiana.

1880–1890: August Weismann proposes the theory of the germ plasm; he disproves Lamarck's theory of the inheritance of acquired characteristics. Defective germ plasm becomes a medical and social problem.

1883: Galton gives eugenics its name. Frank Hamilton ligates the vas deferens as a treatment for masturbation. Joseph Howe publishes *Excessive Venery, Masturbation, and Continence* and offers many surgical and medical approaches to treat masturbation.

1892: Henry D. Chapin argues that vagabonds, tramps, and criminals should be isolated from society. Edward S. Morse condemns congenital criminals and paupers, citing Weismann's germ plasm theory for believing that the unfit have an impaired heredity.

1893: F.E. Daniel recommends sterilization of the unfit as being humane.

1894: Reginald Harrison performs vasectomy for reducing enlarged prostate gland. Martha Clark argues mandatory segregation for life of paupers and repeat criminals.

1895: Charles Dana criticizes Max Nordau's *Degeneration*. He claims that degeneracy is self-eliminating and is not an enduring problem.

1896: Czarist forgery called *Protocols of the Learned Elders of Zion* is released and becomes an international bestseller. It is adopted by Henry Ford's *Dearborn Independent* in 1920 and published by Gerald L.K. Smith as *The International Jew: The World's Foremost Problem*.

1897: Michigan sterilization law fails after passing one house.

1898: F. Hoyt Pilcher castrates 58 retarded boys. Martin Barr advocates sterilization of the unfit; he castrates 2 males and 2 females. Everett Flood castrates 24 epileptics and persistent masturbators.

1899: A.J. Ochsner urges vasectomies for prisoners and other degenerates. Harry Clay Sharp performs first vasectomy to treat masturbation in prisoner in Jeffersonville, Indiana.

1900: Mendel's paper is rediscovered.

1901: David Starr Jordan publishes *The Blood of a Nation* and extols eugenics.

1903: The American Breeder's Association is founded. It creates a committee on eugenics in 1909. In Great Britain, Robert Rentoul proposes sterilization of unfit by vasectomy.

1904: Alfred Ploetz names Race Hygiene (Rassenhygeine) as an extension of Virchow's public hygiene movement. He founds the German Society for Racial Hygiene.

1906: Governor Samuel Pennypacker of Pennsylvania vetoes compulsory sterilization law for feebleminded as a criminal and dangerous act.

1907: Indiana passes first state compulsory sterilization law. Harry Sharp sterilizes by vasectomy 200 to 500 young men.

1913: Mrs. E.H. Harriman provides funds to establish a building and salaries for the Eugenics Record Office in Cold Spring Harbor, New York.

1914–1940: Harry Laughlin serves as Superintendent of the Eugenics Record Office at Cold Spring Harbor. With Charles Davenport, Director of Cold Spring Harbor Laboratory, he becomes the leading promoter of the American eugenic movement.

1916: Madison Grant publishes *The Passing of the Great Race.* He advocates restrictive immigration laws to prevent dilution of the Anglo-Saxon heritage of the United States.

1920: Euthanasia is promoted in Germany by publication of Karl Binding and Alfred Hoche's *The Release of the Destruction of Life Devoid of Value.*

1921: Erwin Baur, Eugen Fischer, and Fritz Lenz publish *Human Genetics* (English edition in 1931), which is read and admired by Adolf Hitler while he is under house arrest after failed putsch. German genetics shifts as new Nazi party endorses race hygiene as its goal.

1924: Johnson Act restricting immigration to ethnic composition of United States in 1890 census becomes law. Harry Laughlin serves as expert witness for the Johnson Committee and provides evidence for the inferiority of southern and eastern Europeans.

1927: Buck v. Bell upholds Virginia's sterilization law by 8–1 vote. Harry Laughlin provided the model eugenic law for the state of Virginia. Justice Oliver Wendell Holmes, Jr., argues "three generations of imbeciles are enough."

1933: Adolf Hitler becomes Chancellor of Germany. His Nazi party advocates a program of harassment of Jews and eventually the establishment of a "Jew-free" Germany. He advocates a widespread state eugenic program favoring the Aryan race and purging it of its alleged inferior strains.

1933–1935: Hitler enacts by decree Enabling Laws that bar marriage of Jews to non-Jews, classify Jews as a biological race, and promote sterilization of the unfit through decisions of eugenic courts.

1939: A secret order, initiated by Hitler with the onset of World War II on September 1, permits Nazi doctors to kill Germany's mentally retarded, deformed, and psychotic through designated centers. Program is stopped after rising protests create a war morale problem in Germany.

1942: Reinhard Heydrich chairs and Adolf Eichmann serves as secretary for a conference on the Final Solution ordered by Goering, Himmler, and Hitler. Death camps are advocated with massive removal of Jews from occupied territories. All activities are disguised with use of a coded language and information passed along a strict chain of command.

1942–1945: Six million Jews are killed primarily by gassing followed by cremation in death camps. Several million Poles, Russians, and smaller numbers of Gypsies, political opponents, and homosexuals are also killed. The event is given its historical name, the Holocaust.

THE UNFIT
A History of a Bad Idea

Introduction

*T*HE *UNFIT: A HISTORY OF A BAD IDEA* explores the sources of a movement that was used to justify, at least among those who had the
authority to implement it, the final solution or Holocaust, which claimed
several millions of innocent lives in World War II. The movement is usually called eugenics, but more accurately it represented a branch of eugenics known as negative eugenics: the study of the deterioration of human
heredity and the means that can restore it. This book is not a history of
eugenics, nor is it a history of the Holocaust, although aspects of both are
components of this broader picture of the idea that there are people who
were made to represent a degenerate or unfit class of humans. Rather, the
title reflects the nearly three centuries of belief that some people are socially unfit by virtue of a defective biology. It also echoes an earlier theory of
degeneracy, dating to biblical antiquity, when some people were deemed
unfit because of some transgression against God or God's law. For these
reasons, I have called these people "the unfit," the term used in the late
19th and early 20th centuries to describe those whose needs may require
intense social and personal attention and expensive investments of society's resources, although some of those who were called the unfit may not
have had any problems at all other than being victims of bias.

My interest in the history of unfit people stems from my association
with H.J. Muller, with whom I received my Ph.D. Muller was the founder
of radiation genetics and received a Nobel Prize in 1946 for having shown
that X-rays induce mutations. He was a leader in advocating protection of
the public from unnecessary exposure to ionizing radiation. For this

stand, humanity should be thankful, because, with rare exceptions, the gene mutations and chromosome breakage induced in reproductive tissue by X-rays are harmful to future descendants who receive them. Muller was also a life-long advocate of positive eugenics. He believed humanity possessed the techniques and had the moral judgment to take control of its own evolution and produce future generations that were wiser, brighter, more talented, and healthier in mind and body than is our own generation. For this stand, however, most of humanity condemned him, shunned him, or ignored him. Muller was a critic of negative eugenics and condemned it as bigoted, sexist, and spurious in his public speeches and writings as early as 1932, but this did not alter the hostility he received from many of his colleagues and the public in general for advocating a voluntary positive eugenics program.[1]

I have also kept a long-standing interest in eugenics and the idea of unfit people because I teach biology to non-science majors. I follow the controversies that surround biology and human affairs. Eugenics in various guises recurs in the debates concerning intelligence testing and an alleged genetic difference in social classes or races in their average intelligence. Equally controversial are those advocating an alleged genetic basis of intellect-based merit or failure in social life, economic class, and the professions. Still others debate the value of human genetics, seeing it as a back-door approach to eugenics through genetic screening, prenatal diagnosis, elective abortion, or gene therapy. The same fears apply to new reproductive technologies such as in vitro fertilization, in which sperm donors as well as egg donors are sought to get around the many ways in which natural reproduction can fail for a couple.

I wrote this book because I followed a lead. In the fall of 1986, I was on sabbatical leave as a Fellow of the Institute for Advanced Study at Indiana University. My only formal duty was a public lecture. I wanted to explore the effects on a scientist's career when that scientist speaks out on controversial social or political issues. Indiana University had several such individuals, and I chose geneticist H.J. Muller (the founder of radiation genetics), Alfred C. Kinsey (the founder of scientific human sex research), and David Starr Jordan (a founder of the American eugenics movement). The lead I pursued was the curious fact that the state of Indiana (in 1907) was the first place in the world to authorize by law the compulsory sterilization of unfit people. The connection I sought—the role of Jordan in formulating that legislation—was wrong. The person who played that key role was, instead, a prison doctor, Dr. Harry Clay Sharp, who in 1899 sterilized by vasectomy a young man for therapeutic reasons, as a treatment to prevent masturbation. It seemed less likely to me that a prison doctor would be crazy or obsessed than educated to believe that what he was doing was medically sound. On that assumption, I used the Kinsey Library for Sex Research to study the history of attitudes to masturbation and discovered that masturbation (or onanism as it was then called) served as the first theory of human degeneracy based on a physiological model.

The history of the unfit in this work is presented in a roughly chronological way. From the discussion of biblical views, I jump to the early 18th century to introduce the first biological theory of degeneracy—masturbation. This is followed by a flowering of degeneracy theory and its applications to a variety of social classes in the 19th century. These include tramps, paupers, and criminals as well as more vaguely defined "dangerous classes." Some key intellectual traditions are introduced, including the ideas of progress and its discontents in the debates of Malthus and Godwin over Condorcet's optimistic view of human reason and human potential, and the concepts of evolution and heredity and how they were applied to social problems.

The rise of sociology in the last half of the 19th century brought with it studies of degenerate families and the struggle between environmentalists and hereditarians in interpreting the causes of their failure. The role of professionals is stressed, especially in their efforts for redefining charity, reforming prisons, establishing asylums, and providing medical technology to remedy social conditions. The connections between masturbation, degeneracy theory, and compulsory sterilization are clearly revealed in the beliefs and practices of Harry Clay Sharp, who successfully gave the world its first eugenic compulsory sterilization law. The rise of the two eugenics movements, positive and negative, is covered, as is their sad history of failures and abuses in the United States and Europe. Racism, anti-Semitism, and the ultimate outrage of misapplied science and technology are explored with the Holocaust as the capstone of this miserable period in 20th-century history. I conclude the study with two chapters. The first explores the future of genetics based on the new technologies and applications of the human genome project. The second reflects on the death of the old eugenics, on the problems that won't go away, and on some of the deeper reasons, I believe, that we are so ambivalent about our own biology.[2]

In following the history of degeneracy theory and its ties to those people later called the unfit, I had to go back to original sources and not rely on secondary sources, which often reflect the values of the writer's generation. To the extent possible I have done so, and I quote liberally, both in the text and in the accompanying notes, from these sources because they illustrate how diverse are the origins of what in the 20th century was called the eugenics movement.

I consider myself fortunate by circumstance and choice to have a background suitable for this interdisciplinary study. I am a geneticist by training, but for more than 20 years I have made the history of genetics a central focus of my scholarship. I was the child of a Lutheran Swedish father (who had become an atheist) and a Jewish-American mother (who was rejected by her family and who abandoned most of her religious practices). My childhood was spent in slums and with a brother whose life was precarious from birth because of a congenital heart defect. My brother and I had to cope with the unpredictable behavior of our mother, who was a paranoid schizophrenic and had been institutionalized before she met our father. By many of the standards of the late 19th and early 20th centuries, my parents, my brother, my half-siblings (raised in an orphan asylum after my mother's first marriage had failed), and I would have been classified among the unfit. My mother, had she lived in another state, might have found herself sterilized against her wishes. Had we lived in Europe during the years of my childhood, all, except perhaps my father, might have perished in concentration camps.

Although many, if not most, of the descendants of those classified as unfit in the 19th century are among today's productive blue collar, middle class, and professional families, there are babies born with physical and mental impairments that require surgery, medical management, special education, physical therapy, and other social services to survive and function as best as those professional services can provide. The fact that many of those labeled unfit were victims of a spurious diagnosis does not negate an underlying biological reality that touches all of humanity and that at least 5% of all families have to face when their children are born with birth defects.

In the last two chapters of this book I have wrestled with this problem. It cannot go away because it is part of the biology of every generation, but it is a problem whose remedies, and the values we bring to it, change with our increasing knowledge of biology and the new technologies that have become available to us. Readers of this book may feel uncomfortable, as I certainly did, when they realize that there is a lot of mythology associated with the origins of the eugenics movement. It is often portrayed as a philosophy of the successful and well-to-do, conservative, and elitist class in which the unfit are the exploitable, repressed, and victimized classes whose failings were largely attributed to innate factors. The story is far more complex, and eugenicists and their predecessors cannot be classified in such simple terms. It is, indeed, embarrassing to see many strange bedfellows in the development of the idea of unfit people, and it should give us pause if we believe that the Holocaust could have been predicted from its earliest roots.

I have documented my sources with scholarly care, and the reader should be able to look up or extend further inquiry into any theme through the references in the footnotes. I also provide a list of very useful books on the history of eugenics and cognate fields for those who wish to read more about this movement. Most of the scholarly and popular writings since World War II are hostile to eugenics in any form. To provide the reader with some guidance, I have given a brief commentary on some of these works. My own bias is tilted toward the environmentalist position (somewhat to the right of Stephen Jay Gould and much to the left of my own mentor, H. J. Muller). I believe that some 95% of all children born are capable of being M.D.s, lawyers, college professors, or other professionals. In about 5% of humanity I believe there may be impediments to learning or behavioral function because of congenital or genetic defects or environmental damage associated with the gestational or birth process. I do not believe that those infants have the capacity to handle higher education, and many (about 0.5%) of all infants will not have the capacity to earn an independent income or live an independent life. There may also be rare talented individuals of exceptional ability whose origins, whether genetic or environmental, have eluded all attempts to identify them.

For the heritability of personality and many of the behavioral traits that attract the attention of parents (natural or adoptive), I can offer very little evidence that meets the standards of good science. At the time of the writing of this book, the evidence is still unconvincing that either an environmentalist or a hereditarian theory of human behavioral traits is primarily true. For physical defects, however, the number of documented familial disorders associated with single genes is in the thousands, and a great deal is known about the molecular biology of birth defects and how they can be detected in the child (or embryo) that expresses it or in the parent or sibling that harbors, but does not express, the defective gene.

FOOTNOTES

[1] Muller's essays on theoretical genetics and the applications of science to society are in two volumes that I edited. See H.J. Muller, *Man's Future Birthright* and *The World View of Moderns* (State University of New York Press, 1972).

[2] The last chapter is meditative and philosophic rather than narrative and interpretive. I chose this format to convey the lack of scientific answers or social consensus for some of our most enduring concerns. I also felt it was a more powerful way to remind us of the ambivalence we feel when confronted with our own imperfections.

Part I: Before Darwin

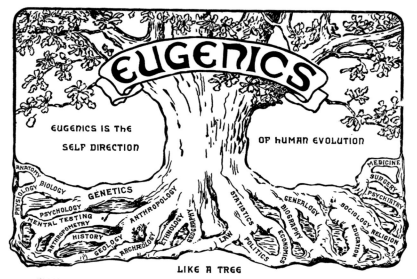

LIKE A TREE
EUGENICS DRAWS ITS MATERIALS FROM MANY SOURCES AND ORGANIZES
THEM INTO AN HARMONIOUS ENTITY.

1

Who Are the Unfit?

ROM THE 1880S TO THE 1940S the use of the phrase "the unfit" was widespread in American culture. It evoked an image of physically and morally weak people associated with society's failures—paupers, criminals, psychotics, the mentally retarded, vagrants, prostitutes, and beggars. From 1910 to 1940 the term coexisted with a variety of technical and semipopular terms associated with the eugenics movement. The term "eugenics" was coined by Francis Galton in England in 1883 as a moral philosophy to improve humanity through selective breeding. Galton was mainly interested in breeding the best of humanity to constantly improve the quality of succeeding generations. He particularly favored intelligence, cultural talents, and physical strength and dexterity. His form of eugenics is now called positive eugenics.[1]

POSITIVE VERSUS NEGATIVE EUGENICS

Positive eugenics never found favor in the United States, but it did have a following in Great Britain, especially among the intellectual class. It hoped to bring about a change through moral suasion—the ablest and the brightest would be educated and urged to have larger families than the average couple. Great Britain had a tradition of social classes based on wealth, property, education, and royalty. The United States did not. The eugenics movement in the United States tried to preserve the basic good-

"That is, with questions bearing on what is termed in Greek, *eugenes*, namely, good in stock, hereditarily endowed with noble qualities. This, and the allied words, *eugeneia*, etc., are equally applicable to men, brutes, and plants. We greatly want a brief word to express the science of improving stock, which is by no means confined to questions of judicious mating, but which, especially in the case of man, takes cognisance of all influences that tend in however remote a degree to give to the more suitable races or strains of blood a better chance of prevailing speedily over the less suitable than they otherwise would have had. The word *eugenics* would sufficiently express the idea; it is at least a neater word and a more generalised one than *viriculture*, which I once ventured to use."

Galton's coining of the term " eugenics" (Courtesy of CSHL Archives).

ness of its people by preventing those deemed unfit from breeding with each other or with essentially decent people. The reproductive isolation of such unfit people is called negative eugenics. This became very popular in the United States, Scandinavia, and Germany during the first four decades of the 20th century. Negative eugenics was given enthusiastic state support in Germany when the Nazis were elected to run the country. One feature of that Nazi eugenic movement was the purging of racial impurity from the German stock, an activity particularly directed at German Jews,

ESSAYS

IN

⚘ EUGENICS. ⚘

BY

SIR   FRANCIS   GALTON,   F.R.S.

London :
THE EUGENICS EDUCATION SOCIETY.
1909.

"EUGENICS: ITS DEFINITION, SCOPE
AND AIMS.

Eugenics is the science which deals with all influences that
improve the inborn qualitites of a race; also with those
that develop them to the utmost advantage."

Galton's definition of eugenics (Courtesy of CSHL Archives).

and later, all Jews as they were arrested in the conquered nations of an
advancing German army during World War II. The Holocaust represented the systematic genocide of a people and its culture by annihilation,
mostly carried out by a coded secret command sanctioned by Nazi ideology and its chief leaders, especially Hitler, Himmler, and Goering. The
overwhelming number of those killed were Jews, although the Holocaust
also included the mass murder of gypsies, homosexuals, and selected
political enemies.

Ever since the revelations about the death camps in which millions of
Jews were gassed, shot, cremated, or buried in mass graves, the word eugenics has connoted an evil doctrine created and fostered by bigots, racists, the
selfish, the uncaring, and those who believe in their own superiority. This
view is comforting because it portrays an identifiable cast of personalities

that can make us watchful and prevent future holocausts.[2] Unfortunately, my research on the history of the idea of unfit people does not support such a simple image of society. Most governments and the people they represent are more complex and do not fall simply into liberal and conservative political thought, into bigots and humanitarians, into those with a negative view of human nature and those with a positive view. Instead, in our own times and in the past, the persons involved with the idea of unfit people and the social philosophies of eugenics that responded to them were much more diverse in their personalities and social outlooks.[3]

HISTORY OF THE IDEA OF UNFIT PEOPLE

Those who read this book will follow the history of the idea of unfit people. It was not an invention of the latter half of the 19th century. It is a story that goes back to antiquity, with many examples in the Old and New Testaments of the Bible.[4] Biological interpretations of the unfit began in 1710, with masturbation as the first alleged cause of physical and mental degeneracy in both the abuser and the abuser's descendants. Degeneracy theory became more encompassing and included many occupational hazards such as tanning, hat-making, and food preparation, as well as social conditions such as poor housing, malnutrition, alcoholism, and deficient hygiene. Many of these observed associations turned out to be correct. Some, like masturbation, proved false.

Masturbation was nevertheless the reason vasectomy was tried as a treatment to stop the habit and as a preventive to the assumed degenerate progeny of the masturbator. The physician Harry Clay Sharp (see Chapter 12), who first performed vasectomies on young men in his prison clinic, successfully lobbied his state of Indiana to pass the world's first compulsory sterilization law in 1907. More than 30 states in the United States passed compulsory sterilization laws. Most students are astounded when they learn that the Supreme Court upheld such laws by an 8 to 1 vote in 1927 and that the court has never fully reversed itself on this issue.[5] They are also astounded to learn that the chief lobbyist for compulsory sterilization laws, Harry Laughlin (see Chapters 13 and 14), whose office was in Cold Spring Harbor on Long Island in New York state was praised and awarded a gold medal in 1936 by the Nazis, who used his model eugenic law for their own eugenics program!

Those who contributed to degeneracy theory in the 19th century were often professional people—charity leaders, sociologists, physicians, and prison reformers. They were often advocates of public hygiene; they fought for slum prevention through changes in building codes; they became friends of the new labor movement, champions of public education, providers of public libraries, creators of visiting nurse associations, promoters of public bath houses, and founders of settlement houses. Many of them were what today would be called liberals in their political philosophy.

Throughout the 19th century, social philosophers who sought ways to address the failings of society relied on science for its theories of the causes of human failure and for its technology to prevent or remedy the social pathology of the times. This was considered an advance over asking the local government, organized religion, or the families of the unfortunates to handle a problem of gargantuan proportions. In prior centuries, these three traditional approaches had often been relied on, but they failed to solve the on-going problem of dealing with at least 10% of the population who could not support themselves or their families. It was hoped that science would be the savior of society.[6]

BIOLOGY VERSUS SOCIOLOGY

Much of the writing on the history of eugenics examines it from two perspectives—the biological validity or spuriousness of the assumptions about the physical and mental qualities of life and the political climate that favored or discouraged applications of eugenic thinking in society, especially that of the United States and Great Britain. Social history is more complex than that. It helps us considerably when we also study the intellectual traditions, the cultural differences, the social structure, the religions, and the networks of communication in different countries where the ideas of unfit people developed. A knowledge of Lamarck's ideas (see Chapter 8) regarding the way heredity can be modified by the environment and his long-lasting influence on French biological thought reveals why an American eugenics movement never found favor in France. American society encouraged social mobility through merit and hard work; British society placed limits on the opportunities of even its most talented individuals if they had the bad fortune of being born into the wrong class.

It is probably not a coincidence that Americans preferred negative eugenics (keeping the basic genetic stock from being corrupted) and the British favored positive eugenics (persuading the elite classes to have more children than the lower classes).

Of particular importance are long-lasting intellectual beliefs in shaping the policies and ideas of many of those who contributed to the idea of unfit peoples and the eugenic or environmental proposals to thin their ranks. These include the conflict between a belief in progress, represented in its extreme form as "the perfectibility of man," an idea shaped during the French Enlightenment, and the belief in degeneracy, an extension of Adam's fall from grace and often associated with a negative view of human nature plagued by original sin, moral error, or passions that cannot be effectively restrained.[7] Equally important and of long-standing debate are assumptions about the innate and malleable aspects of human behavior. Today we see the argument as one of biological determinists and environmentalists on such issues as human aggression, intelligence test scores, talents, and pathological behavior. Other important influences include our mostly erroneous, or at least controversial, beliefs that we can read character through facial appearance or body shape and build or through ethnic and racial status; that the burden of support for those who need assistance should be familial or on the private sector or the public sector; that we have a duty to empathize with others and show a caring concern for the unfortunate or, quite the contrary, that we owe nothing to the downtrodden except tolerance and our good wishes that by their own efforts they may reverse their sad condition.

FOOTNOTES

[1] A good collection of primary historical documents on the history of both negative and positive eugenics is found in *Eugenics: Then and Now*, edited by Carl Bajema (Halstead, 1976). Good histories of the negative eugenics movement include Kenneth Ludmerer's *Genetics and American Society* (Johns Hopkins University Press, 1972); Daniel Kevles' *In the Name of Eugenics: Genetics and the Uses of Human Heredity* (Knopf, 1985); and Philip Reilly's *The Surgical Solution* (Johns Hopkins University Press, 1991). No historical work on positive eugenics has been written. Galton introduced the term eugenics in 1883 in *Inquiry into Human Faculty and Its Development* (MacMillan), pp. 24–25, but it was not until the first decade of the 20th century that societies with the word eugenics in their names appeared.

[2] This is particularly so for such social movements as "Science for the People," which publishes its own alternative press journal. A comparable group exists in Great Britain. They tend to be environmentalists who reject alleged scholarly studies of personality differences among groups as racist, incipiently racist, or politically naive. They often see new technologies in human genetics and reproductive biology as potentially hazardous or as a means to revive failed eugenics programs of the past.

[3] A good example of recent scholarship on this diverse response to eugenic issues is William Schneider's *Quality and Quantity: The Quest for Biological Regeneration in Twentieth-Century France* (Cambridge, 1990), which portrays French eugenics as a coalition of many contradictory political and philosophic outlooks.

[4] I do not imply from this biblical account of unfit peoples that there is a causal connection running from them to the Holocaust. That would be a simplistic interpretation of history. What I do claim is that for various reasons the idea of unfit people is an ancient one. Neither do I imply that there will always be people who may be called the unfit. There are many ideas that persisted for millennia and have now virtually disappeared, including beliefs in the legitimacy of slavery, witchcraft, or mental illness as an outcome of possession by demons.

[5] Fortunately, most of the 30 states that passed such laws have repealed them, or they have been overturned by state courts. Those that still carry these as state laws rarely enforce them because of the negative publicity (and costs in lawsuits) they generate.

[6] It still is perceived that way. We seek technological solutions for environmental pollution, infertility, induced cancers, new infectious diseases, famines, floods, conflagrations, beach erosion, and every possible misfortune that afflicts the individual or society.

[7] Essentially it is a difference in personality. Those who believe in progress are optimists; those who believe in a corrupted human nature are pessimists. These basic personality outlooks surely influence the way one sees human goals and future outcomes.

2

The Unfit in Biblical Times

IT IS NOT EASY TO LIVE AS A HUMAN. We must adjust to the personalities of our parents and siblings while we are children and then struggle to assert our independence in our teens. We try to find a job or a career, raise a family, and hope we will find sufficient support to keep us going when we are too old to work. Most healthy people, as long as there has been a recorded history, have managed with more success than failure to live that life cycle. The Bible is a record of how humanity lived in the past, especially in the 2000 or so years that preceded the Christian era. It should not be a surprise that those who failed to abide by the customs, laws, and faith of their contemporaries were sometimes seen as unfit people.

THE REBELLIOUS SON AS UNFIT

The Bible distinguishes two states of unfit people. Some are condemned forever or without reprieve, and some are condemned for a fixed number of generations. In the first category is an unusual family disturbance of a "rebellious disobedient son" described in Deuteronomy 21:18–21.[1] The cause of the son's troubling behavior is not specified, and talmudic interpreters have not specified either an acquired or an inherited basis for the son's personality. If it were interpreted in genetic terms, the condition would be described as rare, sporadic, and sex-limited to males. Commentators infer that the son was born that way.[2]

Despite the eugenic implication of this category of unfit persons, note the highly legalistic restrictions put on the determination of such a disobedient son. The son has to be of a relatively precise age (before he shows a beard or pubic hair); both parents must condemn him (his entire family is a victim of his personality); witnesses must testify at the gate of his household (there should not be a kangaroo court miscarriage of justice); the elders of the city must determine his guilt (the legal responsibility must rest with those in highest authority to exercise it); and all the adult males of the unruly son's city must stone him to death (everyone in the community must take responsibility for purging the potential evil from that community).[3] All of these limitations on the determination of the rebellious, disobedient son as an unfit person imply that a potential abuse, by a misguided group, of this obligation to purge the community would not be allowed. Only both parents can bring such charges, and only a careful trial can determine the outcome. This is similar to the protections put into the United States Constitution to assure that the charge of treason (often a capital offense) would not be abused by a government eager to purge itself of its most irritating or persuasive critics.

The description of the rebellious son's behavior as evil may be seen as his potential to live the life of a thief, murderer, or abominator of God's laws. The community is threatened by the son's existence, and the seriousness of this threat requires society's response by capital punishment to protect society from the future havoc this unruly son can bring about.[4] The importance of that warning to purge society of its occasional unfit sons is reinforced by the requirement that the execution of the sentence be made publicly known ("all Israel shall hear, and fear") to as many Jews as possibly can be reached.

The "bad seed," possessed child, or child "born to raise hell" is a recurrent theme in literature. The child may be interpreted as an atavism or "throwback" to an ancestral feral condition that is assumed to be brutish and ferocious, unrestrained by moral training. Another mechanism proposed for such children is their deliberate creation as a punishment for a past ancestral sin. A third possible model is their origin as a degenerate offshoot caused by an untimely intercourse or improper living habits by one parent or the other. A fourth explanation may be biological; that is, the child may have a disordered personality because of some physiological defect associated with his genotype.[5]

BASTARDS AS UNFIT PEOPLE

A second category of unfit people applies to males and females who are the products of a specific illegal fornication of a married (or betrothed) woman with a man who is not her husband (or fiancé), or of a relation prohibited by Jewish law from marriage, such as incest. Such a child is a special kind of bastard or *mamzer*.[6] Ordinary bastards, or those born out of wedlock to unmarried and unbetrothed and unrelated couples, have full religious and social rights in the Jewish culture of ancient Israel.[7] The mamzer is a more serious offense than mere out-of-wedlock children, because the patriarchal lineage is threatened for both religious and social standing. There is an implication that it is not just how a child is brought up, but what paternal biological stock the child comes from that matters. The Bible's punishment is explicitly laid out in Deuteronomy 23:2: "A bastard shall not enter in the congregation of the Lord; even to his tenth generation shall he not enter into the congregation of the Lord." This could bring severe consequences, because such persons would be legally denied a Jewish marriage to another Jew, and because of this stigma they might be treated as a pariah caste, excluded (at least by social stigma) from governance, social mobility, and opportunities for work or to own or inherit land, although no such prohibitions were officially specified. Ten generations is at least two and possibly three centuries, enough time to establish an outcast population within the community, shunned or limited in their activities by their fellow citizens who label them as mamzerim. They resemble, superficially, the outcast populations first identified in the late 1860s along the Hudson River Valley in New York State and assigned the name Jukes (see Chapter 10).[8]

THE AMALEKITES AS AN UNFIT POPULATION

A third category of unfit people in ancient times were the Amalekites.[9] They lived in Ashdod, the most important of the five towns that constituted the Philistines in the southwest of Palestine, or what is called, since the creation of the state of Israel, the Gaza Strip, about 7 kilometers (3 miles) east of the Mediterranean and about 40 kilometers (18 miles) north of Gaza. They were at the time of the Exodus from Egypt a Bedouin tribe that harassed the fleeing Jews in their journey across the Sinai desert.

The Amalekites were perceived as a degenerate people with evil habits, who should not only be shunned but exterminated, including their wives and children, and even their cattle. In talmudic interpretation, the Amalekites were believed to have been created evil, but their extermination was never to be complete because they had intermarried with other tribes as well as with Jews, and their seed (hereditary nature) was mixed in undetectable ways. The presence of Amalekite heredity is inferred when particularly evil people arise, such as Haman in the book of Esther or, in more recent times, King Ferdinand of Spain in the 15th century, or Hitler in the 20th century.[10]

The Amalekites may be thought of as a model for racism or even genocide. In this interpretation, the racist often attributes inhuman practices to the offending race, such as the Amalekites with blood in their mouths and abominations between their teeth. These might be examples of their violating dietary laws, Jews being prohibited from eating meat that has not been drained of its blood and Jews being prohibited from eating specified animals or animal parts. The abomination may be more repulsive to the sexual mores of that era, especially if blood as a vital fluid is equated with semen as a vital fluid.[11] In that case, the Amalekites practiced sodomy (fellatio).

Even if one acknowledges the source of God's wrath, the slaying of those weak and infirm Jews least able to defend themselves, it is difficult for contemporary readers of Exodus to imagine why the children and future descendants of the Amalekites, no matter how many centuries pass, should be considered as evil in the eyes of the Jews as they were at the time of their original crimes. An inference may be made that the culture is corrupting, and children raised in it are necessarily going to practice the abominations of the parents. This should not apply to infants or very young children, yet there is no sparing of even these children. Nor are there any careful protections of legal rights as in the determination of the rebellious and disobedient son; the Amalekites are perceived evil as a class, and their death is laid down as a commandment. The miscegenation of the Amalekites may have rendered that commandment moot, but nevertheless it might be interpreted as a warning that unspeakably evil people will occasionally appear because they were not destroyed at the appropriate time, and Jews cannot take their freedom or safety as a people for granted. Scapegoating the Amalekites in this interpretation is analogous to 19th-century ideas of atavisms.

FOOTNOTES

[1] "If a man have a stubborn and rebellious son, which will not obey the voice of his father, or the voice of his mother, and that, when they have chastened him, will not hearken unto them: Then shall his father and his mother lay hold of him, and bring him out unto the elders of his city, and unto the gate of his place; And they shall say unto the elders of his city, This our son is stubborn and rebellious, he will not obey our voice; he is a glutton and a drunkard. And all the men of the city shall stone him with stones, that he die: so shalt thou put evil away among you; and all Israel shall hear, and fear."

The citations are from the King James version of the Bible. I am grateful to Rabbi Howard Diamond for drawing my attention to many of these biblical and talmudic sources on unfit persons and for the informative discussions they provided.

[2] Note that alcoholism and gluttony are specified as early symptoms of the rebellious and disobedient son. Alcoholism was often cited in the 19th century as a major cause of degeneracy in the individual and in his progeny. Also note that the conflict was more often perceived as a father–son strife rather than a mother–son strife, a situation similar to family conflicts in the 20th century, where father–son or mother–daughter conflicts are more often experienced by family counselors.

[3] That is, all the males are responsible to do so because in that ancient Jewish society the governance was assumed by males. If this is interpreted from a legal or religious view, only males have the power to try, convict, and punish an evil that would threaten the welfare of the community. The only role of the woman in this episode is as a witness for or against her son. If she will not condemn her son, the father's displeasure and misery is not sufficient to condemn him.

[4] Gemara: Sanhedrin, 8.

[5] The first three interpretations are based on false assumptions about heredity. The biological theory is unproved but was revived for 47,XYY males (males whose cells have 47 chromosomes, instead of the normal 46, and an extra Y chromosome, which contains the testes-determining factors). The evidence for psychotic or criminal or violent behavior in 47,XYY males is unproved, although such men are overrepresented in prison and in asylums for the insane. All criminal cases using such a defense have failed to convince juries and judges that the accused deserved either freedom or a lighter sentence. There are single-gene (even X-linked) disorders associated with disturbed behavior, such as Lesch-Nyhan syndrome, which can lead to self-destructive behavior or to violent attacks on others.

[6] The term mamzer is also used as an epithet in its Yiddish form, where it may variously be spelled momzer, momser, momza, or momsa.

[7] The mamzer is forbidden to marry a Kahan or Cohen (the highest or priestly caste of Jewish religious authority); the mamzer cannot serve as a witness. These prohibitions did not apply to an out-of-wedlock child. The 10-generation prohibition is sometimes interpreted as permanent. The only escape from the cursed state is for a male mamzer to marry a non-Jewish woman; then their children can be converted

to Judaism and be free of the mamzer stigma. This only applies to males who marry to free the lineage of their mamzer status. In Conservative and Orthodox Jewish law, Jewishness is passed on through a Jewish mother and never through the Jewish father. In that strict Jewish law, a convert has full status as a Jew, and it is considered inappropriate to distinguish between a born Jew and a converted Jew.

[8] The Jukes are discussed in Chapter 10. Like Biblical mamzerim, they had a bastard ancestry and persisted for about ten generations from 1720 to about 1920, after which they disappeared as an alleged kindred of hereditary degenerates.

[9] "Amalekite" *Encyclopedia Britannica* 1: 724; "Ashdod" *Encyclopedia Britannica* 2: 509. In Deuteronomy 25: 17–19 that dismal episode is recounted: "Remember what Amalek did unto thee by the way, when ye were come forth out of Egypt; How he met thee by the way, and smote the hindmost of thee, even all that were feeble behind thee, when thou wast faint and weary; and he feared not God. Therefore it shall be, when the Lord thy God hath given thee rest from all thine enemies round about, in the land which the Lord hath given thee for an inheritance to possess it, that thou shalt blot out the remembrance of Amalek from heaven; thou shalt not forget it." This divine mandate to punish and kill the Amalekites is stated in Exodus 17:14: "And the Lord said unto Moses, write this for a memorial in a book, and rehearse in the ears of Joshua: for I will utterly put out the remembrance of Amalek from under heaven." It is repeated in Exodus 17:16: "For he said, Because the Lord hath sworn that the Lord will have war with Amalek from generation to generation." Later the Amalekites are singled out and cursed as if they were mamzerim in Zechariah 9:6–7: "And a bastard shall dwell in Ashdod and I will cut off the pride of the Philistines. And I will take away his blood out of his mouth, and his abominations between his teeth...."

[10] Gemara: Bruchot, page 28 side A.

[11] In Chapter 2, the "spilling of seed" by masturbation or coitus interruptus is interpreted by some talmudic scholars as a "shedding of blood" or murder. Hence, blood in the mouth may have been deliberately used to express in revolting terms what could not be literally described.

CHAPTER 3

Self-Pollution and Declining Health

L ong before Francis Galton introduced the term eugenics, there was a growing concern during the 18th and 19th centuries that degeneracy was a major problem. The degeneracy might be physiological (caused by masturbation, occupational exposure, or alcoholism), moral (leading to innate criminality), mental (resulting in feeblemindedness or insanity), or economic (in which the pauper lacked the ability to rise out of poverty). There was confusion as to what was the cause and what was the effect, because the prevalent idea of heredity assumed that environments both caused and reversed hereditary behavior, health, and physical structure. In this chapter, I discuss the earliest model of degeneracy based on self-pollution, or masturbation. The moral, mental, and economic theories of degeneracy are discussed in Chapters 4–6.

Masturbation was the first type of degeneracy associated with a physiological mechanism. The loss of semen or the ejaculatory shocks to the nervous system were believed to cause physical and mental illness. This idea appeared in print around 1710 when a pamphlet appeared in England with the title *ONANIA, or the Heinous Sin of Self-Pollution and All Its Frightful Consequences in Both Sexes, Considered.*[1] The author of the pamphlet is unknown but was probably a clergyman.[2] Beginning with the 4th edition, the anonymous author added the unsigned correspondence of his readers, who poured out their confessions of personal ruin and profound guilt stemming from acts of masturbation that commenced during puberty. As

Title page of *Onania* (courtesy of the Kinsey Institute, Indiana University, Bloomington).

a result, from an initial slim 60 pages, *Onania* contained 336 pages in the 18th edition of 1756.[3] By 1778, the book was in its 22nd edition.

In the preface to *Onania*, the author attributes the idea of masturbation as the source of ill health to a *Treatise on Uncleanness* by Ostervald (probably Jean Frederic Osterwald, whose treatise appeared in English translation in 1708), but he points out that Ostervald "passed over masturbation out of decency, only alluding to its evil."[4] The term masturbation is derived from the Latin (*manu* = hand and *stuprare* = to defile). The synonym, onanism, is derived from the title of the book. The author explained in his preface that "The sin of Onan, and God's sudden vengeance upon it, are so remarkable, that everybody will easily perceive that, from his name I have derived the Running Title of this little book: and though I treat of this crime in relation to women as well as men, whilst the offence is Self-Pollution in both, I could not think of any other word which would so well put the Reader in mind both of the sin and its punishment at once, as this."

THE BIBLICAL ACCOUNT OF ONAN

The author of *Onania* associates masturbation with the sin of Onan, described in Genesis 38, who spilled his semen rather than impregnate his dead brother's wife.[5] The Bible does not specifically mention masturbation, and most theological commentary on the sin of Onan assumes coitus interruptus to be the act Onan performed with Tamar. Matthew Henry (1662–1714), a Presbyterian divine and author of a popular and much reprinted *Commentary on the Whole Bible* (1704), does not specify how Onan spilled his seed but remarks that "to the great abuse of his body, of the wife that he had married, and of the memory of his brother that was gone, he refused to raise up seed unto his brother, as he was in duty bound."[6,7] Henry's reference to Onan's "great abuse of his body" implies that the idea of seminal loss resulting in physical damage may have preceded the *Onania*. The Bible is more specific in Leviticus 18–20 on sexual practices that are forbidden.[8,9]

SEMINAL LOSS AND ILL HEALTH

The idea of seminal losses leading to physical weakness is an ancient one, probably arising from the sleepiness many men experience after ejaculation. Such losses were assumed to arise from three sources—masturbation, excessive venery, and spermatorrhoea. All of the books on masturbation as a vice or a disease refer to "involuntary seminal losses" or spermatorrhoea as a consequence of a prolonged habit of masturbation or excessive venery. The earliest descriptions of spermatorrhoea are by Hippocrates and Celsus.[10] Celsus claimed that the loss of semen could occur without conscious pleasure or voluptuous dreams and that this could lead to a fatal consumption. The medical references to masturbation are strangely absent after the fall of Rome and do not recur until the 17th century. Although physicians may not have considered onanism or excessive venery a major medical problem in the Middle Ages or during the Renaissance, these practices were condemned as sins.

Onania cites many physical consequences of masturbation, ranging from stunted growth to abnormalities of the sexual organs and consumption (tuberculosis). A Dr. Etmuller is credited with attributing gonorrhea to "damnable self-pollution." Masturbation is also associated with the acquisition of other sexual vices; it leads "to atheism and loss of Godliness," obscene discourse, puny offspring, sterility, and miscarriages.[11] The author's remedy is primarily religious; the patient must find cure through repentance, but he may also benefit from a lean diet and moderate exercise. Most of all, he urges his reader to stop by the bookseller's shop to purchase some "tinctures" that he had prepared to stem the damage and fortify the sinner's body while he awaits God's grace after repenting.[12]

INFLUENCE OF THE ONANIA

Although the book was immensely successful, serving as the *Psychopathia Sexualis* of its day, it was not cited by reputable physicians. The endorsement of the book's thesis came from Switzerland, where the physician Samuel Auguste Andre David Tissot (1728–1797) in 1758 published in Latin, and thereafter had translated into French (1760) and English (1832), *Onania, or a Treatise upon the Disorders Produced by Masturbation.*[13] Tissot, like his English counterpart, mixed his religious views with his medical views and condemned the practice as a vice. It was Tissot who suggested that Balthazar Bekker was the author of the English *Onania*, perhaps hoping to discredit it. Tissot's book bears the motto:

> When base lust fills thy thoughts
> Let a horrible picture rise before thy mind
> Of withered dead men's bones,
> So let the sensual stimulation be driven away.

Voltaire accepted Tissot's views and included a discussion of the dangers of masturbation in the *Dictionnaire Philosophique*. A few years after Tissot's endorsement of the idea of masturbation as a disease, English physicians adopted the idea and offered such publications as W. Farrer's *A short treatise on onanism, or the detestable vice of self-pollution. Describing the variety of nervous and other disorders that are occasioned by that shameful practice, or too early and excessive venery, and directing the best method for their cure. By a physician in the country.* The second edition was published in London in 1767.[14]

It was not until the 1830s that the religious aspect of masturbation was minimized or altogether removed from the medical literature. Leopold Deslandes, a French physician and a member of the Royal Academy of

Medicine at Paris, published in 1835 *De l'onanisme et des autres abus vene-
riens consideres dans leurs rapports avec la santé (On onanism and other sexual abuses considered in relation to health)*, a book that was popular with
the profession in Europe and the United States. Its American edition
appeared in 1839.[15] Deslandes's approach was physiological. He argued
that there is a normal rhythm of body function associated with the reproductive organs. If they are removed, as by castration, severe effects result;
thus, the gonads serve a purpose other than reproduction itself. If they are
precociously exercised or overstimulated, they produce physical and mental damage to the individual.[16] Deslandes's theory couples the idea of seminal loss as the cause of weakness with the idea of "spinal marrow" being
the primary tissue damaged by bad sexual practices. Among the consequences described by Deslandes are "loss of flesh," "loss of strength," pale
countenance, diarrhea, consumption,[17] and other effects on the heart,
musculature, and skeletal system.[18] Deslandes also describes the psychological effects on the onanist: "His eyes are turned from the gaze of those
around: he loves solitude, avoids the world, and is embarrassed, and almost
as it were, ashamed of himself."[19] Further difficulties are loss of attention
span and memory: "Young men, who previously showed considerable
vivacity of mind and aptitude for study, become, after being addicted to
this habit, stupid, and incapable of applying themselves."[20]

Deslandes uses hundreds of cases from his own practice and from the
medical journals of his day to illustrate these points, building a powerful
case that the confessed onanism or excess venery of the patients was the
basis for the medical complaints that brought them to seek a physician's
help. He is ambivalent about the value of reading books such as *The Onania* or Tissot's treatise. He tells of a case cited by Tissot involving a 40-year-
old male: "In 1762, he procured from Frankfort the remedies mentioned in
the English treatise, *Onania*, which were of no use,"[21] but mentions from
time to time young onanists who were sobered and cured by reading Tissot, whose treatise "is the only one which possesses much reputation."[22]

Spermatorrhoea is perceived by Deslandes as a serious symptom, often
leading to fatal consequences. His predecessors and contemporaries were
debating that view.[23] Deslandes claimed that Herman Boerhaave did not
believe in spontaneous seminal emissions and attributed such discharges
to fluids that were not semen. Jan Swammerdam, John Hunter, and
Albrecht von Haller also were in agreement with Boerhaave's views. It was

Deslandes's hope that the dozens of cases he cited would convince his colleagues that spermatorrhoea was a real disease.

Females also masturbated and they too were at risk to the shocks to their spinal marrow that could cause mental weakness or derangement. Since they did not have the equivalent of seminal losses, their physical defects were not as extensive as those encountered among males. A major effect was on the clitoris: "By too frequent titillation, the clitoris may become enormously large." Deslandes approved of removal of the clitoris (clitoridectomy) for compulsive female onanists. "Levret was, we believe, the first who conceived the idea, curing nymphomania by this operation. Dubois performed it on a young girl, who was so addicted to onanism, that she was almost in the last stages of marasmus."[24] Very likely the marasmus Deslandes cites was anorexia nervosa and not a direct consequence of her sexual practices.

Treatments for masturbation were nonsurgical because Deslandes rejected castration. He suggested instead "cold lotions or applications of ice to the scrotum, and of leeches around it,"[25] as well as cold douches to the perineum, cold hip baths, or cold enemas. More important to Deslandes was prevention of masturbation, and he suggested careful design of boarding schools including open sleeping areas, coarse linens, doorless privys, and cold showers.[26] Boys should also have lots of physical exercise to occupy their time and cause them to sleep readily from exhaustion.[27] Many of the public boarding schools in England still adopt these Spartan measures, but they no longer associate them with their original intent for the prevention of masturbation and, instead, attribute them to the tradition of toughening the body and soul in the development of manly character.[28]

In 1847, Claude Lallemand, a urologist who is frequently cited by Deslandes, published *Involuntary Seminal Losses* and also confined the physical and mental defects caused by masturbation to physiological factors, such as loss of vital fluids and shocks to the nervous system.[29] Lallemand favored acupuncture as a treatment for the spinal shocks.[30] The French physician, Louis Auguste Mercier, in 1841 associated masturbation and venereal excess with the formation of enlarged prostate glands, which he attempted to remedy by castration.[31] In the United States the masturbatory theory of ill health received strong support in 1812, from Benjamin Rush, one of the most prominent physicians in the new nation.[32]

It was widely believed in the last half of the 19th century that masturbation led to involuntary losses of seminal fluid (especially losses other than through nocturnal emissions, or wet dreams). These seminal losses were associated with a depletion of physical strength, making both body and mind subject to illness.[33] Books like the Reverend John Todd's *The Student's Manual*, appearing in 1835, warned of the deathly consequences of masturbation. The book was a bestseller and went through 24 editions by 1854. Male self-sufficiency required will and energy; masturbation dissipated that energy. Masturbation also depleted the potentially good heredity of a robust male, and "runts, feeble infants, and girls would be produced by debilitated sperm, old man's tired sperm, masturbator's exhausted, debaucher's exceeded, contraceptor's impeded, coward's unpatriotic, and newlywed's green, sperm."[34]

MASTURBATION AS A DISEASE IN THE UNITED STATES

Characteristic of the medical books on masturbation in the United States by American physicians is Joseph Howe's *Excessive Venery, Masturbation, and Continence* (1883).[35] Howe, who taught at New York's Bellevue Hospital, urged that physicians, when asked to advise a boy about masturbation, should make "clear to him how terrible the consequences must be if the life be continually flowing away from the body."[36] He acknowledged that not all physicians believed ejaculation itself was harmful; a Dr. Hamilton Mcgraw of Detroit claimed that a person could remain in good health while ejaculating once a day. Howe felt that masturbation created shocks to the nervous system resulting in nervous debility, languor, loss of spirit, feebleness of mind, dimness of sight, loss of manly bearing, and "many cases of the loss of reason and an imbecile and driveling old age."[37] He also attributed to masturbation acne, pallor, lusterless eyes, a coated tongue, constipation, tuberculosis (phthysis), hypochondria, insanity, and epilepsy.

Howe acknowledged that dogs, cats, and monkeys occasionally masturbate, but he attributed these to their mimicry of "depraved beings of the human species."[38] He believed masturbation was caused by mothers who stimulated the genitals of nursing infants to pacify them, to nurses of dubious character who corrupted youth, and to vicious playmates.[39] The mental effects of masturbation were atavistic, imposing behavior like that of "lower animals," including a slouching posture and absence of eye contact.[40]

Treatment of boys who masturbated included low-fat, low-pork diets, with raw beef, few spices, and no fried foods. Cold sponge baths, enemas, normal sleep patterns, and exercise were also part of the therapy. For compulsive, frequent masturbators over the age of 18, he recommended "castration without delay,"[41] and for those with less serious problems he recommended frequent sitzbaths, a morning enema, and three to six small meals a day. "Female masturbators," he noted, "can only be cured by marriage."[42]

TREATMENT OF MASTURBATION AS A DISEASE

The Swiss physician, August Forel, classified five types of onanism in a paper he prepared in Zurich in 1889.[43] These included imaginary masturbators or hypochondriacs who exaggerate the frequency and the consequences of their habit; congenital sexual perverts who devise outlandish ways to masturbate; those who are led on by bad example and who cease to masturbate when they learn it is unmanly; those who are aroused by irritation or excitement; and so-called "nothonanists" such as prisoners who have no natural outlet for their sexual needs. Forel felt that masturbation was not seriously harmful except among the young and those who do it in excess.

Castration for masturbation was first used in the United States in 1843 by Dr. Josiah Crosby of New Hampshire for a young man who claimed he was on the verge of madness from compulsive masturbation. Although Crosby claimed he had completely cured the patient, castration for masturbation was rarely performed.[44]

Until the late 1880s, treatment was essentially benign, with diet, moderate exercise, and proper sleep habits most often prescribed. Marriage was considered a sure cure, but physicians warned worried fathers that "prostitution is no cure for onanism."[45] There were some short-lived treatments. Camphor was applied as a local repellent in 1815 by Schwarz; potassium bromide was suggested in 1869; and occasional physicians would blister the foreskin or attempt an electrical stimulation of the spine.[46] Particularly painful was the procedure of infibulation, a sewing of the foreskin around a small wire ring to prevent its retraction during erection. Ligature of the vas deferens for treating masturbation was attempted in 1883 by Frank Hamilton.[47]

More moderate views emerged in the 1890s. E.L. Keyes, a surgeon specializing in genitourinary defects, claimed in the 1895 text for his specialty that masturbation "does not necessarily produce disease unless it is carried to excess." He denied that it is seminal loss "which is of the first importance in producing disease from sexual excess, but the nervous shock of the oft-repeated orgasm." Most youthful masturbators abandon their habit with maturity, he noted, but "the longer and more frequent they yield to the vicious habit, the stronger does its hold become, so that in case they escape the physical and mental disorders to which excessive venery in extreme cases may give rise, still they may pay the penalty of excess by some diminution of vigor in after-life, by throwing confusion into their sexual hygiene, and establishing sexual necessities which they find it difficult to meet suitably; and, finally, they may continue on through life victims to a perverted sexual sense, shunning women, from whom they aver that they derive no pleasure, totally wrecked as to their morale, often hypochondriacs, physical and intellectual, real and fancied."[48]

Hypnotism was tried in 1887 by Voisin; mechanical appliances to prevent erection were described by Flood in 1888. Females were subjected to clitoridectomy as early as 1858, and the procedure was used by Bloch on a 2-and-a-half-year-old girl in 1894.[49] The revival of clitoridectomy in the late 19th century was associated with the erroneous belief that the sewing machine treadle stimulated the clitoris and led to masturbation.[50] Hutchinson was offering circumcision of the newborn male as a preventive of masturbation in 1890,[51] and by 1896 surgical cutting of the nerves to the genitalia was attempted.[52] The widespread habit of circumcising newborn males for nonreligious reasons in the United States owes its origin, in part, to Hutchinson's belief that it would prevent masturbation.[53]

MASTURBATION THEORY IN THE 20TH CENTURY

The fear of masturbatory insanity and other illnesses associated with the habit persisted until the first decade of the 20th century. Iwan Bloch, one of the early pioneers in the field of sexology, was reassuring when he wrote, in 1908: "At the present day all experienced physicians who have been occupied in the study of masturbation and its consequences hold the view that moderate masturbation in healthy persons, without morbid inheritance, has no bad results at all."[54] Bloch tried to distinguish between harmless masturbation and onanism, which he defined as excessive and prolonged masturbation. He attributed to onanism depletion of vital energy, diminished spiritual and physical activity, cold-heartedness, and minor physical complaints including "masturbator's heart," photophobia, and hypochondria. In females he felt onanism led to frigidity and in males to perversions. He dismissed effects on the intellect. For cures he suggested a light diet, cool clothes, and light bedding.[55] He was not averse to the efficacy of fear: "The methods of the older physicians who appeared before the child armed with great knives and scissors, and threatened a painful operation, or even to cut off the genitals, may often be found useful, and may effect a radical cure."[56] He also cited the virtues of circumcision: "Furbinger cured a young fellow in whom no instruction and no punishment had proved effective, by simply cutting off the anterior part of his foreskin with jagged scissors."[57] Most physicians writing on the subject in the 1920s rejected the notion of masturbation as the cause of physical or mental disease, although they were still ambivalent about its practice in adults. At worst it was regarded as a habit rather than a disease, whose psychic effects offended the moral feelings of its practitioner.[58]

Although it disappeared from the medical literature as an illness, masturbation remained as a vaguely unhealthy practice in Robert Baden-Powell's Boy Scout manual for another 25 years. Baden-Powell, who in 1908 founded the Boy Scouts, felt strongly about the physical and mental damage caused by masturbation. His intent was to convince Scout leaders and their boys that "the result of self-abuse is always —mind you, always —that the boy after a time becomes weak and nervous and shy, he gets headaches and probably palpitations of the heart, and if he carries it on too far he very often goes out of his mind and becomes an idiot."[59] This draft that he submitted to the publisher was rejected because it was thought too direct and indelicate for young boys to read. He got his message across in a more subtle wording, denouncing "indulgence" or "self-abuse" as a contemptible vice avoided by manly men.

Masturbation is still grounds for dismissal from some schools, such as the U.S. Naval Academy. For many religions it is still a vice or a sin, although neither insanity, physical deterioration, nor the prospect of "God's sudden vengeance" is now cited as the reason against it.

FOOTNOTES

[1] The *Onania* is difficult to date because the first three editions are not available, and the first edition may have appeared as early as 1710 or as late as 1716. Robert H. MacDonald in "The frightful consequences of Onanism: notes on the history of a delusion" in the *Journal of the History of Ideas* 28(1967): 423–431, gives "1710?" as the date cited by the British Museum (p. 424). The 4th edition has 88 pages (it is the first edition with letters from the readers, the earliest being dated June 5, 1717), the 9th edition (1723) has 197 pages, the 12th edition (1726) has 214 pages, the 16th edition (1737) has 194 pages, the 18th edition (1756) has 336 pages, the 22nd edition (1778) has 328 pages. A French version, perhaps derived from the *Onania*, was published in English in 1719 entitled *Onanism Display'd, Being, I An Enquiry into the True Nature of Onan's Sin. II Of the Modern Onanists. III Of Self-Pollution, its Causes, and Consequences, IV Of Nocturnal Pollutions. V The Great Sin of Self-Pollution. VI A Dissertation concerning Generation. With a Curious Description of the Parts*. Translated to English from the Paris edition, the 2nd edition (London, E. Curll, 1719). In older books, the distinction between the terms "edition" and "printing" was not always made, and not all 22 editions of the *Onania* differ in content. Until copyright laws secured authors' rights, pirated versions were common, and *Onanism Display'd* was possibly such a work.

[2] Sometimes Charles Corbett, the London publisher of the book, or Balthazar Bekker, a quack who flourished in the 17th century, is erroneously assigned as the author. The Indiana University Research Library lists Bekker as the possible author. Tissot, a later writer on masturbation, also assumes Bekker was the author. Until the late 19th century, it was the custom of authors who wished to remain anonymous not to give their names. In the 20th century, authors protect their anonymity with pseudonyms.

[3] Anonymous, *ONANIA, or the Heinous Sin of Self Pollution and All Its Frightful Consequences in Both Sexes Considered*, 18th edition (Charles Corbett, London, 1756).

[4] Ibid. p. vi.

[5] Genesis 38: 1–10 describes the story of the killing of Onan after he spills his seed. Onan was the second of three sons. Their father, Judah, had married impulsively, at a young age, an unnamed Canaanite woman (later referred to in the Bible as BathShua, daughter of Shua, I Chronicles 2:3). Judah was one of Joseph's older half-brothers and a son of Jacob.

When Judah's oldest son, Er, reached his maturity, Judah arranged a marriage for him with Tamar, another Canaanite. Although the reasons are not provided in the Genesis account, Er was struck dead by God as wicked, leaving Tamar with no children. The custom of levirate marriage, which was not to be made a Mosaic law until the book of Deuteronomy, was used by Judah, who ordered Onan to marry his widowed sister-in-law.

Onan disobeyed Judah's request to "raise up seed to thy brother," and instead "it came to pass, when he went in unto his brother's wife, that he spilled it on the ground, lest that he should give seed to his brother." It is not clear what Onan did, how long he practiced spilling his seed, and what aspect of his behavior is referred to when the narrative states that "the thing which he did displeased the Lord: wherefore he slew him also." Tamar eventually had children, not with the third brother, Shelah, but through deception with her father-in-law, Judah, producing twin boys, Perez and Zerah. "God's sudden vengeance" upon Onan, according to the *Onania*'s author, was not merely for disobedience, a sin of its own accord, but for the inferred "self-pollution" by which Onan carried out the act.

The story of Judah, his three sons, and Tamar has many interpretations. At one extreme, Tamar is seen as an unfit person who repelled both Er and Onan, neither of whom could consummate a marriage with her. Her unfit status may have been associated with a loathsome appearance, an incompatible personality, or a disease that scared off her bridegrooms (Howard Diamond, pers. comm.). At another extreme, Tamar is seen as an unsubmissive victim of male deceit, and she emerges as the first female heroine, tricking Judah, and thus the patriarchal culture, in revenge for his dishonesty.

6 Matthew Henry, *Commentary on the Whole Bible, Volume 1 Genesis to Deuteronomy*. Reprint of 1704 edition (Fleming H. Revell Co., Old Tappan, New Jersey), p. 128. Levirate marriage preserved the law of primogeniture. The first-born son retains the rights to the land. By requiring the brother to marry his widowed sister-in-law and assigning to the first-born son of that marriage all the property that would have been his brother's, the father (the second-born son) becomes a tenant farmer of his brother's estate. The second-born son is also deprived of the full benefit of the dowry his sister-in-law's family gave to his deceased brother.

7 *Onania*, p. 132. The actual act may have been coitus interruptus, as one of the correspondents asserted and the author of *Onania* disputed, maintaining that it was the unnatural ejaculation, resulting in seminal loss, that corrupted Onan and caused his death.

8 These include observing the nakedness of one's parents, sister, granddaughter, aunt or uncle, or daughter-in-law. Fornication with animals and homosexual fornication are banned, as is adultery or fornication with one's mother-in-law. A parent is also forbidden to sell a daughter into prostitution. Spilling one's seed only appears in the story of Onan, and for this reason some commentators on the episode favor the interpretation that Onan was punished for not fulfilling the levirate marriage, thereby jeopardizing the lineage that led to Jesus. The New Testament begins with that lineage, and Judah is cited as a direct ancestor (Matthew 1:2–3), with Perez being the next of some 35 descendants to Jesus.

The stress of the laws laid out in Leviticus is on fornication and nakedness leading to incestuous unions and unnatural unions (such as bestiality). The absence of "spilling seed" is notable in this section. Masturbation is nevertheless considered a sin by Roman Catholics because it is claimed to be an unnatural act, copulation with the potential to procreate being considered the natural act as part of a divine intent for human practice. It is also held to be a sin because it constitutes a sexual activity. It may be thought of as a sin because it is intrinsically self-indulgent, the normal sexual act having as one of its functions the bonding in love of two individuals. Jewish talmudic commentary includes the idea of spilling seed as an act of a murderer or a "shedder of blood" and attributes the practice to idol-worshippers. Julius Preuss (1911) *Biblical and Talmudic Medicine*. Translated by Fred Rosner (Sanhedrin Press, New York, 1978), p. 489. The interchangeable state of body fluids or humors (as they were called in older literature) was possible because they were considered the vital essence of living things.

The failure to cite masturbation as a forbidden act in the Bible may reflect rarity of it in that era when marriages were early and usually polygamous, and prostitution was a tolerated outlet (as Judah demonstrated when he solicited Tamar, believing her to be a temple prostitute of Baal). Some talmudic commentators do try to make a case for a Biblical condemnation of masturbation by citing Isaiah 1:15: "And when ye spread forth your hands, I will hide mine eyes from you: yea, when ye make many prayers, I will not hear: your hands are full of blood." One rabbi interpreted this last phrase as "those who are lewd with their hands" and attributed the Noachian Flood to God's displeasure that "everyone expended his sperm on the ground," a somewhat loose rendering of Genesis 6:13: "And God said unto Noah, The end of all flesh is come before me; for the earth is filled with violence through them; and behold, I will destroy them with the earth." (Preuss, *Biblical and Talmudic Medicine*, p. 489.)

[9] Henry, *Commentary on the Whole Bible*, p. 218. Similarly, talmudic commentators have applied the same argument to David, who is a direct descendant of Perez (Howard Diamond, pers. comm.). According to Henry, the sudden deaths of both Er and Onan following their marriages to Tamar led Judah to fear that Tamar was bewitched. He tried to protect his third son, Shelah, from Tamar, and this forced her to take on the guise of a temple harlot to lure Judah into impregnating her. The story of Onan may also reflect the negative attitudes Jews had for the Canaanites who worshiped Baal and fertility goddesses such as Astarte. The corruption of Er and Onan may have begun with their father's marriage out of faith.

[10] L. Deslandes, *A Treatise on the Disease Produced by Onanism, Masturbation, Self-Pollution, and other Excesses*. Translated from the French, second edition (Otis Broaders & Co., Boston, 1839), p. 143.

[11] *Onania*, pp. 13–17.

[12] *Onania*, p. 26.

[13] Samuel Tissot, *Onania, or a Treatise upon the Disorders Produced by Masturbation*. Translated into English from the Latin 1758 edition.

[14] The Kinsey Library for Sex Research at Indiana University has many of these early books condemning masturbation. The card catalog is at the Kinsey Library and most of their items are not cross-listed at the main library.

[15] Leopold Deslandes, *On Onanism and Other Sexual Abuses Considered in Relation to Public Health*. Translated from the 1835 French edition, 1839.

[16] The idea of a mean or average activity as an optimum was favored by physicians in the early 19th century and popularized in 1835 by the Belgian statistician, Lambert Quetelet (1796–1874). Quetelet extended the concept to the average man (l'homme moyen) for both physical and behavioral attributes. In Quetelet's model, both

departures from the norm are pathological. At the physiological level, infrequent usage is as harmful as excessive usage of a body function or part.

17 Deslandes, *On Onanism and Other Sexual Abuses*, pp. 52–53.

18 Ibid. p. 54.

19 Ibid. p. 114.

20 Ibid. p. 61.

21 Ibid. p. 64.

22 Ibid. p. 230.

23 The idea of spermatorrhoea as a disease may have arisen from the observation of seminal fluid that is excreted after necking and petting or other erotic activities that are not accompanied by intercourse or masturbation. Also, after ejaculation, some seminal fluid may remain in the urethra and leak out later, especially during micturition. A number of infections, including gonorrhoea, can lead to urethral discharges, which may have been falsely assumed to be of the same physiological composition as normal semen.

24 Deslandes, *On Onanism and Other Sexual Abuses*, p. 188.

25 Ibid. p. 201.

26 Despite all the effort to discourage masturbation by these dismal boarding school practices, masturbation flourished among the youngsters. See Julian Huxley's *Memories I*, p. 144 (George Allen and Unwin, London, 1970), for the "discovery" of this habit at Eton.

27 Deslandes, *On Onanism and Other Sexual Abuses*, pp. 222–232.

28 Peter Medawar, *Memoir of a Thinking Radish* (Oxford University Press, Oxford 1986), pp. 28–31.

29 Claude Lallemand, *Involuntary Seminal Losses*, 1847, cited, p. 2, in E.H. Hare. "Masturbatory insanity: The history of an idea." *The Journal of Mental Science* 108(1962): 1–25.

30 Deslandes, *On Onanism and Other Sexual Abuses*, p. 236.

31 The history of medicine is filled with erroneous theories that served as a basis for surgical correction. Treatments for enlarged prostates also included vasectomies in the 1890s. See Chapter 14.

32 Hare, *Masturbatory Insanity*, p. 4.

33 Ben Baker-Benfield, "The spermatic economy: A nineteenth century view of sexuality." *Feminist Studies* 1(1972): 336–372. The idea of a "spermatic economy," as Baker-Benfield calls it, may have originated from the much earlier observation that massive blood loss causes death. Semen in that model would be a vital fluid, renewable like blood, but only when released in biologically tolerable quantities. Again, note the association between blood and semen, as in the talmudic prohibition of masturbation as equivalent to shedding blood.

[34] Ibid. p. 342.

[35] Joseph William Howe, *Excessive Venery, Masturbation, and Continence* (Birmingham & Co., New York, 1883).

[36] Ibid. p. 24.

[37] Ibid. p. 25.

[38] Ibid. p. 63.

[39] Ibid. p. 64.

[40] Ibid. p. 207.

[41] Ibid. p. 265.

[42] Ibid. p. 268.

[43] Anonymous. "A word upon the regulation of prostitution and sexual hygiene." *The Medical Record* 36(1889): 320–321. A translated abstract of the work of August Forel in *Corrpondenzblatt fur Schweitzer Aertze* 1889.

[44] E.L. Keyes, *The Surgical Disease of the Genito-urinary Organs Including Syphilis* (D. Appleton & Co., New York, 1895), p. 439.

[45] Forel, *The Medical Record* (see in 47), p. 320.

[46] Hare, *Masturbatory Insanity*, p. 9.

[47] Cited in Howe, *Excessive Venery*, p. 278.

[48] Keyes, *Surgical Disease of the Genito-urinary Organs*, p. 437.

[49] A.J. Bloch, "Clitoridectomy in a two and a half year old child." *Transactions of the Louisiana Medical Society, New Orleans*, 1894, p. 333.

[50] John Duffy, "Masturbation and Clitoridectomy." *Journal of the American Medical Association*.186(1963): 246–248.

[51] Jonathan Hutchinson, "On circumcision as preventive of masturbation." *Archives of Surgery*, 1890–1892, ii, p. 838.

[52] J.H. McCassey, "Adolescent insanity and masturbation with excision of certain nerves supplying the sexual organs as the remedy." *Cincinnati Lancet Clinic* 37(1896): 341–343.

[53] My father told me that when he was a boy in Stockholm, about 1913, his mother caught him masturbating. She consulted a physician who recommended and performed a circumcision on him.

Jonathan Hutchinson, "A plea for circumcision." *British Medical Journal*, September 27, 1890, p. 769. His views reflect both his personal biases and his broader and more complex perception of health: "It is surely not needful to seek any recondite motive for the origin of the practice of circumcision. No one who has seen the superior cleanliness of a Hebrew penis can have avoided a very strong impression in favour of the removal of the foreskin. It constitutes a harbour for filth, and is a constant source of irritation. It conduces to masturbation and adds to the difficulties of sexual continence. It increases the risk of syphilis in early life, and of cancer in the aged. I have never seen cancer of the penis in a Jew, and chancres are rare."

[54] Iwan Bloch, *The Sexual Life of Our Time.* English translation of the 1908 edition (Allied Book Co., New York, 1928), p. 421.

[55] Ibid. pp. 424–426. American approaches to the treatment of masturbation also included dietary regimens. Graham crackers and corn flakes were both introduced as treatments for masturbation. Eggs and breakfast meats were considered conducive to masturbation. See John Money's *The Corn Flake Wars* (Prometheus Press, Buffalo).

[56] Ibid. p. 427.

[57] Ibid.

[58] John F.W. Meagher, *A Study of Masturbation and Its Reputed Sequelae* (William Wood & Co., New York, 1924), p. 28.

[59] Michael Rosenthal, *The Character Factory: Baden-Powell and the Origins of the Boy Scout Movement* (Pantheon Books, New York, 1924), p.187.

4

Degeneracy Theory: Identifying the Innately Depraved and the Victims of Vicious Upbringing

COEXISTING WITH THE ALLEGEDLY MEDICALLY SIGNIFICANT defects of masturbation was the idea of moral degeneracy, originally stressed by the author of *Onania* but only marginally discussed in the later books on masturbation in the 19th century. As moral degeneracy became separated from onanism, its victims could be characterized into distinctive classes, especially vagrants, paupers, criminals, the insane, and the retarded. Masturbation was practiced among so many adolescents that no one class of defectives could have its problems assigned to excessive masturbation.

In the late 18th century, a theory of medicine devised by the Scottish physician John Brown (1735–1788) was in vogue. Known as the Brunonian system, it was based on a theory of balance or unity. Excess stimulation of the muscles or vascular system led to spasms, disease, and fever; deficient stimulation of these systems led to weakness and loss of normal function.[1] The nervous system was considered involved in unspecified ways, and the preferred treatment was not by proper exercise of the organs but by bleeding, purging, or modification of diet.[2]

MOREL'S THEORY OF DEGENERATION

The first scholarly proposal that degeneration occurs in human populations was offered in 1857 with the publication of *Traité des Dégénérescences Physiques, Intellectuelles, et Morales de l'éspèce Humaine, et des Causes qui Produisent ces Vérites Maladives* (*Treatise on Physical, Intellectual, and Moral Degeneration in Humans and the Conditions Producing these Detrimental States*). The author was Benedict Augustin Morel (1809–1873), a physician who was born of French parents in Vienna. Morel was a friend of Claude Bernard and shared his enthusiasm for seeking physiological explanations for living phenomena. He was also a supporter of Darwin's theory of natural selection. At Nancy he studied the mentally ill and became a pioneer alienist, as psychiatrists were then called, using family histories, social circumstances such as poverty, and organic illnesses as a means of identifying the onset and classification of mental illness. He was the first to describe and name dementia praecox, in 1860, and that name endured until 1908 when E. Bleuler replaced it with schizophrenia.

Morel believed that degeneracy arose from many sources, but chiefly from the toxic effects of poisons, such as alcohol, narcotics, tobacco, tainted bread (ergotism), and organic poisons, as well as chronic diseases such as syphilis, tuberculosis, and goiter.[3] Degenerate individuals might transmit their degeneracy to their descendants, but they were rarely persistent in the population, he believed, because they were defective and died out.[4]

Morel claimed that "the clearest notion we can form of degeneracy is to regard it as a morbid deviation from an original type. This deviation, even if, at the outset, it was ever so slight, contained transmissible elements of such a nature that anyone bearing in him the germs becomes more and more incapable of fulfilling his function in the world and mental progress, already checked in his own person, finds itself menaced also in his descendants."[5] Morel identified three major features in such physiologically caused degenerates: mental impairment such as idiocy and imbecility (described as "debiles"), stigmata of the face and body, and an emotionalism or pessimism of behavior. The deviations from normal behavior could go either way, and Morel recognized both the weak-minded (inferior degenerates) and the geniuses (superior degenerates) as pathological. Morel believed the prevention of degeneracy in society was a function of public health. Altered life styles, occupational safety, education to prevent self-abuse, or prohibition of toxic substances such as alcohol and narcotics were some of his recommendations. He recognized masturbation as a contributor to degeneration, but it was not singled out as a major cause.

Some of the stigmata associated with Morel's degenerative diseases may have been collectively lumped together into a single theory of origin. From his experience as a physician, Morel probably saw children born with Down syndrome (not yet recognized as a syndrome in his day) who would show facial stigmata and rarely survive past their infancy. Children born to alcoholic mothers sometimes show what is today called a fetal alcohol syndrome, and babies born of syphilitic parents often express characteristic dental and physical defects. Mental retardation syndromes associated with chromosomal imbalances are also frequently associated with physical defects. These physical stigmata or associations, whose causes were largely unknown to Morel, may have led him to infer that mental, physical, and moral degeneracy were related primarily to common environmental causes.

Morel's theories were adopted by many social thinkers and physicians in the last half of the 19th century. Like Morel, they believed that acquired characteristics can be transmitted to the offspring. Although the mechanism of inheritance was unknown, there was almost universal agreement, until the 1890s, that a damaging environment would result in defective offspring whose subsequent progeny would receive that physical or mental impairment. There was less agreement on beneficial environments and how effective they can be in reversing the damage of prior generations.

During the 19th century, there was a widespread belief that a person's facial features were tied to personality traits, a field known then as physiognomy.[6] Captain Robert Fitzroy nearly rejected Charles Darwin as a naturalist for the *HMS Beagle* in 1831 because he did not like the shape of Darwin's nose. Also of wide interest and belief was the claim of advocates of phrenology (invented by Franz Joseph Gall) that the bumps and shape of the cranium revealed a person's character, talents, and other behavioral traits.[7] Both views were on the wane as the 20th century began, and psychological and psychiatric approaches replaced physical features of the head as a superior method to interpret behavior.

ENVIRONMENTAL THEORIES OF DEGENERACY

Although theories of degeneracy pointed to a subsequent hereditary class of social failures, not all European scholars were convinced that these failings were biological, no matter how they were produced. The French psychologist Jean Esquirol (1772–1840) in 1835 claimed he had been arguing since 1805 that the particular type of insanity, as well as its frequency, was related to social conditions, including fluctuations in suicide rates, arrests for prostitution, and the incidence of infanticide (based on body counts of infants retrieved from sewers).[8] Pierre Proudhon (1809–1865), the founder of anarchism,[9] asserted in 1851 that poverty leads to crime when the worker is "stupefied by the fragmentary division of labor, by serving machines, and by obscurantist education, ..." The pauper becomes a criminal after he is oppressed "by low wages, demoralized by unemployment, and starved by monopoly. ...When he lacks bread, home, etc., he begs, filches, cheats, robs, and murders."[10]

Many physicians attributed the high criminality among illegitimate children to their "vicious upbringing" and the family environment in which they were raised. In 1836 in Paris, one-third of the births were illegitimate, most of them from working-class women. Foundlings were abandoned in larger numbers during the winter months when the mothers could not feed their children. Those abandoned as infants often died in hospitals, but many were shipped out of Paris to the provinces where they were raised by farmers.[11] The privileged classes often looked upon the less fortunate classes as victims of their own moral failings. They were considered by them to be "branded with the marks of vice and destitution" and "reduced by sheer besottedness to a life of savagery." Their lives inspired "disgust and horror."[12] This popular image was reflected in Balzac's portrayal of the Parisian demi-monde as "a people of ghastly mein, ...[whose] contorted, twisted faces exude at every pore of the spirit the desires and poisons teeming in their brains...."[13] They were depicted as deteriorating, the damp and darkness in which they lived affecting them like plants cut off from sunlight.

LOMBROSO'S THEORY OF INNATE CRIMINALITY

Among those who extended these biological theories of degeneracy was Caesar Lombroso (1836–1909), who was born in Verona. After his medical studies in Padua, Vienna, and Paris, he served as an army surgeon and began classifying the personalities and behavior of soldiers he

PROFESSOR CÆSAR LOMBROSO,

TURIN.

DEAR AND HONOURED MASTER,

  I dedicate this book to you, in open and joyful recognition of the fact that without your labours it could never have been written.

  The notion of degeneracy, first introduced into science by Morel, and developed with so much genius by yourself, has in your hands already shown itself extremely fertile in the most diverse directions. On numerous obscure points of psychiatry, criminal law, politics, and sociology, you have poured a veritable flood of light, which those alone have not perceived who obdurately close their eyes, or who are too short-sighted to derive benefit from any enlightenment whatsoever.

  But there is a vast and important domain into which neither you nor your disciples have hitherto borne the torch of your method —the domain of art and literature.

  Degenerates are not always criminals, prostitutes, anarchists, and pronounced lunatics; they are often authors and artists. These, however, manifest the same mental characteristics, and for the most part the same somatic features, as the members of the above-mentioned anthropological family, who satisfy their unhealthy impulses with the knife of the assassin or the bomb of the dynamiter, instead of with pen and pencil.

  Some among these degenerates in literature, music, and painting have in recent years come into extraordinary prominence, and are

Dedication pages of Max Nordau's "Degeneration," citing his debt to Caesar Lombroso and Benedict Morel for their contributions to degeneracy theory. (Reprinted, with permission, from M. Nordau [1968] *Degeneration,* Howard Fertig, New York.) (*Continued on following pages.*)

observed and treated. By relating these traits to their physical features, he hoped to establish a biological basis for mental health and disease. Like Morel, he chose to become an alienist (psychiatrist) and, after holding appointments in Padua and Pesaro, he was appointed to the medical faculty in Turin, a position he held for the rest of his life.

  Lombroso's fascination with the biological determination of behavior led him to seek an effective model. On one occasion, while he was still a young physician, he performed an autopsy on an infamous criminal and

viii   DEDICATION

*revered by numerous admirers as creators of a new art, and
heralds of the coming centuries.*

*This phenomenon is not to be disregarded. Books and works of
art exercise a powerful suggestion on the masses. It is from these
productions that an age derives its ideals of morality and beauty.
If they are absurd and anti-social, they exert a disturbing and
corrupting influence on the views of a whole generation. Hence
the latter, especially the impressionable youth, easily excited to
enthusiasm for all that is strange and seemingly new, must be
warned and enlightened as to the real nature of the creations so
blindly admired. This warning the ordinary critic does not give.
Exclusively literary and æsthetic culture is, moreover, the worst
preparation conceivable for a true knowledge of the pathological
character of the works of degenerates. The verbose rhetorician
exposes with more or less grace, or cleverness, the subjective impressions received from the works he criticises, but is incapable of
judging if these works are the productions of a shattered brain,
and also the nature of the mental disturbance expressing itself by
them.*

*Now I have undertaken the work of investigating (as much as
possible after your method) the tendencies of the fashions in art
and literature; of proving that they have their source in the degeneracy of their authors, and that the enthusiasm of their admirers is for manifestations of more or less pronounced moral
insanity, imbecility, and dementia.*

*Thus, this book is an attempt at a really scientific criticism,
which does not base its judgment of a book upon the purely accidental, capricious, and variable emotions it awakens—emotions
depending on the temperament and mood of the individual reader
—but upon the psycho-physiological elements from which it sprang.
At the same time it ventures to fill a void still existing in your
powerful system.*

*I have no doubt as to the consequences to myself of my initiative.
There is at the present day no danger in attacking the Church, for
it no longer has the stake at its disposal. To write against rulers
and governments is likewise nothing venturesome, for at the worst
nothing more than imprisonment could follow, with compensating
glory of martyrdom. But grievous is the fate of him who has the
audacity to characterize æsthetic fashions as forms of mental decay.*

noted an unusual depression in the occiput of the skull. He was struck by
this similarity to more primitive animals and then realized he had discovered "the nature of the criminal—an atavistic being who reproduces in his
person the ferocious instincts of primitive humanity and the inferior animals. Thus were explained anatomically the enormous jaws, high cheek
bones, prominent superciliary arches, solitary lines in the palms, extreme

DEDICATION   ix

The author or artist attacked never pardons a man for recognising in him a lunatic or a charlatan ; the subjectively garrulous critics are furious when it is pointed out how shallow and incompetent they are, or how cowardly when swimming with the stream ; and even the public is angered when forced to see that it has been running after fools, quack dentists, and mountebanks, as so many prophets. Now, the graphomaniacs and their critical body-guard dominate nearly the entire press, and in the latter possess an instrument of torture by which, in Indian fashion, they can rack the troublesome spoiler of sport, to his life's end.

The danger, however, to which he exposes himself cannot deter a man from doing that which he regards as his duty. When a scientific truth has been discovered, he owes it to humanity, and has no right to withhold it. Moreover, it is as little possible to do this as for a woman voluntarily to prevent the birth of the mature fruit of her womb.

Without aspiring to the most distant comparison of myself with you, one of the loftiest mental phenomena of the century, I may yet take for my example the smiling serenity with which you pursue your own way, indifferent to ingratitude, insult, and misunderstanding.

Pray remain, dear and honoured master, ever favourably disposed towards your gratefully devoted

MAX NORDAU.

size of the orbits, handle-shaped or sessile ears found in criminals, savages, and apes, insensitivity to pain, extremely acute sight, tattooing, excessive idleness, love of orgies, and the irresistible craving for evil for its own sake, the desire not only to extinguish life in the victim, but to mutilate the corpse, tear its flesh, and drink its blood."[14]

Lombroso was convinced that he had defined a criminal type, a separate race, or *Homo delinquens*, who was a "ghost from the past" or a "relic of a vanished race." He borrowed from Darwin's evolutionary views on the *Descent of Man* (1871) and believed his "criminal man" was a throwback (atavism, as they called it then) to this past line of descent from ape to man.[15] In 1876 he summed up his views in a book composed chiefly of his essays, *L'Homo Delinquente [Criminal Man]*. He classified 7,000 criminals into subtypes; some were "epileptoid" and showed many of the behavioral, physical, and moral traits of the epileptic. He classified others as "mattoid," attributing to them an imbalanced mentality that allowed them to function in society and often hide their pathology. Still other categories of hysterical, inebriate, and morally insane criminals revealed their own combinations of physical and behavioral traits.

GENIUS AS A PATHOLOGY

Besides the criminal type, Lombroso investigated the "man of Genius" and noted the similarities of works of art by insane patients and those who were capable of functioning in society.[16] He believed these were variants of a common epileptoid personality defect. Among his epileptoid geniuses were Caesar, St. Paul, Mohammed, Petrarch, Swift, Peter the Great, Richelieu, Napoleon, Flaubert, and Dostoyevsky, all of them victims of the "morbid rage" encountered by epileptics. Other geniuses, including Marlborough, Faraday, and Dickens, he assigned to a pathology characterized by "vertigo." Lombroso's approach was one of correlation biased by his selection of examples that fitted his types. In its day it was an immensely popular theory, and the criminal was often perceived to be physically defective in a recognizable way.

The work inspired many physicians on both sides of the Atlantic to seek examples of criminal types. In 1891, the *Journal of the American Medical Association* promoted Lombroso's views as two physicians, G. Frank Lydston and E.S. Talbot, presented pictures of prisoners as well as the deformed skulls and jaws of deceased prisoners. They were satisfied with their survey: "As far as our observations go, they tend to show that a degenerate type of skull is common among criminals, and that the assertion of Lombroso that the deviation of type, as far as the index is concerned, is toward brachycephaly, is correct."[17]

Many other physicians and criminologists in the last half of the 19th century disputed Lombroso's claims and found more than enough evidence to convince themselves that almost all criminals were the products of bad upbringing and social neglect (see Chapter 5).

NORDAU'S THEORY OF DEGENERACY AND CULTURE

Also influential toward the closing decades of the 19th century was Max Nordau (1849–1923), born Max Simon Sudfeld in Budapest. He started out as a physician, receiving his medical degree in 1873, but he switched to journalism and became the Viennese correspondent for his newspaper. Later, he moved to Paris, changed his name to Nordau (i.e., he translated his name from the abandoned and German "south field" to the newly adopted and French "north meadow") and established his reputation as a critic of popular culture. He became internationally known through his scathing criticism of contemporary society, *Conventional Lies of Our Civilization* (1884). In 1892 he published in German, *Entartung*, a two-volume analysis of cultural degeneracy which he associated with the physiological models of degeneration of Morel and Lombroso. The English translation, *Degeneration*, appeared in 1895. Nordau dedicated the book to Lombroso and acknowledged his debt to Morel as the first to introduce the idea of degeneracy. Nordau, like Lombroso, was Jewish, but whereas Lombroso kept his religious beliefs as private matters, Nordau felt obliged to make his religion a central theme, and much of his later career was devoted to the Zionist movement. This may have stemmed from his reaction to the growing anti-Semitism in the closing decades of the 19th century, especially the Dreyfus case, in which French defeat by Germany in the Franco-Prussian War was attributed to Jewish treason by those seeking a scapegoat for France's failure on the battlefield. He joined Theodor Herzl in 1892 and helped organize the international Zionist congresses, serving as vice president until 1911, when he broke with the movement. Nordau would not compromise on what he felt was needed for the political guarantees of sovereignty for a Jewish state in Palestine.

Nordau's social philosophy was classical, conservative, and idealistic. He believed in a "spiritual hygiene," consisting of altruism, an ardent work ethic, the dominance of reason over instinct, and a reverence for truth. Any artistic or literary departure from this norm he classified as pathological. He used the theme of biologically impaired genius, reflected by terms such as moral insanity (H. Maudsley), higher degenerates (proposed by V. Magnan, after Morel's superior degenerates), and Lombroso's mattoids (especially his writing mattoids or "graphomaniacs"). In degenerate literature, Nordau specified three forms, those stressing mysticism, egomania, or prurient realism. Among the "pseudogeniuses" whose writings or works of art were degenerate, he included Ibsen, Nietzsche, Wagner, Rodin, Verlaine, and Mallarme. He described them as "lunatics and madmen, ... the cripples and clowns of art and literature."[18] Among his contemporaries whose art he held as exemplary was Sully-Prudhomme.[19] He rejected impressionism in art, romanticism in music, and the psychological and realistic novels of Balzac, Flaubert, and Zola.

Nordau's interpretation of degeneracy included familial dwarfs, idiots, cretins, and other medically impaired individuals, as well as many of the weak-minded and criminal classes of society. Most of his attention was directed at the destroyers of culture and civilization; the intellectuals and artists who depicted vice, drug addiction, prostitution, and criminality in their novels or paintings, who distorted reality, or who chose for literary subjects the least of civilization's accomplishments. He railed against their publishers for selling out for money, and he condemned galleries and museums that exhibited the degenerate paintings and sculptings that corrupted public taste.

Nordau accepted the physical causes of degeneration cited by Morel, but he considered the vast increase in numbers of degenerates in his day to have arisen from the tempo of 19th century living, especially in the period of rapid industrial growth and urbanization after 1840. The railroad, the telegraph, a surfeit of newspapers, mass literacy, and mass consumption of manufactured goods, he believed, left much of humanity fatigued and exhausted. For Nordau, life was too frenzied to be appreciated, and the overstimulation of the senses was producing mass degeneracy. "All these new tendencies, realism or naturalism, 'decadentism,' neo-mysticism, and their subvarieties, are manifestations of degeneration and hysteria and identical with the mental stigmata which the observations of clinicists have unquestionably established as belonging to these."[20] According to critic Charles Dana, Nordau was less concerned about the long-range hereditary harm from his degenerates than he was about their immediate effect on his generation of susceptible youth. Like Morel, Nordau believed degeneracy worsened each generation and was self-eliminating.[21]

ZOLA'S THEORY OF FAMILIAL DEGENERACY

Emile Zola (1840–1902), although vilified by Nordau for writing about the seamier side of French society, strongly supported the theory of degenerate classes. Twenty of Zola's novels (from 1871 to 1893) followed the histories of two fictional families, the Rougons and the Macquarts.[22] Zola believed degeneracy found its outlet in the reinforced bad habits and

MACQUART AFTER HIS RETURN FROM THE ARMY.

Emile Zola's Rougon-Macquart series of 20 novels described heredity across four generations in 19th century France. Zola believed that personality, intelligence, and talent were inherited. (*Left*, Reprinted, with permission, from F. Brown [1995] *Zola: A Life*, Farrar, Straus, & Giroux, New York [Cliché Bibliothèque nationale de France, Paris]; *right*, reprinted from E. Zola [1888] *The Fortune of the Rougons: A Realistic Novel*, Vizetelly & Co., London.)

petty corruptions that characterized the failures of each social class. Degeneracy ran erratically and unpredictably in families and formed a class or fixed type that would eventually become extinct. This, in part, is an extension of Morel's theory of degeneracy, but Zola had read extensively about heredity as medical scholars and biologists of his day perceived it. Zola's theory of heredity is a more sophisticated amalgam of these ideas and recognizes many of the apparent inconsistencies that Mendelism later resolved. He recognized the skipping of generations, the disappearances and reappearances of characters, the amalgamation of traits, and the strange combination of traits that prevail in a careful study of a kindred. "What an immense fresco there is to be painted," he exulted in *Dr. Pascal*,

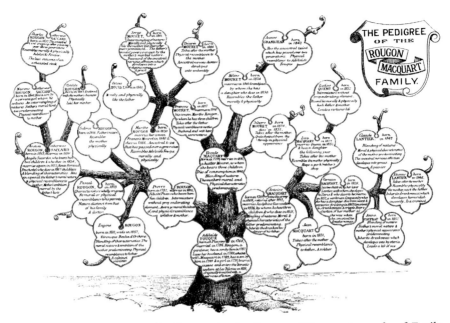

Pedigree represented as a family tree in the Rougon-Macquart novels of Emile Zola. (Reprinted from E. Zola [1888], *The Fortune of the Rougons: A Realistic Novel*, Vizetelly & Co., London.)

"what a stupendous human tragedy, what a comedy there is to be written with heredity, which is the very genesis of families, of societies, and of the world!"[23] Nordau believed, erroneously, that Zola's ideas of degeneracy were based on the work of Saint-Brieve, who described a degenerate family in Brittany, the Kerangal family, who between 1830 and 1890 produced seven murderers, nine prostitutes, and one each of painters, poets, architects, actresses, and musicians, as well as several blind individuals. [24]

Although Zola develops the character of a particular genius or madman, his human beings, according to a contemporary critic, Vernon Lee, writing in the *Contemporary Review* for 1893, "are but mediocre creatures, lacking all strength and newness." In *La Terre* (*The Earth*), a novel of peasant life, "the chief character is a born criminal, what Lombroso and his school would call a *moral idiot*; and, as such, his feelings and doings must be deducted from the frightful bill brought against the normal peasant." In *L'Assomoir* (*The Dramshop*), "a generation of gluttons, tipplers, and good fellows who amuse themselves, is preparing in densest moral darkness, a race of paupers, prostitutes, and criminals...This marvelous study of the gradual degradation of a family of respectable and well-to-do artisans is not really, as the title implies, the novel of Drink." Instead, Lee attributes to the inhabitants of this Parisian street, a "slow destruction, accomplished by a number of bad habits and small vices, which arise out of emptiness and idleness of life."[25] In a similar manner, *Pot Bouille* (*Piping Hot*) deals with the degeneracy of the bourgeoisie; *Nana* illustrates the effects of a charismatic slut who leads the upper class to degradation or crime through sex; and *L'Argent* (*Money*) follows the path of destruction of the wealthy who become obsessed with their material possessions. Perhaps the most pessimistic of Zola's novels is *La Bête Humaine* (*The Human Beast*), a symbol of the political degeneracy of France during the Second Empire. The novel concludes with a fight to their death of a train engineer and boiler stoker; the former, the lover of the mistress brutally murdered by the latter, a homicidal maniac afflicted with atavistic lust. They leave the train leaderless as its packed passenger cars, filled with drinking and carousing soldiers, hurtle into the night in anticipation of engaging the Germans in the Franco-Prussian War.

The American biologist and educator David Starr Jordan was not convinced that degeneracy was as sweeping and widespread as Nordau claimed. Nordau's degenerate literature he classified as just "unwholesome fashion."[26] He agreed, however, with the thesis of Morel and Nordau that degeneracy is a biological process arising from bad environments. Using his perception of Darwinian evolution, Jordan claimed degeneration "takes place whenever a relation of the struggle for existence permits life on a lower plane of activity or with less perfect adaptation to conditions..." to prevail. Applying this biological lesson to the human species, he concluded that "inactivity and dependence, protection in idleness, bring about deterioration and end in weakness, incapacity, and extinction."[27]

DEGENERACY AND SOCIAL CLASS

The idea of degeneracy led to the belief that three social classes—tramps, paupers, and criminals—were the products of generations of neglect and abuse and that two types of defective individuals—the feeble-minded and the insane—could appear as degenerate offshoots from any social class in society. In 1892, after a generation of suggestions for environmental solutions to the growing problem of unfit people, Henry Chapin, a physician writing in the *Popular Science Monthly*, raised the question his readers had also been framing, "What is society to do with its horde of defectives?[28] Chapin lamented that society has done little to check their production and he suggested, instead, that "we can cure by preventing." By now the census data had reinforced the opinions of clergymen, physicians, and sociologists that "it is a fact proved by statistics that a large percentage of criminals are defective either physically or mentally, and have had an unfavorable heredity and environment."[29] Chapin offered a medical model, "When such a class is formed, it should be permanently isolated from the rest of society...Such a permanent quarantine should be applied to all tramps, cranks, and generally worthless beings. Society must do this for protection, not punishment, to avoid their contamination; and above all to prevent the propagation of their kind. Advanced sociology will devote its principal energies to avoiding the production of the unfit, and then see to it that they do not survive beyond one generation."[30]

By the beginnings of the 20th century there was a common consensus that the unfit were the products of environments that brought about a hereditary degeneracy. Dr. G. Frank Lydston, in *Diseases of Society and Degeneracy* (1905), enumerated 20 different causes of degeneration, leading to criminals, prostitutes, the insane, paupers, and inebriates. The causes included heredity and habit, defective physique, neglect of children, acquired diseases, brain injuries, alcohol, herding of criminals resulting in vicious examples to the naive, defective moral training, lack of education, unjust dispensation of laws, marriage among criminals, menopause, sexual perversions, anarchy, poverty, idleness, gambling, high cost of living, the stress of urban living, and the immigration of the "criminal refuse of the old world."[31]

FOOTNOTES

[1] C. Keith Wilbur, *Revolutionary Medicine 1700–1800* (The Globe Pequot Press, Chester, Connecticut, 1980). Almost all medical practice in Europe and North America until the mid-19th century was based on the idea of body fluids (humors) that occasionally became corrupted. The detoxifying of these humors was accomplished by induced vomiting, by applications of enemas, and by cutting veins to

remove blood. No doubt some conditions were relieved by these measures (such as edema from congestive heart failure), and those successes may have become the basis for their continued use on other patients.

2 Brown's system is a variation of what became Quetelet's theory of the "homme moyen" (average man). The optimum was the mean, and departures from the mean were considered pathological. Francis Galton changed this perception, although, like Quetelet, he tried to quantify human behavior. In Galton's system only the departures in the degenerate or lower direction were pathological, and the optimum shifted to the extremes of physical talent, intellectual talent, and artistic talent. Those who followed Quetelet believed, for example, that ideal Greek sculpted bodies and faces were those that approached the mathematical mean for height, shape, and facial features of the "homme moyen." For the Queteletans mediocrity was excellence.

3 B.A. Morel, *Traité des Dégénérescences de l'éspèce Humaine* (Chez J.B. Bailliere, Paris, 1857). The work is cited in Max Nordau's *Degeneration* (D. Appleton & Co., New York, 1895), p. 34. Most of Morel's agents of degeneracy are considered valid today as teratogens or agents that adversely affect the embryo or fetus. There is a fetal alcohol syndrome, babies can be born with narcotic withdrawal symptoms, babies are of smaller birth weight if their mothers are smokers, ergot (the cereal mold toxic alkaloid produced by *Claviceps pupureum*) induces abortion, goitrous women can produce athyrotic infants (cretins), and syphilitic babies may have numerous anomalies, such as congenital deafness.

4 M. Nordau, *Degeneration* (D. Appleton & Co., New York, 1895), p.16.

5 Ibid. p. 16.

6 "Physiognomy," *Encyclopedia Britannica* 17: 886–887, p. 886. Bell (1774–1842) was a Scottish anatomist who pioneered studies of the nervous system. He was the first to distinguish sensory nerves from motor nerves. He attempted to relate specific nerves to their muscle functions.

7 Robert M. Young, "Franz Joseph Gall (1758–1828)" in *Dictionary of Scientific Biography* 5: 250–256. Gall is a generation older by birth than F.X. Bichat (1771–1802), the discoverer of tissues and one of the founders of the mechanistic approach to biology. Yet Bichat believed that passionate emotions had their origin in the thorax and abdomen. Gall claimed that all emotions, wherever they may be localized in the body, have their origin in the brain. See p. 253.

8 Esquirol pioneered the asylum movement for the study and treatment of the insane. He designed and established asylums at Rouen, Nantes, and Montpellier. He was the founder of the French psychiatric school that later attracted Sigmund Freud as a young physician to come to Paris. Esquirol also produced the first classification of the insane based on scientific or rational characteristics.

9 Proudhon was largely self-taught. He became a printer and taught himself Latin, Greek, and Hebrew, publishing a comparative grammar of these languages and his native French. He believed in the idea of justice, liberty, and equality as the basis for an ideal society. From these principles he concluded that "property is theft" because it denies people access to land that can be used productively. In Proudhon's ideal

society there would be less need for government if all work were to be equally reimbursed for the time put in and if private property were to be abolished. The ideal state, or anarchy, was the outcome of prolonged freedom. Marx strongly opposed Proudhon's views, which are closer to Libertarian thought than to socialism. Herbert Spencer used some of Proudhon's ideas in developing his ideal society in *Social Statics* (see Chapter 8).

[10] Louis Chevalier, *Laboring Classes and Dangerous Classes in Paris during the First Half of the Nineteenth Century*. Translated by Frank Jellink (Howard Fertig, New York, 1973), p. 269.

[11] Ibid. p. 314.

[12] Ibid. p. 359.

[13] Ibid. p. 386.

[14] Gina Lombroso-Ferrero (1872–1944), *Criminal Man, According to the Classification of Cesare Lombroso*. Reprint of the 1911 edition with introduction by Leonard D. Savitz (Patterson Smith Publishing Company, Philadelphia, 1972). Quoted on p. xxv of Savitz's introduction.

[15] The popularization of Darwin's scholarly works, especially his *Origin of Species* (1859), the *Descent of Man*, and his two-volume *Variation of Animals and Plants under Domestication* (1868), led to exaggerated theories of human inheritance. Darwin gives immense detail on dogs, cats, horses, pigs, sheep, and other animals as well as medical disorders and behavioral traits in these works, especially his 1868 effort to come to grips with the problem of heredity. Although atavism was recognized by Darwin as an inheritance resembling an ancestral form at least as far back as the grandparents and often more remote, it was not a central theme of Darwin's theory of heredity. The public was titillated by atavism, and R.L. Stevenson's *Dr. Jekyll and Mr. Hyde* reflected the fear and fascination the lay public had with "feral people" displaying savage ferocious behavior. Lombroso legitimized for his followers this popular view by projecting it onto the criminal class. The genetic basis for atavism was the reduction in the offspring of the number of homozygous selected traits present in the two parents. The highly heterozygous offspring from very long-selected strains resembled the wild-type or feral organism from which the domesticated breeds arose.

[16] Lombroso's perception of genius as a pathology was rejected by two very different movements that were developing contemporaneously. Galton's perception of genius as an ideal or optimal state of potential appealed to the talented and successful who did not think of themselves as being pathological. Galton even rationalized the peculiar quirks of his eminent peers and himself as of little consequence to their overall accomplishments and reputation (his autobiography *Memories of My Life* [Methuen & Co., London, 1908], describes many of them). Unrelated to Galton's celebration of hereditary genius was the development of the romantic movement, especially through Friedreich Nietzsche, with an emphasis on the romantic leader using his genius to capture the national or racial spirit of a people. The romantic movement competed with Galton's eugenics movement for the sympathies of the elite, but both denounced Lombroso's and Nordau's characterization of genius as a diseased state.

[17] G. Frank Lydston and E.S. Talbot, "Studies of criminals" in *Journal of the American Medical Association* 17(1891): 903–923, pp. 904–905.

[18] Nordau, *Degeneration*, pp. 31–32.

[19] Rene Sully-Prudhomme (1839–1907), a French poet, was the first recipient of the Nobel Prize in Literature, 1901. Some literary critics castigated the selection of Sully-Prudhomme, who has since sunk into obscurity as a world figure, and the failure of the Nobel literary committee to select Leo Tolstoy (1828–1910).

[20] Nordau, *Degeneration*, p. 43.

[21] Charles Dana, "Are we degenerating?" in *Forum* 19(1892): 458–465.

[22] Zola read extensively on heredity in 1868 at the Bibliotheque Nationale in Paris and he was impressed most by the work of Prosper Lucas (*Traite Philosophique et Physiologique dans les états de Santé du System Nerveux*, Paris, 1847–1850, 2 volumes). Zola used this model of heredity, based in part on Morel's theory of degeneracy, to sketch a family tree of descendants of Adelaide Fouque, an unstable personality who eventually went mad. Her first husband was a Rougon, and after his death she had a common-law husband, a Macquart. Both liaisons produced children who would play prominent roles in a series of novels relating the individual to French society. The first novel, *The Fortune of the Rougons*, follows the children of these two unions. The series concludes in the 20th novel, *Doctor Pascal*, whose leading character is a fourth-generation descendant who contemplates the history of his family and uses it as a study of the heredity of all humanity. He leaves his niece pregnant and a widow as she worries about the family's proclivity for producing alcoholics, prostitutes, thieves, murderers, and fools scattered among the more successful middle-class financiers, priests, politicians, and landowners. Zola's naturalistic (his term for literary realism) assessment is expressed by Dr. Pascal: "Heredity [is] life itself, which produces imbeciles, madmen, criminals, and great men. Certain cells collapse, others take their place, and a rascal or a raving lunatic appears instead of a genius or a mere honest man. And meantime mankind continues rolling onward, carrying all along with it." (*Dr. Pascal*, MacMillan, New York, 1898, pp.121–122).

[23] Zola, *Dr. Pascal*, p. 109.

[24] Nordau, *Degeneration*, p. 496.

[25] Vernon Lee, "The moral teaching of Zola" in *The Contemporary Review* 63(1893): 196–212, p. 203.

[26] David Starr Jordan, *Footnotes to Evolution* (D. Appleton & Co., New York, 1898), p. 291.

[27] Ibid. p. 279.

[28] Henry Dight Chapin, "The survival of the unfit" in *The Popular Science Monthly* 41(1892): 182–187.

[29] Ibid. p. 184.

[30] Ibid. p. 187.

[31] G. Frank Lydston, *Disease of Society and Degeneracy* (J.P. Lippincott, Philadelphia, 1905), p. 91.

5

Dangerous Classes and Social Degeneracy

Breaking the laws of God and society is as old as human consciousness. From the expulsion from Paradise of Adam and Eve, the murder of Abel by Cain, and the cheating, theft, and deception among Joseph and his brothers, the Old Testament portrays a humanity potentially corruptible and ever in need of reform. The varieties of criminals are legion and include prostitutes, pickpockets, murderers, burglars, gamblers, cheats, robbers, embezzlers, and con artists. These criminals of antiquity are individual transgressors whose disregard for moral or social law made them sinners in the eyes of their peers. The oldest view of crime is intensely personal, with the criminal a victim of his or her own greed, a weak will, corrupted values, or bad judgment. The fault is in moral reasoning or selfishness, and the criminal who is responsible for an illegal act must pay in some way, usually through punishment, to satisfy both the wronged and society itself.[1]

THE TRAMP AS FREE SPIRIT

Every age has encountered homeless people who travel from place to place. They may be romanticized as sturdy vagabonds, valiant beggars, or hobo kings, but more often they are looked upon as public nuisances, petty thieves, or social parasites. In the 15th century the English vagrant "was so objectionable then and later that the whipping-post, ear-slitting, and hanging were his legal portion, and a fine was the reward of the man who harbored or helped him." The term "tramp," which appeared in the periodical literature in the United States in the late 19th century, appeared as early as 1760 in English literature in the longer form, "tramper." The term "bum" was also in use in post-Civil War America. Characteristic of the criticisms of these vagrant classes[2] was the belief that they were lazy or indolent and that they were alcoholic. They were usually young men, unmarried and uninterested in civic virtues.

The ambivalence about the tramp in the 1870s was reflected in the periodical literature in England and the United States. An article on "Tramps and pedestrians" appeared in *Blackwood's Magazine* in 1877.[3] The tramp was accepted as "an essentially English type, as England is the country of pedestrianism, par excellence."

During the 1870s in England, a contrary view was also held. The *Contemporary Review* in 1870 featured an article by E.W. Hollond on "The vagrancy laws, and the treatment of the vagrant poor."[4] Hollond claimed that the amount of vagrancy was "not so slight an evil as might be at first supposed." He cited a Mr. Hardy, who estimated that "there were no fewer than from forty thousand to fifty thousand vagrants known to be tramping about the country from one end to the other." Historically the English people engaged in a "continual struggle that has been going on between the two opposite principles of the punitory and the compassionate treatment of vagrants." They were sent to jail or to "the eating-house." Elizabeth I had set up Houses of Correction for "rogues" and the poor, which were made into reformatories by James I "for the keeping, correcting, and setting to work of the said rogues, vagabonds, sturdy beggars, and other idle and disorderly persons."[5]

AN AMERICAN CAMPAIGN AGAINST THE TRAMP

In the United States a much more systematic campaign against vagrants was initiated by the editor and publisher of *Scribner's Monthly*, J.G. Holland (1819–1881), not to be confused with England's Hollond.

Holland's campaign against tramps began in 1876, when he described "the dead-beat nuisance."[6] Holland extended his ideas about tramps and dead-beats to society as a whole. He warned of "the pauper poison," which is the habit of getting something for nothing.[7] He criticized public hospitals for giving in too readily to the claims of the patients that they couldn't pay for their medication. He condemned the practice of tipping waiters in restaurants because this makes the waiter "paid twice" for the same service or, even worse, forces him into begging for tips by an employer who does not pay him. A third victim of his criticism was the clergy. He claimed that a seminary student on scholarship was corrupted into pauperism because he was receiving money he had not earned. Similarly, ministers who were supported by their congregations' weekly donations were practicing the same habit of dependency that he condemned in tramps and deadbeats. To prevent the "moral poison" of almsgiving, he asserted that one must give in work or goods the equivalent of what one receives; giving in to the "pauper poison" only leads to a loss of manhood because the "poison is in his soul."[8]

By 1880, Holland had come up with a solution. He blamed the educational system for imposing a liberal arts education on all the students in the cities.[9] It was an education that was unsuited to the world most of the poor would live in. They would work as laborers and in services where the liberal arts education was unrelated to their lives. Holland urged the cities to introduce Industrial Schools to teach practical skills, which, if not immediately suited for the job market, would be of value at home. He cited the benefits to the girls of Boston who were assigned to sewing classes.[10]

Holland's further influence ended with his death in 1881, but the battle continued. At the Annual Conference of Charities and Correction, held in Boston in 1881, Levi L. Barbour discussed the issue of "Vagrancy."[11] He repeated the charge that "a vagrant is the sturdy beggar of today and the criminal of yesterday and tomorrow." He claimed that "the suppression of vagrancy and street begging is probably the most important work undertaken by those engaged in organizing charity, for this is the tap-root of pauperism." This theme would echo throughout the closing years of the 19th century. In 1893, John J. McCook, discussing "Tramps" in the *Charities Review*, passionately argued against almsgiving. "The person who will give any beggar a coin just because it seems too hard to refuse him, ought on similar grounds to give razors and guns to madmen and children."[12] Originally a distinct class, with some claims to grudging admiration for their freedom to move on to more pleasant climate and new adventures, the tramp or vagabond had become synonymous with the unwanted and feared pauper and thief.

CRIMINALS AS A SOCIAL CLASS

In large cities like Paris and London, the criminal population and concern about crime varied with natural disasters and severe climate. When times were stable and the poor could afford the price of bread, as in the Paris of 1783, crime was essentially petty, rare, and picturesque, but after the restoration of the monarchy, it became "commonplace, anonymous, impersonal, and obscure."[13] Diseases such as smallpox and cholera would reduce the number of urban poor, and the immediate survivors would be described as savages. Their shift to crime was more organized, and Honoré de Balzac depicted them as a class, forming a "republic with its own law and its own manners and customs."[14] The criminal was primarily a swindler rather than a robber or killer, and his individuality was stressed during the early years of the 19th century. Crime was still an aberration, alien to the poor and the wealthy alike. Toward mid-century, the criminal was more often identified with a mass or anonymous class and less often described as an individual. Executions after 1832 were shifted from the traditional public view in the afternoon at a busy location in the city to a secret dispatch at dawn in a more obscure neighborhood, the abode of tanners and ragpickers.[15]

In Victor Hugo's era, the nature of crime was changing rapidly. Wandering hordes of children were possible because mass vaccination greatly reduced smallpox, and infant mortality was in decline, creating an illusion of increased fertility among the lower classes. The burgeoning population of Paris was supplemented by the immigration of laborers and peasants from the countryside looking for opportunities to work in the city. This changed the social structure of the city, the older established residents regarding the newcomers as less capable, less restrained, and potentially dangerous. Surveys of the poor were not sympathetic. They were regarded as a "degraded and corrupted people" whose misery was caused by their own inadequacies.[16]

THE SHIFTING PERCEPTION OF CRIMINAL BEHAVIOR

Each generation has its own perception of criminal behavior. What is rudeness to one age may be considered libel to another. The schoolmaster with a switch in the early 19th century would in late 20th-century American culture face not only dismissal, but also a costly lawsuit, if not criminal charges of assault, if he beat a student who didn't prepare an assigned lesson. Almost every crime is, under some condition or some culture, not a crime.[17] Killing in self-defense or carrying out capital punishment may not be murder to the same person who believes that ingesting estrogen to prevent implantation of a fertilized egg is an act of murder. Yet all cultures are concerned about criminal behavior and worry about its causes and consequences. The poor are not normally looked upon as dangerous or a criminal threat, but when food is scare or expensive, when jobs are not to be had, and when society fails to provide hope for a daily living, the poor become transformed into dangerous classes. They are seen as perpetual paupers, or vagrants and tramps, or beggars and thieves. The poor through history have been victims of failed crops, economic collapse, malnutrition, high childhood mortality, visitations of pestilential epidemics, exploitive employers, and inadequate housing. Those who relate the plight of the poor to their consequences see "crime as the expression of a sick society."[18]

CRIMINALS AS A SEPARATE NATION

The transformation of individual criminals into a criminal class was a perception accepted by the urban middle class during the first decades of the 19th century. As society shifted its perception of crime from individual aberration to the activity of a dangerous class, it changed its habits of managing crime. Historian Louis Chevalier, in his account of the dangerous classes of Paris in the first half of the 19th century, describes crime after 1832 as a by-product of the city, like its garbage.[19] In 1845, the attitudes had hardened and Eugene Buret, after surveying the misfortunes of the laboring classes in England and France, concluded that "thousands of human beings [were] reduced to a state of barbarism by vice and destitution."[20] These "degraded and corrupted people" were ignored, and their misery was regarded "only as something regrettably exceptional, the cause of which lies wholly with the poor themselves."[21]

In that same year, M.A. Fregier warned that "the poor and vicious classes have always been and will always be the most productive breeding ground of evildoers of all sorts; it is they whom we shall designate as the dangerous classes."[22] Chevalier attributes the rise in urban crime to a sudden rise in population. The city fails to assimilate the newcomers who cannot find employment or displace higher-paid workers with their desperate need to work at any wages. At the same time, the older residents view the new arrivals with suspicion and expect this class to carry out criminal acts. In the third component of Chevalier's transformation of the poor into a criminal class, the city's politics, religion, and values become rigidly defined, and departures are then perceived as criminal acts.[23]

The causes of these dangerous classes were still environmental. Air, water, and food pollution were identified with poverty in the 1780s. The immigration of beggars, ragpickers, invalids, and criminals took place during epidemics, but these were assimilated in the early 1800s as manufacture increased and urban improvements kept up with the needs of trade. If housing was substandard and poorly lit or poorly ventilated, the inhabitants were thought to be like wilting or etiolated plants deprived of sunlight.[24] By 1838 the crowding of the slums exceeded the growth of new industry, and the poor lived in sluggish air, the streets muddy, damp, polluted, and crowded, making the inhabitants "like reptiles in a marsh."[25]

THE ENGLISH PERCEPTION OF DANGEROUS CLASSES

In England, that same perception of the poor as a dangerous class grew from the concerns of the new middle class. James Greenwood in the 1870s identified "the seven curses of London," singling out paupers, professional thieves, professional beggars, fallen women, inebriates, gamblers, and mismanaged charity as the problems in want of effective solution.[26] The pauper children swarming in London, about 100,000 in his estimate, were the offspring of "felons, cripples, and idiots, or orphans, bastards, and deserted children." Many were forced to feed on garbage and few received the protection of the law that was their due. Instead, Greenwood lamented, they were "in fair training for the treadmill and the oakum shed, and finally for Portland and the convict's mark."[27] He deplored their exploitation as cheap labor by greedy businessmen.

London's criminal class in the early 1870s was estimated at 20,000, many of them having started life as "gutter children." Greenwood cites Lord Romilly's observation: "There is a great disposition on the part of children to follow the vocation of their father, and in the case of the children of thieves, there is no alternative. They become thieves because they are educated in their way, and have no other trade to apply themselves to."[28]

Lord Romilly's draconian solution was to have convicted felons forfeit their children to the state for adoption. The emergence of a criminal class, Greenwood believed, followed the cessation, in 1853, of the deportation of criminals to Australia.[29] The 22,000 thieves were supplemented by 33,000 vagrants and tramps, 3,000 receivers of stolen goods, 27,000 prostitutes, and 29,000 "suspect persons."[30]

CRIMINALS AS VICTIMS OF SOCIETY

In the United States, as in France and England, there were those who believed criminals to be victims of their circumstances and those who interpreted criminal behavior as a volitional moral flaw meriting no sympathy. In his address to the National Prison Association at Baltimore in 1873, Horatio Seymour clearly saw the problem as society's. "We are apt to look upon the inmates of prisons as exceptional men, unlike the mass of our people. We feel that they are thorns in the side of the body politic which should be drawn out and put where they will do no more harm. We regard them as men who run counter to the currents of society, thus making disorder and mischief. These are errors."[31] Instead, Seymour claimed that "prisoners are men like ourselves, and if we would learn the dangers which lurk in our pathways we must learn how they stumbled and fell."[32] For Seymour, prisons were inappropriate because they assumed the criminal's personal failing should merit punishment. Seymour wanted the prisons turned into "moral hospitals" where criminals would be treated and not punished. "Crimes always take the hues and aspect of the country in which they are committed. They show not only guilty men but a guilty people."[33] Seymour expressed an opinion not often heard among the educated professionals of his day, "Many of the transactions of our capitalists are more hurtful to the welfare of our people than the acts of thieves and robbers."[34] Although he offered no specific training program for the moral defects of those convicted of crimes, Seymour expressed a pious hope, "I have never yet found a man so untamable that there was not something of good upon which to build a hope. I never yet found a man so good that he need not fear a fall."[35]

Professional thieves, such as burglars and bank robbers, emerged in the post-Civil War era. They generally shared the same values as their victims and took some pride in their skill at planning and carrying out their robberies. Some of the more notorious thieves wrote autobiographies, blaming their choice of a criminal life on some false accusation that branded them thereafter as criminals. Others attributed their criminal life-style to their dependence on alcohol or drugs, and a few rationalized their criminal behavior as a response to an unjust world.[36]

Although almost no criminal claimed to have always felt destined to live a life of crime, that view developed among interpreters of crime after the mid-19th century. The older theories held that crime was a moral error for which the criminal was responsible and accountable to society or that criminal behavior was a defect of society which failed its neglected citizens. Many adherents of these two views (what would later be classified as conservative or liberal views on crime) found more attractive a new theory of criminality that assumed the behavior was innate. Although such persons could not be blamed for their criminal acts, they could be identified and separated from the rest of society so they would not commit any crimes.

The biological theory of crime developed in the latter half of the 19th century in response to scientific studies of animal behavior and the attempts by physicians to replace religious ideas of spiritual possession with medical classifications of behavioral pathology. Although masturbation was the first of these behaviors to become medicalized and removed from a purely moral realm, theories of personality, disordered behavior, and criminality soon emerged among physicians hoping to turn social problems into medical problems that would yield to scientific study.

THE CRIMINAL TYPE AS INHERITED PATHOLOGY

Henry M. Boies was persuaded by the biological theories of criminality and pauperism of his day. He was a member of the Board of Charities of the State of Pennsylvania, the National Prison Association, and the

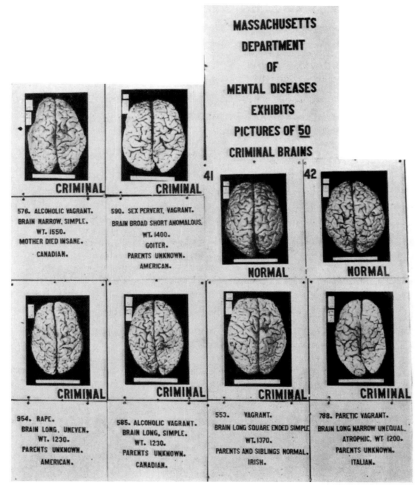

Medical support for criminality as a consequence of brain patterns. Exhibited at the Second International Congress of Eugenics, New York City, 1921. Shown is part of an exhibit of the Massachusetts Department of Mental Diseases prepared by Dr. Myrtelle M. Canavan. (Reprinted from H. Laughlin[1923] *Second International Exhibit of Eugenics*, courtesy of CSHL Archives.)

Committee on Lunacy in his home state, where he resided in Scranton. In his *Prisoners and Paupers* (1893),[37] Boies presented the evidence for an alarming increase in the criminal population of the United States. Although convictions actually decreased from 15,033 in 1868 to 9,348 in

1889, the frequency of reported crimes increased from 1 in 3,500 in 1850 to 1 in 786 in 1890. He attributed most of the increase to immigration, alcoholism, and laws that unfairly discriminated against blacks. He acknowledged that one-third of the convict class was "colored" in contrast to their comprising one-eighth of the population. He blamed their high crime incidence on their ignorance, the prejudice of white officials, their lack of voting rights, and the profitable convict labor system where the convicts were "leased out to the highest bidder."[38]

There was a real criminal class, however, that did not find Boies's sympathy. He cited his colleagues in Europe, Herr Sichart, the Director of Prisons in Wurtemberg; Dr. Vergilio in Italy; and Dr. H. Maudsley in England; as well as his colleague at the New York State Reformatory, Dr. H.D. Wey; who all agreed that there was a recognizable criminal type. "It would seem that a composite photograph of a hundred or so of them might produce the typical criminal. ...These are they who have inherited criminality from the parents, who are the products of generations of vice and crime, or who have slid down the plane of transgression and excess to the very bottom of degradation. ...They are human deformities and monstrosities, ill-shapen, weak and sickly, with irregular features. They bear a sinister, ignoble, and furtive expression."[39] Most alarming of all, they have a "depraved if not utter absence of moral sense or conscience."[40] Boies's animalistic image of the criminal class made it easy for him to favor a harsh solution: "As soon as an individual can be identified as an hereditary or chronic criminal, society should confine him or her in a penitentiary at self-supporting labor for life. ...No pardon, no hope of liberty, should be possible except in a clear and positive case of mistake in the character, or where, after indubitable reformation, the convict should be made incapable of reproduction."[41] He attributed their hereditary criminality to "natural incapacity from deterioration."

To those who indicted society and not the criminal, he claimed, "most of those characteristics which have hitherto been treated as causes, such as ignorance, intemperance, poverty, disease, and defects, are symptoms indicating a social state or condition of crime or pauperism, rather than the causes of them."[42] He claimed that "there is never found in the criminal or pauper class, except by accident, a normal, well-developed healthy adult."[43]

A similar view was expressed by Nathan Oppenheim in his article "The stamping out of crime" (1896), which appeared in the *Popular Science*

Monthly.[44] Oppenheim was pleased that after centuries of misconduct deemed part of human nature, slavery was abolished. He also felt that the fight against alcoholism was making headway, and this left criminality as the major social issue for science and the professional intellect to address. Oppenheim summed his findings: "This much, at least, we have learned: that the criminal forms a class by himself, no matter whether he is born so or grows into vice; that not only in his acts but likewise in mind and body does he vary from the healthy individual."[45] He cited Dr. Ogle, a British authority on criminology. The literacy rate in England had increased to a degree where 90% were literate, and yet crime had actually increased. Want of education, he claimed, could not be a cause of crime. Similarly, he claimed real wages in England had increased without causing a diminution of crime; instead he argued, crime actually rises as wages go up. Like Boies, he urged "right methods of development based upon normal marriages and normal breeding."[46]

Conflicting Views of Social Workers and Social Reformers

Although the biological view of criminality was embraced by many physicians and social commentators, the sociologists and charity workers, in general, were not prepared to abandon the criminal. Adelle Wright, writing in the *Arena* in 1902, acknowledged that "we should treat crime as we would disease."[47] Her approach was curative and reflected the rest and nursing that went with the process of repair during recovery from illness. She advocated a "Christian socialism" that would include reformatories for youthful offenders in which "sympathy and encouragement, self-support-ing work, reading and study" would be stressed.[48] "The most important steps ... necessary for the extermination of the criminal classes are as fol-lows: first—the establishment of homes for convicts, second—the educa-tion of the young, third—the providing of proper dwellings for the poor, fourth—the prohibition of the sale of intoxicating liquors, fifth—the establishment of the curfew, or its equivalent."[49]

Lombroso's theory of criminal anthropology appeared in 1876 and sup-plied the biological arguments for those who believed in a fixed criminal class (see Chapter 4). Atavisms were seen as primitive ancestral types, fre-quently reflecting a more savage than cultivated state, lurking in the hereditary fabric of persons whose outward appearance provided no evidence of such a fearful past. Degeneration, epilepsy, and "moral imbecility" were also associated with the criminal pathology, a view popularized by novelists from Dostoyevsky to Zola. Smerdyakav, the epileptic half-brother in the *Brothers Karamazov*, reflected the popular image of the innate criminal.[50]

The clash between supporters of the free-will or criminal by moral choice and the biological or born-criminal schools of thought took place in Europe during the mid-1880s.[51] The sociological model of the environmental origins of criminal behavior developed about the time of the debate. The oldest of these models, the free-will model, was the traditional religious view and had both popular and legal sympathy, law having been built on the assumption of right and wrong behavior. The biological model got its boost from Darwinism, although Darwin himself cautiously kept out of the multitude of attempts to use his ideas for interpreting human society. The French, loyal to the theory of the inheritance of acquired characteristics promulgated by their own Jean Baptiste Lamarck, favored the environmentalist school of thought. To Lacassagne, "Societies have the criminals they deserve."[52] To the followers of Lombroso, the criminal problem was solved through emigration, perpetual imprisonment, and capital punishment to protect the present and to prevent the genetic spread of crime.[53] Unlike their American counterparts, who often felt sympathy for the criminals as victims of society's neglect, the French environmentalists were also free-will advocates and held the criminal responsible for individual acts but believed in the potential to reform criminals and to raise their children so that they would exercise better moral choice than their parents. By 1889, criminologists and sociologists meeting in Paris had routed Lombroso's anthropological school of criminology. Repeated studies by English and French criminologists of photographs of prisoners of varying degrees of danger to society showed no criminal type or characteristic stigmata.

FOOTNOTES

[1] This is still widely regarded as correct. It is the basis for criminal justice and assumes people have the capacity to distinguish right from wrong and must be held accountable for their wrongs. The debate is at the scientific level. Psychologists usually reject free will and can sometimes demonstrate effectively the components that lead

to behavioral actions. Even those who accept the validity of behavior being determined by experience will not use that as a basis for exonerating the wrongdoer from
accountability or punishment.

[2] John J. McCook, "Tramps," *The Charities Review* 3(1893): 57–69. Within the culture
of vagrants, a hierarchy of skills was recognized. There were saltigrades or train
jumpers, pike bums, shovel bums, mission bums, bay cats, and, most respected of
all, the hobo or railroad tramp who rode the rails from city to city.

[3] Anonymous, "Tramps and Pedestrians," *Blackwood's Magazine* 122(1877): 325–345.
The tramps of the first half of the 20th century in the United States were, like the
ones described in this English article, treated kindly in popular culture. Tramps
were perceived as rascals with a generous heart and quick wit in the comic strips.
During the Depression of the 1930s, the tramp was portrayed with a broomstick
carrying a cloth bundle of personal goods at the distal end, and at his footsteps a
faithful dog frequently followed.

The anonymous author expressed his own repressed desires for the freedom of
the vagabond. "We do not say that the professional tramp is a creditable member of
society; but we must own to a certain sympathy with him, not altogether untem-
pered by envy." The author characterizes the tramp as "a practical philosopher," and
admires him for having "the faculty of throwing off worries as a Newfoundland dog
shakes himself dry after a header into broken water."

Among the vices cited by the author as habits of the tramp were stealing clothes
off the line and robbing chickens from their coops, "but it by no means follows that
the vagrant need be a scoundrel, although he almost necessarily has a dash of the
scamp in him" (p. 326). There is a social value in the odd jobs that the tramp does,
such as raking leaves, chopping wood, and other seasonal tasks that many would
welcome, including *Blackwood's* author, who summed him up as a cheery, friendly,
and "penniless bachelor with a light pair of breeches" (p. 329).

[4] E.W. Hollond, "The vagrancy laws and the treatment of the vagrant poor," *The Contemporary Review* 13(1870): 161–176, p. 162.

[5] Ibid. p. 170.

[6] J.G. Holland, "The dead-beat nuisance," *Scribner's Monthly* 12(1876): 592–593. Holland's editorials are unsigned, but at his death his obituarist noted that Holland
wrote these "Topics of the Times" editorials.

Holland's views are difficult to classify. He was sympathetic to the freed black
slaves and wanted them to have an opportunity to establish themselves in American
society ("The shadow of the negro" *Scribner's Monthly* 20[1880]: 304–305). He dis-
liked gratuitous habits like tipping in restaurants and any form of private charity
that did not involve some direct return for the money spent. His magazine, like the
Atlantic Monthly, supplied the middle class with entertainment and informative
articles.

Holland's views echoed those of the "new charity movement" which arose in
England during the early 19th century as a reaction against the dole provided by
taxpayers for the poor. The new charity movement approved of personal involvement in helping the destitute, especially by funding jobs for the able-bodied. It
rejected almsgiving as corruptive to self-worth and blamed the extent of poverty on

the long persisting belief that almsgiving was a moral act. See Chapter 8 for a more extended discussion.

[7] J.G. Holland, "The pauper poison," *Scribner's Monthly* 14(1877): 399–400.

[8] Ibid. p. 400.

[9] J.G. Holland, "Industrial education again," *Scribner's Monthly* 20(1880): 464–465.

[10] Holland's views should be contrasted with those of his predecessor, Richard Dugdale (Chapter 10). Dugdale sought the development of a liberal arts culture and rejected vocational training because he believed paupers needed the aesthetic, intellectual, and moral enrichment that the liberal arts provided. Dugdale's views are based on his environmentalist position that pauperism is an acquired trait. Holland's views are based on an intrinsic difference between the middle-class persons like himself and paupers, whom he perceives as a degenerate class.

[11] Levi L. Barbour, "Vagrancy," *Proceedings of the Eighth Annual Conference of Charities and Corrections Held at Boston, July 25–30, 1881*, pp. 131–138.

[12] John J. McCook "Tramps," p. 69.

[13] Louis Chevalier, *Laboring Classes and Dangerous Classes in Paris During the First Half of the Nineteenth Century*. Translated by Frank Jellink. (Howard Fertig, New York, 1973), p. 61.

[14] Ibid. p. 61.

[15] Ibid. p. 83.

[16] Ibid. p. 139.

[17] This is a consequence of the ease with which we organize what I call "constellations of values" that have numerous inconsistencies and an intrinsic arbitrariness, but which hold together like the patterns uniting stars in a constellation. One could argue that we are forced to create such constellations because any single principle of consistency (such as not killing any human for any reason) is difficult or impossible for a society to abide by.

[18] Chevalier, *Laboring Classes and Dangerous Classes*, p. 48.

[19] Ibid. p. 87.

[20] Ibid. p. 139.

[21] Ibid.

[22] Ibid. p. 141.

[23] Ibid. p. 258.

[24] Ibid. p. 151.

[25] Ibid. p. 152.

[26] James Greenwood, *The Seven Curses of London* (Stanley Rivers & Co., London, 1870).

[27] Ibid. p. 2.

[28] Ibid. p. 173.

[29] Ibid. p. 186.

[30] Ibid. p. 189.

[31] Horatio Seymour, "On the causes of crime," *Popular Science Monthly* 2(1873): 589–596, p. 590.

[32] Ibid. p. 590.

[33] Ibid. p. 591.

[34] Ibid. p. 592.

[35] Ibid. p. 593.

[36] Larry K. Hartsfield, *The American Response to Professional Crime: 1870–1917* (Greenwood Press, Westport, Connecticut, 1985), p. 151.

[37] Henry M. Boies, *Prisoners and Paupers* (G.P. Putnam's Sons, New York, 1893).

[38] Ibid. p. 71.

[39] Ibid. p. 172.

[40] Ibid.

[41] Ibid. p. 178.

[42] Ibid. p. 263.

[43] Ibid. p. 266.

[44] Nathan Oppenheim, "The stamping out of crime," *Popular Science Monthly* 48(1896): 527–533.

[45] Ibid. p. 528.

[46] Ibid. p. 532.

[47] Adele Williams Wright, "The extermination of the criminal classes," *The Arena* 28(1902): 274 – 280. p. 276.

[48] Ibid. p. 276.

[49] Ibid. p. 279.

[50] Although epilepsy is of physiological origin, usually from brain injury, the epileptic was often singled out as a dangerous person. This may have been because the seizures fit the stereotype of a person possessed by the devil. Prayer and religious conversion would not have helped an epileptic needing pharmacological regulation of this disorder. Dilantin and other prescription drugs have greatly relieved the suffering and have diminished the public witnessing of epileptics undergoing seizures. Few seizure disorders are inherited, but epilepsy was frequently thought to be inherited in the late 19th and early 20th centuries.

[51] Robert A. Nye, "Heredity or milieu: The foundations of modern European criminological theory," *Isis* 67(1976): 335–355.

[52] Ibid. p. 339.

[53] Ibid. p. 340.

6

Poor Laws and the Descent to Degeneracy

A LMSGIVING AND CHARITY ARE OLDER than the generosity of Job in an agrarian or nomadic tradition, and they became a major focus of Christian teaching for the urban multitudes in the Roman empire.[1] Giving alms to the poor is a religious tradition in both Christian and non-Christian cultures. The assumption of almsgiving was that some needy individuals would survive through the generosity of acts of charity and eventually find themselves and their families back among productive citizens. The other assumption was that other people, such as the blind, the crippled, and the elderly were too frail or impaired to ever return to productive self-sustaining lives, but their support was a test of the spirit, like that described in the parable of the good Samaritan in St. Luke 10: 25–37.[2] The obligation to perform Christian charity was subsequently reinforced, and children and adults alike were exhorted with aphorisms celebrating behavior in which "it is better to give than to receive."

Almshouses or refuges for those whose needs could not be met by the family developed in the 3rd and 4th centuries. There were Roman gerontochia for the aged, ptochia for the poor, and nosochomia for the sick.[3] In Anglo-Saxon times, the Church had exclusive control and administration of poor relief. Almsgiving was considered a religious duty for all, from the king to his least affluent subject. This tradition was temporarily set aside following the years of the Black Death in the 14th century, when almost one-third of Europeans died from the disease. The shortage of labor in Great Britain led to local wage controls, the banning of almsgiving to the able-bodied poor (many were professional beggars), and less successful attempts to bar laborers from moving to districts where wages were higher. The travel ban of 1388 was circumvented by many laborers who joined religious pilgrimages (as in Chaucer's *Canterbury Tales*) in the hopes of finding towns desperate for laborers.[4]

The Church administered the almshouses or "houses of pity" in medieval times, but after the Protestant reformation, both the philosophy of caring for the poor and the administration of the support for them were transformed. In Germany, Martin Luther urged the abolition of begging and shifted poor relief to the parishes or local communities. The breaking up of baronial control over local inhabitants and the development of national governments with a single king led to the dismissal of private soldiers once employed by the barons. They swelled the ranks of the unemployed and were depicted by contemporaries as a "rowsey ragged rabblement of rakehelles."[5]

Punishment of beggars was sometimes severe.[6] First offenses usually merited a whipping; a second offense could result in the loss of an ear; and a third offense might lead to several years of slavery or even death. Branding of beggars was not uncommon, either on the shoulder or the cheek.[7] An able-bodied man who refused work was subject, in the law of 1572, to a whipping and then "burned through the gristle of the right ear with a hot iron of the compass of an inch about."[8] Failure to support the poor, however, was not as harshly treated. In 1551 collectors were to "gently ask every man and woman, that they of their charity will give weekly to the relief of the poor."[9] Persistent refusal could lead to the summoning of the reluctant donor to court with a tax set at the magistrate's pleasure.

Henry VIII's break with the Church over the divorce of his wives led to his takeover of the Church's property. He closed down the monasteries and disrupted the tradition of the Church looking after the poor. The city of London, in 1536–1547, had to demand poor rates (taxes) to care for its unemployed and unemployable.[10] Accompanying the disruptions caused by Henry VIII's troubles with the Church were the changes in England's economy. Where once the peasants found security in working on the estates of their landowners, they now found themselves dispossessed from their farms. The wool industry proved more profitable than food crops, and sheep grazing demanded much of the land that supported the peasantry. By the end of the 16th century, rebellions broke out over food shortages, the high price of grain, and the dislocation of the peasants from their land. As the 17th century commenced, the economic burden proved too great for private charity to bear.

THE ELIZABETHAN POOR LAWS

In 1601, the Poor Relief Act was passed in Elizabethan England, and the administration of poor relief was assigned to local appointed "overseers" who set up poorhouses, raised local poor rates, apprenticed children to skilled workers, provided loans, sent able-bodied workers to other parishes needing laborers, and made sure that the deserving poor were not abandoned. The system remained in force for more than 200 years and helped promote a social harmony among the various classes of British society.[11] The local commitment was extensive, requiring the "taxation of every inhabitant, parson, vicar, and other, and of every occupier of lands, houses, tithes, mines, etc., such sums of money as they shall require for providing a sufficient stock of flax, hemp, wool, and other ware or stuff to set the poor on work, and also competent sums for the relief of the lame, blind, old, and impotent persons."[12]

As a consequence of the act of 1601, "poor houses" were secularized. The poorhouse was a building in which "the young and the old, the sick, the infirm, the blind and the mentally ill, the homeless and the destitute were all given asylum."[13] Until 1834, the poorhouse was the social structure for caring for the needy and those who could not participate effectively in the community. In the mid-18th century, considerable criticism of the Elizabethan poor laws was raised. The rising middle class in Great Britain objected to the inadequate administration of the poorhouses. Workhouses for the able-bodied were notorious for their low wages and exploitation of the unemployed. Overseers were uncomfortable raising taxes from local landowners who felt neither a religious nor a personal duty to those they did not know or who did not live on their land. In 1767 George III tried to protect the poor from the spreading parsimony of the overseers, but the clamor to reduce both the taxes and the obligations to aid the poor continued to grow.[14]

Criticism accompanied these Poor Laws from the start. Some argued that they prevented public disturbance by preventing riots for food and reducing the more likely crime that would accompany the roving habits of unsheltered paupers. In Victorian England they were openly acknowledged by W.E. Foster, Minister of Education, as a "safeguard against revolution."[15] Others argued that these laws "fell into the mistake, even as far back as the reign of Elizabeth, of giving the English working classes the feeling that they had 'a right to relief' on the part of the governing classes; or, in other words, that 'charity was a fund' on which they 'could confidently depend.'"[16]

A major overhaul of the Poor Laws was considered in the first third of the 19th century. The Elizabethan Poor Laws broke down particularly during the reign of George III. In 1767 Parliament increased the proportion of outdoor relief, a dole given to families who lived in their own homes, or to laborers living in depressed areas where the wages were too low to support a family. The intent of the act of 1767 was to protect the poor from the parsimony of the overseers, who often denied the needy the relief they required. Since travel from one parish to another by unemployed laborers was prohibited by the Elizabethan Poor Laws, the overseer's decision not to support a family could be devastating. Critics of the law of 1767 claimed it created a dependent class of paupers who lost incentive to work and who looked to outdoor relief as a substitute for hard work.[17]

The plight of the poor and the fate awaiting them in the poorhouse in the late 18th century was depicted by poet George Crabbe (1754–1832) in *The Village* (1783):

> There children dwell who know no parents' care,
> Parents, who know no children's love, dwell there!
> Heart-broken matrons on their joyless bed,
> Forsaken wives, and mothers never wed.
> Dejected widows with unheeded tears,
> And crippled age with more than childhood fears.
> The lame, the blind, and, far the happiest they!
> The moping idiot, and the madman gay.
> Here too the sick their final doom receive,
> Here brought, amid the scenes of grief, to grieve.[18]

Although generosity for the needy was a civic as well as a religious virtue, many able-bodied, working families resented the idea of giving alms to those who used it to buy alcoholic beverages or whose livelihood was that of being a professional beggar. Since the needy do not always present themselves in a way that those who are worthy beneficiaries of charity can be distinguished from those who are abusers of public and private philanthropy, the tradition through the ages has been to assume that abusers are few and that it would be unfair to punish the many for the selfishness or criminal exploitation of the few. Agitation for reform came from many sources, including the clergy. A Dr. Thomas Chalmers in Glasgow studied the history of charity given to the poor during the years 1790–1820.[19] He found that while the paupers were supported by the Poor Rates (tax) distribution of public funds amounting to 1400 pounds per 10,000 poor, the number of poor in Glasgow during this same time interval increased, and "crowds poured along the highways, and bands of idlers, with sullen and scowling visages, filled the streets."[20] Chalmers disapproved of a government that was "taxing the industrious to support the indolent." Chalmers devised a plan that would stem the unbroken cycle of subsidized paupers perpetuating new generations of subsidized paupers. He asked the government not to give his parish further aid. Instead he divided the parish into 24 districts. He appointed 24 deacons, mostly merchants and property owners active in the church, to be in charge of these districts. The poor in each district came individually to the deacon for an interview. If they were needy they were helped with employment or private funds donated by the church. Drunkards were excluded from charity. The poor recognized that this was true charity, he claimed, because it was not a handout from the government. The encouragement and personal interest of the deacons gave the poor a sense that the community and the church cared about them, and this motivated many of the able-bodied poor to work rather than to abandon hope that no one would hire them. What would have cost the government 1400 pounds was accomplished, Chalmers was happy to report, by a private expenditure of only 280 pounds per annum.

Although the Poor Law was enacted in Elizabeth's reign and served erratically ever since, there were episodic attempts to modify it. In 1723 only the destitute were supposed to be supported; but enforcement and

effective means for determining who was able-bodied and who was destitute were lacking. In 1782, this means test was repealed and the poor were
allowed to work in their own houses rather than in parish-supported
almshouses. By 1795 crop failures and inflation combined to make the
price of bread beyond the means of the poor. A mob of unemployed
stopped the King's carriage and cried out "Bread, bread!" In response, Sir
William Pitt amended the Poor Law, permitting the unemployed poor to
leave their parishes in search of employment.[21] Pitt hoped to educate the
poor in new skills and to restore their loyalty and patience. "Let us make
relief a matter of right and honour, instead of a ground for opprobrium
and contempt. This will make a large family a blessing and not a curse; and
this will draw a proper line of distinction between those who are to provide for themselves by their labour, and those who, after enriching their
country with a number of children, have a claim upon its assistance for
their support."[22]

The Reform of the Poor Laws

A commission to investigate the effectiveness of the Poor Laws was
established by Parliament in 1832, and 2 years later the commission presented its findings. A typical poorhouse maintained about 60–80 paupers,
including 12 children, 30 able-bodied adults, 30 aged adults, and a smat-
tering of prostitutes, mothers of bastards, inmates from jail, poachers,
vagrants, beggars, the blind, one or two idiots, and several lunatics. It took
another 20 years before the poorhouses were broken up and the sick were
separated in hospitals independently of the poor.[23]

The Commission reported that "the poor man who once tried to earn
his money, and was thankful for it, is now converted into an insolent, dis-
contented, surly, thoughtless pauper, who talks of rights and income, and
will soon fight for these supposed rights and income, unless some step is
taken to arrest his progress."[24] Despite the unsympathetic attitude
expressed by the Commission, the Poor Laws Act of 1834 did not abolish
the practice of outdoor relief.

The new poor law reforms of 1834 reflected the interests of the new
middle class. Critics of the poor laws pointed out that the costs of maintaining the poor had risen from 2 million pounds in 1785 to 8 million

pounds in 1817. The middle class insisted on cutbacks in such expenditures, much of which they felt was unnecessary because it perpetuated pauperism. The criticisms of poor relief led to the Poor Law Amendment Act of 1834. A central authority was established to oversee the local management of the poorhouses; vagrants who were able-bodied were required to support themselves by working on chores that reduced the costs of the poorhouses; fathers were to be made responsible for their illegitimate children; and laborers who were unemployed were permitted to travel to find work in other communities. Workhouses were henceforth to be deliberately designed to discourage the poor from working there. Parishes were grouped into districts for a more centralized administration with a more uniform wage policy. Poverty was effectively made a crime that merited a stigma; and the poor were believed to have failed not through acts of providence but through their own ineptitude.[25]

Popular interest in the Commission's findings was fueled by the publication of Charles Dickens's *Oliver Twist*, which first appeared in serial form in 1837 in *Bentley's Miscellany*.[26] The orphan, Oliver, suggests the prevailing view that an innate goodness cannot be corrupted by a perverse environment. Oliver's evil half-brother, Monks, hopes to corrupt Oliver so that he will be caught and hanged as a thief. He arranges for Oliver to be transferred from the orphan asylum and delivered to the custody of Fagin, who will then train Oliver in his notorious "academy of crime." Fagin is depicted as debauched and cowardly; Sikes, a graduate of Fagin's school, is the hardened criminal, sullen and violent. Oliver's fellow students in crime, Dawkins (the Artful Dodger) and Master Charley Bates, in their comic roles take the edge off the depraved villainous image of the criminal. Of particular interest is the idea of criminals as a community or corrupted class. Fagin, Monks, and Sikes are varieties of hard-core, pathological criminals. Dawkins and Bates are those vulnerable youths whose propensity for good or evil is malleable, and thus society has hopes for the potential pupils of the academies of crime. Oliver satisfies the wishes of humanity that there is a grace, perhaps in the form of heredity, that spares many of us from the viciousness of bad environments.[27]

By 1837 the poor bore the brunt of the new law. Relief to the poor was down by 36%. Children were farmed out as chimney sweeps; indigent families were broken up if the father was unemployed; the dole was abolished; and little distinction was made between the helpless and the "willful

OLIVER AMAZED AT THE DODGER'S MODE OF "GOING TO WORK"

Illustration of Boz in Charles Dickens's "Oliver Twist." Here the Artful Dodger instructs Oliver how to pick a pocket. (Reprinted from *The Works of Charles Dickens,* vol. 5, *Oliver Twist,* Books, Inc., New York.)

pauper." Dickens's biographer, Edgar Johnson, describes the rage Dickens felt which led him to embark "on a scathing denunciation of the new Poor Law and...a lurid and somber portrayal of London's criminal slums." It was Dickens's conviction that "the cold-hearted cruelty that treated pauperism as a crime brought forth its dreadful harvest of criminality and vice."[28]

Nor did the reform act of 1834 diminish the problem of poverty. In the decade 1850–1860 some 7.8 million pounds was raised annually through the poor rates and 5.4 million pounds was spent annually on the poor. In

1868 the poor rates collected 11 million pounds, of which 7.5 million pounds was distributed among the poor. Those condemning these taxes usually paid lip service to the deserving poor but condemned "those male and female pests of every civilized community whose natural complexion is dirt, whose brow would sweat at the bare idea of earning their bread, and whose stock-in-trade is rags and impudence."[29]

The reform law of 1834 was inadequate in many ways. It did not provide for education of the poor, and they remained without marketable skills. By shifting the burden of child support to the unmarried father, it left women who could not identify or find the fathers of their children without support and thus permanently tied to the poorhouse. Finally, it had no means of distinguishing between laborers seeking work elsewhere and vagrants who did not seek permanent employment.[30]

CONTINENTAL EUROPE AND THE CARE OF THE POOR

While the English parliament provided poorhouses and an occasional dole to help the poor, the French and most of Europe did not. The care of the needy was the province of the Church, although some exceptions were made. Charles IX in 1561 ordered the towns to care for their poor. The

Blind beggars of the 16th century as portrayed by Pieter Bruegel, The Elder, in "The Parable of the Blind" (1568). (Reprinted, with permission from AKG London/Galleria Nazionale di Capodimonte, Naples.).

Church administered hospitals and hospices throughout France without a poor law. Despite this governmental neglect, there were fewer paupers in France than in England. One in thirteen French citizens were paupers in the late 18th century, at a per capita monthly maintenance cost by Church and local community of $2.64, while in England there was one pauper for every ten citizens at a per capita cost of $16.[31] In the city of Cologne, Germany, a different plan, similar to Chalmers's, developed. In 1853 there were 4,224 paupers, representing one-twelfth of the population. The city was divided into 18 districts, and each district subdivided into 14 sections. For each section one visitor, a civic-minded and prosperous or middle-class citizen, was assigned. This visitor would drop in twice a month on the two to four families that were impoverished and help the family plan a budget, discuss its expenditures, and refer members of the family to places seeking employees. By 1870 this Elberfeld Plan, as it was called, had reduced the incidence of paupers to one in eighty.[32]

The Treatment of the Poor in North America

In Colonial North America, poor laws were enacted within a generation after the Pilgrim settlement, beginning in 1642 at Plymouth itself. The Dutch in New Amsterdam, however, followed the continental model and assigned to the Dutch Reformed Church the responsibility of caring for the poor. In the North American tradition, hard work was identified with morality, and failure to abide by this work ethic was interpreted as a vice.[33] Periodic economic depressions, such as that following the economic crash of 1819, caused increased migration of the poor into the cities. A New York state commission to study the problem, headed by J. V. Yates, led to a poor relief act. Yates included idiots, lunatics, the blind, the old and infirm, the lame, and the able-bodied among the state's poor. About one-third each were children, alcoholics, and the able-bodied among almost 7,000 indigent New York residents.

Yates proposed a county poorhouse system, each county attaching a farm to the poorhouse enabling the able-bodied poor to help sustain the costs of administering the program. This was approved in 1824, and the number of counties adopting the program rapidly rose, leaving only three counties without poorhouses by 1841, and only Suffolk County remaining

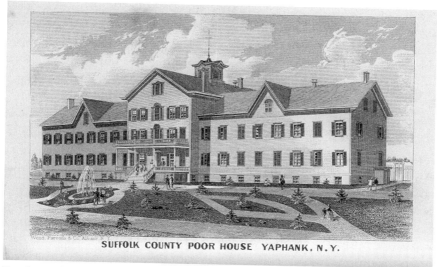

SUFFOLK COUNTY POOR HOUSE YAPHANK, N.Y.

Poorhouse in Yaphank, Long Island. (Reprinted from L.B. Homan [1875] *Yaphank as It Is and Was*, John Polhemus, New York, courtesy of the Thomas R. Bayles Collection, Longwood Public Library.)

as a holdout until 1871.[34] Suffolk had held out against a system it considered a "disgraceful and illy-managed hot-bed of County corruption, poverty and expense."[35] In preparation for caring for its poor, Suffolk County purchased an 80-acre farm with 90 additional acres of woodland in Yaphank, where it established a "model home for the poor." The farm was self-supporting, with the able-bodied working eight hours per day for six days a week making baskets, barrels, and wagons when the weather was bad and outdoors on farm chores during good weather. Paupers were responsible for their rooms and prepared their own breakfast. The sane and the insane were kept in separate wings of the estate, and cripples were housed on the ground floor with their own bathroom. Raving maniacs were kept in cells in the basement. The director of the poorhouse was resented by some because "he failed to kill off the paupers as a miserly element wished" and by others who attributed to him the false remark that "paupers were better dead than alive."[36]

In the Indiana Territories and in the early years of Indiana statehood, adult paupers were offered at public sale to the counties, who employed them at low wages for public works. Children were apprenticed to farmers or businessmen who provided them with a trade and during the years of their apprenticeship they did not have to pay for more than their room and board. The public sale of paupers continued in Indiana until 1897, although this practice was not vigorously enforced after the Civil War.

The cause of pauperism was much debated. In Great Britain, landowners who dispossessed the peasants were blamed by those who found themselves in the poorhouses. Landowners defended themselves by attributing unreliability, laziness, petty theft, poaching, and intemperance to those they forced off their property. Both the desire for the more profitable wool market and the hope to avoid paying taxes by driving the least able-bodied from their properties motivated the selfish. The middle-class tradesman blamed both the landowners and Parliament, the former for exploiting the low wages they could give to poorhouse laborers and the latter for instituting the corrupting influence of the poorhouse. Francis Peek, in his *Our Laws and Our Poor*, summed up these grievances in 1875 and agreed that "these crimes, it must be acknowledged, are the source from which a large proportion of our existing pauperism has sprung."[37]

Peek's sympathetic treatment of the origins of pauperism from the days of Henry VIII to the early 19th century was at odds with his conviction that pauperism can no longer blame these past errors. The new pauper class "arises, to a large extent, from reckless marriages and subsequent improvidence."[38] He deplored the lack of "self-denial" among the poor and recommended "a legal recognition of the indisputable truth, that improvidence is a crime, that it is a dishonorable thing to marry without a reasonable prospect of being able to maintain a family, and that it is still a greater wrong to squander on self-indulgence those resources which should be husbanded against times of stress and difficulty...."[39] Peek also accused landlords of exploiting the poor with overpriced and wretched housing. He also berated the "dram-shops" for fostering drunkenness among the poor. He urged adoption of orphans rather than asylums for their care.

Paupers were frequently associated with alien elements who settled into the villages or cities. The unemployed armies were one source. In London, in 1869, a more convenient scapegoat was the Irish; "They are haunting the metropolis, nearly three mendicants hailing from the Emerald Isle to one of any other nation."[40] A similar outcry against the Irish in the post-Civil War years in the United States, and then against the Eastern and Southern European immigrants, was widely voiced. Opposing this distrust of the foreigner were the voices of social reformers like John Weber, who argued that immigrants create wealth, that very few are paupers, that most want to assimilate into American culture, and that they usually work hard. He had no objection to screening immigrants and sending back those who lacked a sponsor in this country or who had no means of support.[41]

The United States also followed the British model for its poorhouses and used the county to distribute funds for maintaining the needy. The growth of the cities throughout the 19th century increased the number of poor and destitute individuals dependent on public aid. The poorhouse was recognized as a failure. Reformers urged counties not to put pauper children in almshouses because of the bad influence there. "There can be no question to anyone familiar with the influences of almshouse life, that no pauper child, of sound mind and body, should be kept longer than a few weeks in such an asylum. It is of the first importance to the state that pauperism should not be inherited and transmitted, from the familiar scientific principle that inherited evils are intensified in each new generation."[42] Another reformer, in Worcester, Massachusetts, noted that among the paupers "not a few had parents who received aid; and at least one person is the grandchild in a line of unbroken pauper descent in the city records." To the believer in the new charity, based on Chalmers's successful program, the new charity should replace the old charity. The old charity "rapidly turns industrious and self-respecting poor people into greedy beggars, who, as they see their neighbors receiving public aid, demand the same for themselves."[21] Against this system of poor relief that served as "an active school of pauperism and vice" were the aims of the new charity. Henry A. Stimson, an ardent advocate of the new charity, writing in the 1885 *Andover Review,* made a plea to keep the poor out of pauperism by making charity an individual effort, each impoverished family being visited by a person of upright character and concern. What was needed, Stimson claimed, was not support for the body in need, but a "reform of the head and soul."[22]

Throughout the mid-19th century in the United States, experiments in social reform were tried out. In 1844, Dorothea Dix examined the poorhouses in New York state and criticized them as ineffective and harmful to the inhabitants.[43] A committee of the New York State Senate in 1857 reviewed these charges and again looked into the poorhouses, condemning them as a disgrace. Despite this concern, a commission of physicians in 1865 reported virtually no improvement in the care of the poor. After the Civil War, several administrative changes improved the situation. A Board of Charities was established in New York state in 1867, and those who served in the Sanitary Commission during the Civil War helped organize a national network of state Charities Aid Associations to improve the mental, physical, and moral condition of paupers in state institutions. Many important reforms were achieved. Before the 1870s physicians did not have nurses; they used instead what were called "ten days women," mostly alcoholics who were assigned to hospitals instead of to prisons and there they worked off their sentences by attending to the sick. When Dr. W.G. Wylie visited Florence Nightingale and suggested opening a nursing school at Bellevue Hospital in New York City, many of his colleagues opposed this professionalization of nursing because they felt it would create a competitive medical health service in which nurses would usurp the roles of physicians.[44]

In 1885, poorhouses were still operating, and "criminals convicted of minor offenses, tramps, paralytics, imbeciles, lunatics, abandoned women, were found mingled in a hideous confusion with the respectable poor and with children of all ages." This led the *Andover Review* to comment: "Had these institutions been maintained for the purpose of breeding a race of hereditary paupers and criminals, diseased in body, mind, and soul, they could not in many instances have been more successful."[45] Although some progress was made, the editor of the *Andover Review* was not satisfied and hoped for legislative reforms that would guarantee that children would no longer be detained in poorhouses; that criminals would not, out of convenience or cost savings, be sent to poorhouses; that hospitals would be constructed so that those with infectious diseases and mothers who gave birth would not have to be living in the poorhouses with paupers and the aged; and that the poor would be separated from the insane.[46]

Many Americans were opposed to these social reforms. William Graham Sumner, an economist and Episcopal priest, wrote *What Social Classes Owe to Each Other* (1883), a book that took to task the social reformers. "Who are the reformers," he asked, and "Why are they setting problems for others to solve?" Sumner answered his own questions: "So far as I can find out what the classes are ...they are as follows:...the rich, comfortable, prosperous, virtuous, respectable, educated, and healthy," who should solve the problems set by and for "those who have been less fortunate, or less successful, in the struggle for existence."[47] Sumner, who spent most of his career as a professor of political science at Yale, admired the wealthy industrialists of his age, opposed those who wanted to obtain wealth and status without having earned them or, even worse, those who wanted to bring the wealthy and virtuous down to their own misery by bloodshed. "Are there any classes that have rights or demands on the rest of us?" asked Sumner. He thought not, because he believed society is organized through a principle of contractual liberty in which we owe nothing to each other but courtesy and good will. Sumner said it was a virtue to try to become rich, to use nature to increase one's own wealth, and to look after one's own family. This virtue of the wealthy Sumner attributed to foresight and industry. The poor, he claimed, are foolish, not wise; extravagant, not thrifty; negligent, not prudent; and thus should not be helped; "Such classes always will exist." Instead of giving to charity and distributing wealth, Sumner asked society only to "increase, multiply, and extend the chances," for anyone to apply skill and industry for virtue.[48]

The conflict between the new and the old charity was often debated at the social level. It was a problem too vast and too expensive for private charity to solve, and thus it fell upon local and state governments to address the problems of ill health, insanity, poverty, and crime. The clergy in both England and the United States usually participated as advocates of government reforms. An exception to this was the movement initiated by William Booth, founder of the Salvation Army and author of *In Darkest England*.[49] To General Booth of the Salvation Army, there was in the slums of London a "submerged tenth" in which the "poor and the vicious" lived out their lives in poverty. A major cause of their degradation, Booth believed, was intemperance, and he roused his recruits in the Salvation Army to use religious faith and good works as the way to rescue the fallen from their slums. He sought donations from both public funds and private charity to set up farm colonies in England and overseas to empty out the slums so that the new colonies of paupers and former alcoholics could be revitalized through the work ethic and spiritual rejuvenation that the Salvation Army imposed on them. Although the Salvation Army movement was warmly greeted by editorials and commentators in popular periodicals both in England and the United States, it failed to achieve its larger objective of helping the poor and gradually became identified in the public image as a religious refuge for the alcoholic.[50]

PAUPERS AS DEGENERATE STOCK

In contrast, Henry Boies's view on *Prisoners and Paupers* (1893) was not very sympathetic and reflected the biases of a growing number of professionals.[51] He believed paupers were a degenerate stock or a degenerate culture. The poor laws in our various states were "framed to maintain the needy, rather than to assist and enable such to maintain themselves."[52] He believed all able-bodied poor should be forced to work. The paupers should be rounded up and removed from the cities, he argued, and placed in county houses, set up in the farmlands to supply all the needs of the inhabitants through their labors at little cost to the state. Boies deplored "effete races" such as Russian Jews, Moslems, Romanists, and adherents of the Greek Church, "speaking strange tongues, with prejudices, habits, customs, and religions of the inbred strength of centuries." These "wretched people from all lands fly toward 'Liberty Enlightening the World,' like the insects of a summer night, without purpose, without thought, without care, save to bask in the beams of this new sun."[53] While he berated Europe for making "America its almshouse," Boies had no such objection to the "colored people," whose welfare he designated "a National obligation." In a tone remarkably different from his sentiments about the foreign-born pauper, Boies urged unprejudiced acceptance of America's freed slaves. "Let every colored man be treated, everywhere in public and in private, exactly as if he were white; according to his deserts as an individual, his merits as a man."[54]

Boies classified paupers as falling into three groups. The first were the mentally, morally, and physically defective; the "cripples, deformed, deaf, blind, imbecile, weak minded, diseased, insane, or criminal" defectives. The second were beggars, vagrants, tramps, idlers, and criminals who preyed on others in preference to working. These two classes he defined as hereditary paupers. Only the third class, to which he gave the name "heteronomic paupers," deserved charity or the efforts of public good will. These included the victims of adversity, old age, sickness, and accident who were forced into poverty; most of whom, he acknowledged, were capable of resuming productive lives in an improved environment. When Parliament established a Commission in 1832 to study the Poor Laws and make recommendations for their revision, Chalmers's views were widely discussed. Many who were sympathetic to this new approach failed to implement it and it remained impractical. The middle class instead shifted to blaming the poor and embraced the Commission's report. Despite the unsympathetic attitude expressed by the Commission, the Poor Laws Act of 1834 did not abolish the practice of outdoor relief.[55]

Boies's pessimism was characteristic of the social thought of many intellectuals in the 1890s. He accepted with full confidence as correct and proven that "criminals and paupers, both, are degenerate; the imperfect, knotty, knurly, worm-eaten, half-rotten fruit of their race." Only their physical or reproductive isolation would reduce or eliminate such classes from society: "Here is a gangrened member of the body politic; the question is not how it came to be so, but what shall be done to stop the spread of the poison, and save the life of the patient."[56]

In the 1890s, the social problems of the cities and the rural poor continued to plague the efforts of legislators and the clergy. By this time the word "pauper" had taken on the image of a fixed degenerate class, and many reformers were willing to abandon them as a people who would not respond to either the old or the new charity. Margaret Andrews Allen was not impressed by the pessimism of her fellow reformers in Boston. In an account of the work of Jennie Collins appearing in the *Charities Review* for 1892, she quoted Collins's remarks about describing people as paupers: "I have been told that when a woman is sent to Deer Island as a criminal, a kindly interest is felt in her welfare as she passes through the dignity of a court trial, no matter how many times she returns as an offender of the law; but when she sinks to the level of a pauper, she passes out of sight and is only known by 'it' and 'its' number. In view of that state of feeling I would suggest a different mode of helping extreme cases; call them pensioners, wards, proteges of the State; but the word 'pauper' should be expunged from our statute books." Collins had opened a working women's house in Boston that befriended immigrants, deserted wives, and inexperienced teenage girls by obtaining employment for them and serving as a temporary shelter until they established themselves. Collins called her settlement house "Boffin's Bower" after a sanctuary for the poor described in Dickens's *Our Mutual Friend*.[57]

Mrs. E.C. Bolles also championed the new charity movement. She deplored a clergyman's "exhortation to his flock to devote a tenth of their income to the poor." If this were put into practice she admitted, "the increase of pauperism would soon be appalling...." She attributed the causes of pauperism in the cities to "excessive immigration of the worst elements of the Old World," crowded tenements, the fertility of the lower classes, "drunkenness, indolence, ignorance, and inefficiency." Although she did not have any practical remedy to suggest, she wished that "if criminals, paupers, idiots, and incurables were prevented from self-multiplication the next generation of benevolent workers would meet with less discouragement."[58] In the meantime, she suggested direct personal influence through "volunteer visiting" and advocacy of kindergarten classes and industrial schools to improve the skills of the children of paupers.

As the 1890s closed out the 19th century, charity workers were in a dilemma. There was agreement that public relief had not worked in the past to eliminate poverty and that neither good times nor increasing opportunities for mass education had decreased society's troubled classes. The most pessimistic of the social workers were contemplating ways to cut off the reproduction of the unfit. Those with a belief in their reason and good will still held out hope. Francis Peabody, writing on "The problem of charity," expressed this hope: "The old charity satisfied the feelings of the giver of alms; the new charity educates the receiver to do without alms. The old charity was temporary relief; the new charity is continuous education."[59]

MENTAL ILLNESS AND ITS TREATMENT

The paupers and the criminals were not the only classes of a defective humanity. Martha Louise Clark, writing in 1894 for the *Arena*, presented "The relation of imbecility to pauperism and crime." She listed 90,000 insane, 75,000 imbeciles, and a "countless army" of tramps and beggars in the United States. "Does it ever occur to us that their increase might to a certain extent be averted; that crime, imbecility and insanity are hereditary diseases of the mind, and that so long as we allow them to go on breeding their kind, we can expect nothing but constant additions to the burden which we must bear as a nation?"[60] This too, was a shift from the optimism of the 1870s when the insane were considered treatable. Charles D. Robinson discussed the history of the management of the insane in an 1876 article for *Scribner's Monthly*. About one in 2000 Americans was insane, yet "our insane hospitals are merely repetitions of those dungeons of a thousand years ago, with their grim array of dark and filthy cells."[61] The insane were removed from the view of the world by such isolation, but they were not being treated. Robinson opposed this, "insanity means all unhealthiness of mind" found in society, including mania, melancholy, and epilepsy. Educated people no longer accepted, as the ancients did, that insanity was a visitation from an offended deity; they no longer accepted the biblical belief that insanity was evidence of possession by devils. They rejected practices, such as that practiced in Bedlam from its founding in 1246 to 1675 when the insane were let out to beg for their upkeep.[62] Robinson acknowledged that older methods of bleeding and purging did not work, but he believed that the insane were victims of diseased brain tissue and that the skills of 19th-century medicine would eventually lead to successful treatments.

The perception of the failures in American society shifted from one of pity and charity to one of fear, disgust, and rejection in less than one generation. Every generation has some citizens who blame the environment for the predicament of its unfortunates and other citizens who feel these are people who are victims of their own failings. What made the last quarter of the 19th century so different was the rapid growth of science with theories to support both views and the increasing sympathy of intelligent people to reject the views of the environmentalists.

FOOTNOTES

[1] It is not easy for a 20th-century reader to realize how difficult life was for the greater duration of human society in historical times. One's family was the primary source of support in illness or old age, a situation that did not change for any significant portion of humanity until the 20th century.

[2] The Samaritans were one of the early Jewish tribes that lived near Nablus in Palestine, Samaria extending north of that city. They were a schismatic Jewish group, rejected by their fellow Jews at the time of Ezra, and later sometimes allied with the Romans and at other times with the Jews. As a surviving Jewish ethnic group, they now consist of a few hundred individuals, mostly in Nablus, who accept the Torah (the pentateuch or five books of Moses in the Old Testament) but have no talmu-

dic tradition. They differ from other Jews in not wearing prayer shawls or using phylacteries during worship. Their language was Aramaic rather than Hebrew, but after the diaspora they used Arabic as a public language. Jesus singled out the Samaritans because of their pariah status in the Jewish community.

[3]   Peter Townsend, *The Last Refuge: A Survey of Residential Institutions for the Aged in England and Wales* (Routledge and Kegan Paul, London, 1964).

[4]   E.M. Leonard, *The Early History of English Poor Relief* (Cambridge University Press, 1900), pp. 4–5.

[5]   Although the poor laws as well as laws meting penalties for beggary and vagrancy were passed by Parliament, they were not always enforced, or only unevenly enforced.

[6]   James Greenwood, *The Seven Curses of London* (Stanley Rivers and Co., London, 1869), p. 213.

[7]   Ibid. p. 216.

[8]   Ibid. p. 214.

[9]   Leonard, *Early History of Poor Relief,* p. 26.

[10]   Ibid p. 302.

[11]   Greenwood, *Seven Curses of London,* p. 216.

[12]   Townsend, *The Last Refuge,* pp. 12–13.

[13]   Ibid. pp. 12–13.

[14]   Henry Fawcett, *Pauperism: Its Causes and Remedies* (MacMillan, London, 1871), p. 14.

[15]   Charles L. Brace, "Pauperism," *The North American Review* 120(1875): 316–334, p. 317. Brace was the first to give publicity (in this article) to the study of a degenerate family later called the Jukes.

[16]   Ibid. p. 316.

[17]   Fawcett, *Pauperism,* p.14.

[18]   George Crabbe (1754–1832) started as a surgeon and, after failing in that career, became a minister. He found his calling in writing poetry and fiction, but his popularity faded by mid-19th century. Crabbe rejected the romantic view of the village, and his biting description of it as a poorhouse was widely cited and influential on realistic writers. Crabbe's poem was reprinted in Robert Pashley's *Pauperism and the Poor Laws* (Longman, Brown, Green, and Longman, London, 1852), p. 366.

[19]   Thomas Chalmers (1780–1847) was a Scottish minister with talents in mathematics, astronomy, and economics whose scholarly works made him an influential intellectual. While he was minister of the Tron Church in St. John's parish in Glasgow, he carried out the reforms, starting in 1819, that led to the new charity movement. He was later a professor at St. Andrews.

[20]   Henry A. Stimson, "The new charity," *The Andover Review* 3(1885): 107–120, p.108.

[21]   Ibid. p. 116.

[22] Ibid.

[23] Townsend, *The Last Refuge*, p. 13.

[24] Stimson, *The New Charity*, p. 111.

[25] "Poor Law," *Encyclopedia Britannica* 18: 215–224.

[26] Edwin D. Whipple, "Oliver Twist" [Commentary], *Atlantic Monthly* 38(1876): 474–479.

[27] Ibid. p. 474.

[28] Edgar Johnson, *Charles Dickens: His Tragedy and Triumph* (Simon and Schuster, New York, 1952), p. 274.

[29] Greenwood, *Seven Curses of London*, p. 431.

[30] The history of welfare legislation may be read as an ambivalence of legislatures that seek to protect the poor from neglect as a national obligation and legislatures that seek to protect the taxpayers from supporting an unworthy poor who should learn to be more frugal or to plan ahead.

[31] Charles Brace, "Pauperism," *North American Review* 120(1875): 316–334, p. 324.

[32] Ibid. pp. 327–328. This Elberfeld Plan was developed by Herr von der Heydt. It is very similar to a plan developed earlier by Chalmers in Edinburgh. An American attempt to mimic these plans was introduced in Germantown, Pennsylvania. See Joseph Cook, *Boston Monday Lectures* (Houghton, Osgood & Co., Boston, 1879), pp. 217–243.

[33] Robert W. Rightmire, *Suffolk County's Adoption of a County Alms House*. Ph. D. thesis, State University of New York, Stony Brook, 1980.

[34] Ibid. p. 7.

[35] L. Beecher Homan, "The Suffolk County Alms-House," *Yaphank As It Is* (John Polhemus, Printer, New York, 1875), p. 209.

[36] Ibid. p. 212.

[37] Francis Peek, *Our Laws and Our Poor* (John B. Day, London, 1875), p. 9.

[38] Ibid. p. 10. Peek is also influenced in his modern views by the ideas of T.R. Malthus (see Chapter 7).

[39] Ibid. p. 14.

[40] Greenwood, *Seven Curses of London*, p. 218.

[41] John B. Weber, "Pauperism and crime," *The Charities Revue* 3(1894): 117–126.

[42] Brace, *Pauperism*, p. 321.

[43] Dorothea Dix (1802–1887) started out as a teacher and ran her own school in Boston where she developed stories and lessons that were widely published. In 1841 she began her visits to jails and almshouses that led to her later publications. Her documentation proved persuasive and she benefited the pauper insane who were poorly treated. Dix carried her campaign to other countries and then headed the

nursing staff during the Civil War, returning to her campaign for the humane treatment of the insane when her war duties ended.

[44] B.G. Wylie, "Relations of hospitals to pauperism" *Popular Science Monthly* 9(1876): 738–743. See my discussion of Wylie in "R.L. Dugdale and the Jukes family: A historical injustice corrected." *Bioscience* 30(1980): 535–539. Also see reference 20, p. 225.

[45] Anonymous. "Private Aid to Public Charities," *Andover Review* 4(1885): 220–230, p. 227.

[46] Ibid. p. 225.

[47] Rowland Hazard (book review): *What Social Classes Owe to Each Other, Andover Review* 1(1884): 159–174, p. 160.

[48] Ibid. pp. 161–166.

[49] William Booth (1829–1912) grew up in modest circumstances in Nottingham and became a pawnbroker. He had a religious conversion in his youth and became an itinerant preacher. In 1878 he founded the Salvation Army with two major premises: a belief that all will be damned to eternal punishment unless converted to Christianity and that the fallen could be saved by example of the virtuous who cared for them. William Booth is unrelated to Charles Booth (1840–1916), a Liverpool steamship owner who used his wealth to study the conditions of the poor, especially the elderly. Charles Booth's efforts (mostly in the 1890s) led to old-age pensions for the poor, a forerunner of Social Security.

[50] Anonymous. 1891. "The problem of the slums" in *Blackwood's Edinburgh Magazine* 149: 123–136, p. 125.

[51] Henry M. Boies, *Prisoners and Paupers* (G.P. Putnam's Sons, New York, 1893).

[52] Ibid. p. 38.

[53] Ibid. p. 49.

[54] Ibid. p. 83.

[55] Ibid. p. 206.

[56] Ibid. p. 267.

[57] Margaret Andrews Allen, "Jennie Collins and her Boffin's Bower," *The Charities Review* 2(1892): 105–115.

[58] Mrs. E.C. Bolles, "Would direct personal influence diminish pauperism?" *The Charities Review* 2(1893): 410–419, p. 413.

[59] Francis G. Peabody, "The problem of charity," *The Charities Review* 3(1893): 1–16, pp. 5–6.

[60] Martha Louise Clark, "The relation of imbecility to Pauperism and crime," *Arena* 10(1894): 788–794, p. 788.

[61] Charles D. Robinson, "Insanity and its treatment," *Scribner's Monthly* 12(1876): 634–648, p. 634.

[62] Ibid. p. 637.

7

The Perfectibility of Man Confronts Vice and Misery

U NTIL THE EARLY 19TH CENTURY, infant mortality was high among the nobility as well as the paupers, and distinctions between constitutional weakness and strength were not very pronounced. There were fixed groups, such as the nobility and the laity, and clergy chosen by their inner conviction or calling. Class distinctions were given that name in the 18th century but unlike the fixed groups, the status of rich or poor, bourgeois or laborer was potentially mobile, with many nouveau riche arising from the lower classes and many a prodigal son or bankrupt merchant descending back to paupery.

The first recognition of the poor as an imperfect class that does and should bear a disproportionate share of suffering and mortality was developed by Thomas Robert Malthus (1766–1834).[1] Malthus was the sixth of seven children and the son of a well-to-do scholar who was a friend of Jean-Jacques Rousseau and an admirer of the French Enlightenment that had overthrown the monarchy. He grew up in an intellectually stimulating household in which his father used the educational philosophy of Rousseau's *Emile* to educate him, fostering the independence of his mind and the beneficial reflections from hands-on experience rather than memory by rote and harsh discipline. At the age of 10, he was sent to tutors who also used Rousseau's philosophy of education. Malthus was recommended for entry at Cambridge University, where he studied mathematics.

Thomas Robert Malthus, 1766–1834. (Reprinted, with permission, from L. Eiseley [1979] *Darwin and the Mysterious Mr. X*, E.P. Dutton, New York [courtesy of Department of Library Services, American Museum of Natural History].)

Although mildly handicapped with a cleft lip and palate, he managed to communicate his ideas orally and chose to be ordained in the Church of England. His friend, Harriet Martineau, adjusted to his initially inarticulate speech and noted that "his vowels at least were sonorous, whatever might become of the consonants."[2] He held a country curacy for a few years and then traveled in Europe with friends. He married his cousin and began to attract attention through his writings, securing a position as a professor of history and political economy at the newly founded college of the East India Company. He enjoyed intellectual company and was considered a "warm, charming, and lively companion."[3]

MALTHUS RESPONDS TO GODWIN'S OPTIMISM

Malthus's father was of more optimistic temperament than his son and told him to read an essay by William Godwin (1756–1836) on the perfectibility of humanity.[4] Malthus did so and doubted its validity. He had seen far more births than deaths among the parishioners in his church, and he noted, "I had for some time been aware that population and food increased in different ratios; and a vague opinion had been floating in my mind that they could only be kept equal by some species of misery or vice."[5]

What set Malthus on his life career as a political economist was the high regard that his father had for the optimistic philosophy of William Godwin. Godwin's life and ideas were profoundly different from those of the young curate, Malthus. Godwin's parents were nonconformist Calvinists who steered their son into the ministry. As Godwin read about the French philosophers whose passion for the rights of man he shared, he became a nonviolent philosophic radical and left the ministry for a career as a writer. He published several biographies, novels, and histories. He believed in the perfectibility of the human race, the necessity of individual freedom, and a society without government. He married a feminist, Mary Wollstonecraft (1797–1851), author of *A Vindication of the Rights of Women.*[6]

Godwin's philosophy was first published in *The Inquiry Concerning Political Justice, and Its Influence on General Virtue and Happiness* (1793). The radicalism expressed was dangerously like that of the Jacobins of the French Revolution, and might have been interpreted as seditious to the British government. Godwin wisely protected himself by requesting a sale price of three guineas to make sure he would not be convicted of selling his book to the dangerous classes. Godwin believed in human reason, the equitable distribution of the wealth of the world, and Thomas Paine's credo that "Society is produced by our wants, government by our wickedness."[7]

Four years later, Godwin produced a collection of essays, *The Enquirer: Reflections on Education, Manners and Literature.* It was this work that irritated Malthus. Godwin denounced the wealthy as selfish and not meriting their riches. "There is no wealth in the world except this, the labour of man," he claimed. What is falsely called wealth, in Godwin's interpretation, was the power of compulsion over others, especially in the exploiting of the laborer. Since rich and poor alike have the same basic biological needs, he felt that the proper use of riches was "to cheer the miserable, to relieve the oppressed, to assist the manly adventurer, to advance science, and to encourage art." The avaricious rich who hold on to their wealth he condemned as misers who contribute little to society.[8]

INFLUENCE OF CONDORCET ON GODWIN

Much of Godwin's thinking came from the writings of the Marquis de Condorcet, Marie Jean Antoine Nicolas Caritat (1743–1794), a mathematician, philosopher, and prolific writer who served as Secretary of the Academy of Sciences.[9] He was the first to apply mathematics to political and social institutions, but he was not an original mathematician of first rank. Condorcet supported the revolution and served in the legislature, drafting the legislation for public education in France. He opposed the execution of King Louis XVI, and his views were too independent for the more extreme faction of the revolutionists. He fled and, while in hiding, worked on his theory of the perfectibility of humanity. He was recognized and imprisoned. His last work, completed just before his arrest, was *The Future Progress of the Human Mind*. He was found dead the day after his imprisonment, the cause of his death unknown. Condorcet believed that the revolution was the beginning of the eventual end of all political repression. He extolled progress which introduced new ideas and techniques that made society more productive and efficient. He considered the possibility that the supply of food might not suffice for a growing population but argued that "even if we agree that the limit will one day arrive, nothing follows from it that is in the least alarming as far as either the happiness of the human race or its indefinite perfectibility is concerned." Condorcet had faith that medicine, art, and science would banish disease and prolong life indefinitely.[10]

MALTHUS'S IDEAS ON THE POOR

Malthus disagreed with Pitt's support for the poor law and his specific proposal to find employment for the poor and raise their subsidies. He argued in an unpublished essay, *The Crisis*, which he showed to his father and friends in 1796, that increasing the subsidy to the poor only caused increases in prices for everyone during a time of scarcity and that giving employment to the poor when there was a recession only took away jobs from those who were already employed. In his essay he did not refer to population as the root cause of the economic distress of the poor. Malthus felt the children should be supported, because of their innocence, but not the adult poor.

After reading Godwin's views, however, Malthus turned his attention to the problem and read several works that contradicted Godwin's and Condorcet's optimism. From David Hume (1711–1776) he learned in *Political Discourses* (1752) that population declines were temporary and short-lived, as after plagues, when survivors quickly repopulated the settlements that had been devastated. Hume's view on population was essentially an optimistic one: "If everything else be equal, it seems natural to expect, that wherever there are most happiness and virtue and the wisest institutions, there will also be most people."[11]

In 1753 Robert Wallace (1697–1771) published *A Dissertation on the Numbers of Mankind in Antient* [sic] *and Modern Times*. He claimed that on the same plot of land a "rude and barbarous" people living by hunting and gathering can never be as populous as an agricultural people. He also felt that cultures reach an affluent peak and then decline because debauchery and luxury delay marriages and lead to a decline in the population.

The third and most important influence in shaping Malthus's views on population was the popularly acclaimed *An Inquiry into the Nature and Causes of the Wealth of Nations* written by Adam Smith in 1776. Smith saw people as commodities and subject to the same economic laws; but because they were also human beings carrying out their human needs and behaviors, there were added consequences: "...the demand for men, like that for any other commodity, necessarily regulates the production of men."[12] Smith recognized that poverty does not prevent marriages among the poor and he believed that "barrenness, so frequent among women of fashion, is very rare among those of the inferior station." Another influential point that Malthus absorbed was the high incidence of mortality among the newborn and infants. "In some places one half the children born die before they are four years of age; in many places before they are seven; and in almost all places before they are nine or ten. This great mortality, however, will everywhere be found chiefly among the children of the common people who cannot afford to tend them with the same care as those of better station."[13]

Malthus's initial *Essay on the Principles of Population* was a statement of his views with little documentation.[14] His father, while disagreeing with it, was impressed by its logic and felt it should be published, but Malthus was concerned about the effect it would have on his reputation because he was still beginning his career. He published the essay anonymously. The response after its publication in 1798 was rapid and prolific. Malthus's biographer describes well the avalanche of criticism: "He was the 'best-abused man of the age.' Bonaparte himself was not a greater enemy to his species. Here was a man who defended small-pox, slavery, and child-murder; who denounced soup- kitchens, early marriage, and parish allowances; who 'had the impudence to marry after preaching against the evils of a family'; who thought the world so badly governed that the best actions do the most harm; who, in short, took all the romance out of life and preached a dull sermon on the threadbare text— 'Vanity of vanities, all is vanity'."[15]

MALTHUS'S REFUTATION OF HUMAN PERFECTIBILITY

Malthus acknowledges Godwin's essay on "Avarice and profusion" in *The Enquiry Concerning Political Justice, and Its Influence on General Virtue and Happiness* as the stimulus for his essay and asserts that he can rebut the perfectibility of man with two postulates: "First, That food is necessary to the existence of man. Secondly, That the passion between the sexes is necessary and will remain nearly in its present state."[16] What impedes mankind's progress to that perfectibility? Can these impediments be removed now or in the future? Malthus claims that the task is impossible and the cause cannot be removed: "The facts which establish the existence of this cause have, indeed, been repeatedly stated and acknowledged; but its natural and necessary effects may be reckoned a very considerable portion of that vice and misery, and of that unequal distribution of the bounties of nature, which it has been the unceasing object of the enlightened philanthropist in all ages to correct. The cause to which I allude, is the constant tendency in all animal life to increase beyond the nourishment prepared for it."[17]

Malthus used a biological observation to support this inference. "Throughout the animal and vegetable kingdoms Nature has scattered the seeds of life abroad with the most profuse and liberal hand; but has been comparatively sparing in the room and the nourishment necessary to rear them. The germs of existence contained in this earth, if they could freely develop themselves, would fill millions of worlds in the course of a few thousand years. Necessity, that imperious all-pervading law of nature, restrains them within the prescribed bounds. The race of plants and the race of animals shrink under this great restrictive law; man cannot by any efforts of reason escape from it."[18]

Malthus had no direct evidence that the rate of population growth is geometrical in humans, increasing by the square each generation if left to its natural rate. Nor did he have evidence that the rate of increase of food is linear. He inferred the geometrical growth from the rapid colonization of North America and the very large family sizes reported there. He assumed that any acre of farmland had a fixed yield and thus only increases of land acreage would lead to corresponding increases in food. To support the thesis that the North American growth rate was geometric and would have predictable consequences, he claimed "if the United States of America continue increasing, which they certainly will do, though not with the same rapidity as formerly, the Indians will be driven farther and farther back into the country till the whole race is ultimately exterminated."[19]

For the Europeans, there were, Malthus believed, no lands to exploit and turn into new farmland. The population was saturated and any increase would have to lead to misery and vice. Any effort to increase food production by import or other measures was doomed to failure by the dramatic difference between the geometrical and linear growths taking place. "Supposing the present population equal to a thousand millions, the human species would increase as the numbers 1, 2, 4, 8, 16, 32, 64, 128, 256; and subsistence as 1, 2, 3, 4, 5, 6, 7, 8, 9. In two centuries the population would be to the means of subsistence as 256 to 9; in three centuries as 4096 to 13; and in two thousand years the difference would be almost incalculable."[20]

Starvation, except during rare periods of famine, was not what kept the population in check. Malnourished populations, however, were very common and in this weakened state the body was vulnerable to attacks of disease and the behavior of individuals was deflected to satisfying the hunger and poverty of their lives. From this stress, Malthus identified two kinds of checks to the population. "Positive checks" included unwholesome occupations, severe labor, exposure to harsh weather, poor housing, improper clothing, bad nursing practices, and the more familiar horsemen of the Apocalypse, "diseases, epidemics, wars, plagues, and famine." "Preventive checks" were those circumstances that caused a delay in marriage or fostered a celibate state, including the social obligation, if heeded by males, not to marry until they can afford to support a wife and have children. Thus, worry, income, the cost of living, foresight, and care about the quality of life were influential on those persons who were willing to accept a "certain degree of temporary unhappiness" in return for a later peace of mind. Malthus warned, however, that preventive checks could also lead to vice, as in illicit affairs and prostitution, from which misery follows because these practices "poison the springs of domestic happiness."[21] Malthus does not mention masturbation as a preventive check in his essay.

The brunt of the population check, Malthus pointed out, would fall disproportionately on the poor. He recognized that there would be economic oscillations where low wages would bring forth more provisions because the cost of production would decrease; as the price of labor would go up, the amount of land farmed and manufacturing accomplished would diminish and fewer provisions would be available. He believed that these cycles were not noted in the past by historians because "the histories of mankind which we possess are, in general, histories of only the higher classes."[22]

Malthus denounced the poor laws because he believed they do more harm than good. The system of public subsidies to the poor "may have alleviated a little the intensity of individual misfortune, (but) it has spread evil over a much larger surface."[23] He used as an example the price of meat and the availability of meat during a time of scarcity, as during the debate over Pitt's reform of the poor law. If the poor are given more money to buy meat, he claimed, the price of meat will rise and both the poor and the general public will be harmed. A scarce item, he claimed, does not become plentiful when the price for it is increased. The effort to rid the country of poverty by raising wages or donating to the poor was logically false, he claimed, and "we act much in the same manner as if, when the quicksilver in the common weather-glass stood at *stormy*, we were to raise it by some mechanical pressure to *settled fair*, and then be greatly astonished that it continued raining."[24]

MORAL RESTRAINT AS THE REMEDY FOR THE POOR

In the first edition of Malthus's essay there were no specific recommendations to address the problem that he raised except to abolish the poor laws and allow nature to take its course through the misery and vice of the checks on population. In subsequent editions, Malthus added a third component to offset the harshness and apparent cruelty of "misery and vice." Malthus suggested "moral restraint" through the process of delayed marriage, especially among the poor. This would require public education of the poor, with a heavy emphasis on moral virtues, to make them aware of their duty not to procreate beyond the means of subsistence. He also felt that the poor should not be held as objects of pity or sorrowful need, rather, "dependent poverty ought to be held disgraceful."[25]

Malthus feared that the ignorant poor were encouraged by improved wages or public charity to spend their money foolishly rather than save for the future; he was certain that the primary beneficiary of such largesse to the poor would be the proprietors of ale houses.[26] Malthus's formal recommendation was harsh and divisive, generating contradictory sermons on the sagacity or savagery of his values. "To this end, I should propose a regulation to be made declaring that no child born from any marriage taking place after the expiration of a year from the date of the law, and no illegitimate child born two years from the same date, should ever be entitled to parish assistance."[27] He wanted a national campaign by the churches to publicize the coming law, warning men of their responsibility to support their children, stressing the immorality of a marriage without the proper capacity to support a wife and children, the evils of receiving public funds, and the need to abandon the poor laws. He did not feel it necessary to enact a law forbidding foolish marriages because the laws of nature will punish those who enter them. He saw no reason to support illegitimate or abandoned children. "At present the child is taken under the protection of the parish, and generally dies, at least in London, within the first year. ...The death passes as a visitation of Providence instead of being considered as the necessary consequence of the conduct of its parents, for which they ought to be held responsible to God and society."[28]

BLAMING THE POOR FOR THEIR PREDICAMENT

The poor, ultimately, are to blame for their own condition, although they have been misled. If Malthus's proposal were to be enacted, the informed poor would no longer search for blame and would join the ranks of the responsible. If the law Malthus proposed were enacted, he was confident that the widespread expectation of dependency would change. "A man who might not be deterred from going to the ale-house from the consideration that on his death or sickness he should leave his wife and family upon the parish, might yet hesitate in thus dissipating his earnings if he were assured that in either of these cases his family must starve or be left to the support of casual bounty."[29]

Malthus realized that there would be times, even after his proposals were enacted, when temporary economic recessions would lead to unemployment. The solution here was not public funds in lieu of a salary but a tax to provide public works. Since these would be noncompetitive jobs, they would not harm the economy by putting others out of work. "Such are the public works of all descriptions, the making and repairing of roads, bridges, railways, canals, etc."[30]

The response to the first edition was so overwhelming, in both its attacks and praise, that Malthus felt obliged to issue a second edition. He changed it from a statement of belief to a scholarly work, adding more than 200 pages of tables and historical studies of the various countries in which population rises and declines could be documented. He also added his name to all subsequent editions. They appeared in 1803, 1806, 1807, 1817, and 1826. A posthumous reissue of the sixth edition (called the seventh) appeared in 1872. Although the book sold well, it was not uncommon for his ideas to be attacked by those who did not read it, leading his biographer to comment: "Adam Smith has left us a book which 'everyone praises and nobody reads,' Malthus a book which no one reads and all abuse."[31]

GODWIN'S ATTACKS ON MALTHUS

William Godwin was pleased when the first edition of Malthus's essay appeared because it gave him a recognition for stimulating so provocative a work. As the editions continued coming out and Godwin's own work fell into eclipse, he became embittered, and finally in 1820 he published a lengthy response (600 pages) to Malthus, *Of Population: An Enquiry Concerning the Power of Increase in the Numbers of Mankind, Being an Answer to Mr. Malthus's Essay on That Subject.*[32] Godwin's effort to undo Malthus's thesis was marred by his use of a hostile, polemical, ad hominem attack written in repetitious and hasty prose.

Just as Malthus used thin evidence to prove his geometrical and linear laws, Godwin was equally sweeping in his generalizations, but he made no pretense of being scholarly; he used few tables (he considered it pedantic and misleading) and focused on inconsistencies and contradictions in Malthus's different editions. Godwin claimed that the earth could easily support 20 times its present population (then estimated at 1,000 millions by Malthus). His estimate was based on inspection during his travels; "The first thing therefore that would occur to him who should survey 'all the kingdoms of the earth' and their state of their population would be the thinness of their numbers, and the multitude and extent of their waste and desolate places."[33] He denied that the poor were subject to checks by a law of nature and instead asserted that it was a "law of very artificial life" that heaped luxury for the few and condemned most of humanity to want. He successfully rebutted Malthus's claim that the average number of births in Europe was eight, with four dying at birth or in infancy. He pointed out that not all women marry, some marry after their 20th birthday, some women die of childbirth and other causes every year throughout their marriage, some women abort rather than give birth to liveborn, some have only one or two children because their husbands die or become ill, and that even among those not afflicted with some sort of misery, the sexual passion diminishes in many marriages and pregnancy becomes less frequent. Godwin estimated that these factors made it more likely that the average number of births was four, not eight. Malthus used the estimate of eight births from the North American growth rate, one that had fewer checks because of the immense availability of land that could be converted to agriculture. Since both Godwin and Malthus agreed that half the children born died "in their nonage" or years of immaturity, there was only a very slow growth of population during the centuries past and the burden fell less on the poor than on the infants of rich and poor alike. The poor laws would neither increase nor decrease this infant mortality, and thus Malthus's attacks on the poor laws were without foundation. Godwin also considered diminished fertility to be the major reason that the average number of births was four and not eight as in Malthus's thesis. He did not consider these reasons for diminished fertility as vice and misery and, indeed, Malthus did not mention these in his essay. "Nature takes more care of her works, than such irreverent authors as Mr. Malthus are apt to suppose," he scoffed.[34]

Godwin attacked Malthus's arithmetic law of subsistence and denied nature set such a limit. He asserted there were "no assignable limits" to subsistence, only unfavorable seasons and political institutions seriously affected the yield of cultivated land. Godwin was confident that human ingenuity and energy were sufficient to meet the modest growths of population that existed in his day. He denounced Malthus's essay as an example of "what extravagant and monstrous propositions the human mind is capable to engender, when once men shall be prompted, upon a fable, to build a system of legislation, and determine the destiny of all their fellow-creatures."[35]

Footnotes

[1] For a full biography of Malthus, see James Bonar, *Malthus and His Work* (MacMillan & Co., London, 1885). The intellectual influences on Malthus are nicely presented in Philip Appleman's introductory essay for the reprint edition of Malthus's *Essay on the Principles of Population* (Norton Critical Edition, Norton, New York, 1975).

[2] Bonar, *Malthus and His Work*, p. 419. Surgical repair of cleft lip and palate began in 1813 in the United States, but prior to that clips were commonly used to hold the abraded lips together forcing them to heal. Ambroise Paré had much earlier used sutures to repair cleft lips but this was not as often used until the early 19th century. Babies were bottle-fed with a nipple having a shield to prevent milk from being squirted into the infant's lungs. Until the 19th century cleft palate (but not cleft lip) added to infant mortality, and Malthus was fortunate he had a family wealthy enough to care for him. It is ironic that a person with so serious an infirmity should have had so little sympathy for those who were physically healthy but impoverished. See Blair O. Roberts "History of cleft lip and palate treatment" pp. 142–169, in *Cleft Lip and Palate: Surgical, Dental, and Speech Aspects,* W.C. Grabb, S.W. Rosenstein, and K.R. Brock, editors, (Little, Brown, & Co., Boston, 1971).

[3] Diana M. Simpkins, "Malthus, [Thomas] Robert (1766–1834)." *Dictionary of Scientific Biography* 9: 67–71, p. 67.

[4] "William Godwin," *Encyclopedia Britannica* 10: 465–466.

[5] Simpkins, *Malthus,* p. 67.

[6] The Godwin-Wollstonecraft-Shelley relation is one of the more curious and morbid relations in literary history. Percy Shelley was, if not psychotic, at least "strange." He was expelled from school for his defense of atheism (he had originally hoped to become a chemist); he attempted suicide several times; and he had an obsession that he had elephantiasis. Whether his premature death by drowning was a suicide or an accident is not known.

Both Godwin and Wollstonecraft believed in free love and were philosophically opposed to marriage; indeed, Mary had an illegitimate daughter from an American lover she met in Norway who later abandoned her. Despite their beliefs, the Godwins were married so as not to flout convention in Great Britain. From this marriage a daughter, also named Mary, was born, but this birth led to the death of her mother. Mary Wollstonecraft Godwin later eloped with Percy Shelley (he was married at the time), and she is remembered today as the author of *Frankenstein*.

[7] Bonar, *Malthus and His Work*, p. 9.

[8] Appleman, Introductory essay, p. 9.

[9] Gille Granger, "Condorcet, Marie-Jean-Antoine-Nicolas Caritat, Marquis de (1743–1794)," *Dictionary of Scientific Biography* 3: 383–388.

[10] Appleman, Introductory essay, p. 9. Also see Bonar, *Malthus and His Work*, p. 23.

[11] Appleman, Introductory essay, p. 3.

[12] Ibid. p. 7.

[13] Ibid. p. 6.

[14] Simpkins, *Malthus*, p. 471. The first edition appeared in 1798 and the seventh edition appeared in 1872. Thomas Robert Malthus, *An Essay on the Principles of Population* (Reaves and Turner, London).

[15] Bonar, *Malthus and His Work*, p. 1.

[16] His acknowledgment in his preface to the first edition of the *Essay on the Principles of Population* is gracious, "The following Essay owes its origin to a conversation with a friend, on the subject of Mr. Godwin's Essay on avarice and profusion in his *Enquirer*." The friend of course, is Malthus's father, Daniel.

[17] Thomas Robert Malthus, *An Essay of the Principles of Population*, Seventh edition, p. 1.

[18] Ibid. p. 7.

[19] Ibid. p. 4–5.

[20] Ibid. p. 6.

[21] Ibid.

[22] Ibid. p. 10.

[23] Ibid. p. 294.

[24] Ibid. p. 301.

[25] Ibid. p. 303.

[26] Ibid. p. 304.

[27] Ibid. p. 430.

[28] Ibid. p. 431.

[29] Ibid. p. 304.

[30] Ibid. p. 312.

[31] Bonar, *Malthus and His Work.* p. 3.

[32] Richard Godwin, *Of Population: An Enquiry Concerning the Power of Increase in the Numbers of Mankind, Being an Answer to Mr. Malthus's Essay on that Subject* (Longman, Hurst, Rees, Orme, and Brown, London, 1820).

[33] Ibid. p. 15.

[34] Ibid. p. 219.

[35] Ibid. p. 509. The resolution of the debate on Malthus's theory has never ended. Almost all scientists agree that population growth cannot continue indefinitely, but no one can identify an optimal human population size that does not reflect political or personal bias. The zero population growth movement believes much of the tension in the world is caused by overpopulation. Paul Ehrlich (*The Population Bomb*) and Jonas Salk and his son Jonathan Salk (*World Population and Human Values,* Harper and Row, New York, 1981) are the chief advocates among prominent scientists who hold that view. Critics of it include Alan Chase (*The Malthusians*) who provides a history of those neo-malthusians who have used the population issue as a way of dodging responsibility for caring for the world's poor, for addressing urban slums, or for numerous social failures that have existed for millennia and may not simply be a question of population size. The issue is complicated because there are scientists who care about social injustice but who also believe population is a major issue not limited to the amount of food available to populations. Also complicating the problem are those whose position is distorted by religious prohibitions on family planning and the various contraceptive techniques, including elective abortion, that are used throughout the world to achieve birth control.

Evolutionary Ethics before Darwin

U NTIL RELATIVELY RECENT TIMES, apparently healthy-looking infants did not fare well after birth, and it was not uncommon for them to die before they reached their first birthday. In the absence of modern medicine and public health measures, infants were vulnerable to dying from infectious diseases, especially when malnutrition was present. Parents were painfully aware that their infants, no matter how robust and normal at birth, might be carried away by unknown diseases. When children were born with birth defects, past societies, in general, did not regard them with much sympathy. It is very likely that the frequency of infants born with noticeable birth defects has not varied much for our species.[1] In Japan, deformed children were killed or reared according to the wishes of the father.[2] In Roman civilization, Seneca observed that "we drown the weakling and the monstrosity. It is not passion, but reason, to separate the useless from the fit."[3] Much earlier, in Sparta, infants with weak constitutions and birth defects were abandoned to die.

CAUSES OF HUMAN IMPERFECTIONS

The causes of human variation were poorly understood until the 20th century. For most of human history some vague ideas of the inheritance of acquired characteristics were favored. There was a belief in maternal impressions, as in the appearance of a port wine birthmark on a child whose mother was frightened by a fire while she was pregnant. Many acknowledged a paternal transmission of mutilations, such as scar-like lesions on the face of a child whose father, many years earlier, had been slashed in a duel. There was little doubt that alcoholic, malnourished, or tubercular individuals would produce children with similar weak constitutions. The public accepted the belief that well or poorly exercised organs would lead to corresponding development of those organs in the offspring.[4]

Ambroise Paré (1510–1590), a French physician, defined monsters as unnatural births serving as omens for the community.[5] He included among monsters such births as a "child who is born with one arm, another who will have two heads, and additional members over and above the ordinary." Among the causes of monsters he cited: "The first is the glory of God. The second, his wrath. The third, too great a quantity of seed. The fourth, too little a quantity. The fifth, the imagination. The sixth, the narrowness or the smallness of the womb. The seventh, the indecent posture of the mother, as when, being pregnant, she has sat too long with her legs crossed, or pressed against her womb. The eighth, through a fall or blows struck against the womb of the mother, being with child. The ninth, through heredity or accidental illnesses. The tenth, through rotten or corrupt seed. The eleventh, through mixture or mingling of seed. The twelfth, through the artifice of wicked spital beggars. The thirteenth, through Demons and Devils."[6]

Most of the Western world believed, at least through the mid-19th century, that all of humanity descended from Adam (including Eve, who shared his genotype by virtue of being a clonal twin of his rib). Human variation involved not only the occurrence of the healthy and the frail, but also the presence of essentially neutral traits such as skin color, hair texture, body build, and height. These varied among the inhabitants of any nation and even more so when the populations were far removed from one another as among different races. The racial mixture of blacks, whites, or Asians produced children of an intermediate skin color as well as a blending of bodily and facial features. The Reverend Samuel Stanhope Smith, at a meeting of the American Philosophical Society in Philadelphia, shortly after the colonies won their independence from Great Britain, asserted that variations or traits arise from exposure to the "climate and conditions of life." The variations accumulated slowly through generations of living under similar conditions, but they were reversible. He cited tanning and its loss during the winter as an example of cumulative change. In more southern climates the tanning would not be lost but persist all year round and might, in the generations to come, lead to a darker complexion of the children born from such parents. Smith suggested that "color may be justly considered as an universal freckle," subject to the modifying effects of the environment.[7]

LAMARCK'S THEORY OF ACQUIRED CHARACTERISTICS

Although the theory of the inheritance of acquired characteristics goes back to antiquity, the first scientific treatment of that doctrine was developed by Jean Baptiste de Lamarck (1744–1829).[8] Lamarck was a self-taught scientist who made contributions as a botanist, zoologist, chemist, meteorologist, and geologist. He gave the field of biology its name (1801) and is best known for the first systematic theories of heredity and evolution. Lamarck was the youngest of 11 children and was sent by his father, a military officer, to a seminary to become a priest. Instead, Lamarck left the Jesuit school shortly after his arrival and enlisted in the army, where he fought in the Seven Year's War and later was assigned to the Mediterranean. He took up botany and shell collecting as a pastime and returned to Paris, where he hoped to study medicine and science. His knowledge of botany was extensive and he worked in the Jardin du Roi classifying all the flora of France. He wrote in French rather than Latin and followed his *Flore Francaise* (1779) with a *Dictionnaire de Botanique*, these works establishing his reputation as a talented naturalist.

Lamarck observed the difference in appearance of plants when seeds of the same specimen were planted in dissimilar environments. They sometimes resembled different species. He also noted that seeds from plants he collected in the wild produced cultivated plants with noticeably different traits. He called these variations "degradations," and later applied this term to his theory of evolution. The term evolution, however, was not used by Lamarck, nor was it used later by Charles Darwin when they wrote their theories on the origin of life and its diversity. Lamarck initially believed, as did virtually all his contemporaries, that Linnean species were fixed; that is, they did not evolve into new species. His ideas on evolution emerged some 20 years later, in 1800, when he had switched from botany

Jean Baptiste Lamarck was a largely self-educated scholar whose work in botany and zoology launched the fields of biology and evolution. (Reprinted from G.R. D'Allonnes [1911], *Lamarck*, Louis-Michaud, Paris [courtesy of CSHL Archives].)

to zoology. He supported the French Revolution, and since the botanical appointments to the newly reorganized National Museum of Natural History were already filled, he accepted an appointment as Professor of Zoology of Insects and Worms. Lamarck introduced the term invertebrates for these organisms and studied their classification. He made use of the shells he had collected in his youth when he was stationed in the Mediterranean and he now contrasted them to fossil forms. If fossil forms had no modern counterparts they would support the idea of a flood or catastrophe that caused their extinction. But if they included forms that were comparable to surviving species, this would imply a transformation of species rather than their sudden extinction in a flood.

LAMARCK'S EVOLUTIONARY THEORY

Lamarck's theory of evolution was based on a unitary theory of nature. He believed in the classical and then still-extant four-element theory. All matter was thought to be composed of earth, air, fire, and water. Living things, too, he reasoned, were composed of these same elements, but living things make more complex organizations of them. Death permits the degradation of life into the four elements, and much of the mineral world, such as sedimentary rock, has its origins from once living things. Lamarck rejected the idea of a vital force and inferred that fire (energy) somehow maintained the elements in the living state. By adding the dimension of time to the descriptive morphology of organisms, Lamarck explained how gradual changes would lead to new forms. In 1800 his *Systeme des Animaux sans Vertebres* introduced his theory of the "path" or "order of nature," or what would later be called evolution. He assumed that the simplest life arises by spontaneous generation, and as it becomes more complex there are degradations leading to multiple varieties. He assumed that in the natural order of the universe, things tend to become more complex; this arises, among living things, from the special circumstances that species encounter. If they must exert themselves to find food, as by stretching their necks to reach leaves, then the accumulation of the exercised necks over numerous generations will lead to longer necks. Lamarck's theory of use and disuse in the accomplishment of morphological change was not based on a response to a simple desire by the organism, but by the behavioral response the organism made to an altered environment.[9] For the giraffe's neck, that included the stimulation of the neck muscles by the "nervous fluid" that flowed toward them and modified them by repeated attempts to gain food slightly out of reach.

Lamarck's theory of evolution was based more on philosophic principles of how the universe worked than on scientific evidence. He had no evidence that the universe becomes more complex, that the environment directly or indirectly alters heredity, that the habitual use or disuse of an organ produces corresponding hereditary changes in that organ, or that cumulative changes result in the formation of new categories of organisms. Indeed, Lamarck did not emphasize the origin of species; rather, he looked upon evolution producing higher categories, such as Classes or Orders, which then underwent degradation to form clusters of species or genera.

CRITICISM OF LAMARCK BY HIS CONTEMPORARIES

Unfortunately for Lamarck, his reputation as an unlettered scholar and his belligerence toward his critics made him an unpopular colleague in the academies and museums. His work was considered outdated, idiosyncratic, and too philosophical to be of use for further inquiry. This was partly because he stubbornly clung to the four-element theory of matter long after it had been abandoned in favor of the new chemistry based on experimentation. He was also difficult to approach because his life was embittered by the death of two of his wives and three of his eight children, and the failure of all save one to have a career and children (one was deaf, one insane, and two daughters were never married). He also had the infirmity of progressive blindness that limited his activities in his later years. The eulogy for Lamarck was assigned to the Secretary of the Academy, Georges Cuvier (1766–1832), a longtime foe of evolutionary theories and a champion of catastrophe theories, who condemned and ridiculed most of Lamarck's views. Cuvier may have intended his eulogy as a warning to colleagues not to speculate about the mutability of species; he may also have belittled Lamarck out of many years of frustration dealing with him as a colleague. The attack proved too strong for the Academy and was not published until Cuvier's death.[10]

Lamarck's views were debated in France, but almost universally condemned elsewhere. Evolution was known to contemporary English scholars as the development hypothesis. It appealed to Erasmus Darwin (1731–1802), Charles Darwin's grandfather, as a satisfying philosophy suitable for poetic treatment; his views, however, were based on the writings of Democritus and Lucretius rather than Lamarck, and they were not taken seriously by contemporary naturalists. A more serious effort to promote the development hypothesis appeared in 1844.

CHAMBERS'S EVOLUTIONARY THEORY

The *Vestiges of the Natural History of Creation* was anonymously published, its author fearful of personal and professional abuse if his name were associated with it.[11] Robert Chambers (1802–1871) did not acknowledge his authorship while he or his wife was alive. He and his brother were self-taught scholars who became publishers and magazine editors as well as writers of local note in Edinburgh. Chambers had a passion to learn, and he was stimulated by the newest developments in science. He had a gifted amateur's interest in geology and published some professional articles that gained him election as a Fellow of the Royal Society of Edinburgh. He wrote histories, biographies, encyclopedias, and weekly articles on science and invention. *The Edinburgh Journal,* jointly owned and edited by the Chambers brothers, was designed to entertain the newly arising middle class; topics were chosen to "elevate and instruct" and offend no one. When

VESTIGES

OF

THE NATURAL HISTORY

OF

CREATION.

WITH A SEQUEL.

NEW YORK:
HARPER & BROTHERS, PUBLISHERS,
329 & 331 PEARL STREET,
FRANKLIN SQUARE.
1862.

Robert Chambers published the best-selling *Vestiges of the Natural History of Creation* anonymously in 1844, which presented the entire evolutionary history of the earth as he saw it. (*Left,* Reprinted from E.M. Ward [1924] *Memories of Ninety Years*, Hutchinson and Co., New York; *right*, courtesy of CSHL Archives.)

he was approaching 40 years of age, Robert Chambers began forming an all-embracing theory of the world that used the development hypothesis to account for the origins of the stars, the earth, its physical features, and the life upon it. Unlike Lamarck, whose theory of evolution avoided religious or vitalistic interpretations, Chambers adopted a philosophy intended to praise the wisdom of a Creator who used natural laws rather than miracles to create his universe.

In 1841, Chambers moved his family to St. Andrews, where for the next three years he put together the ideas of Comte, Laplace, Lamarck, von Baer, Quetelet, and other contemporary and influential scholars. The resulting book had the distinction of being a runaway best-seller (eleven editions and almost 25,000 copies sold) and a universally condemned work. Chambers accepted the nebular hypothesis for the origin of stars and planets from gases. He used the 55 known elements as the material basis for all matter. He believed that life was diffused throughout the universe and that the fossil record showed an ascending complexity. Unlike Lamarck's gradually accumulating changes through use and disuse, Chambers's new life-forms arose through gross embryonic changes resulting in "sports," as the horticulturists and farmers described them.[12] The preference for an embryological model came from two sources. Karl Ernst von Baer (1792–1876) had firmly established the epigenetic model of development in which new organs appeared from simple rudiments rather than by an enlargement of smaller, preformed organs. Von Baer noted the similarity of all vertebrate embryos in their earliest stages and proposed a "law of corresponding changes" that took place whether the embryo was that of fish, chicken, pig, or human. Although later evolutionists seized on von Baer's observations, he rejected evolution as the explanation. He believed in an archetypal model of a universal form shared by the vertebrates. "Are not all animals," he pointed out, "in the beginning of their development alike, and is there not a primary form common to all?"[13] Also supporting a sporting mutability of life in Chambers's theory of evolution was a family trait. Both Robert and his brother William were born with six digits on each hand and foot (hexadactyly). Although the extra digits were surgically removed in early childhood, Robert Chambers suffered from slightly deformed feet that gave him pain and limited his physical activities.

Chambers's *Vestiges,* despite its hostile reception by the clergy and scientists who read it, challenged the prevailing view established by the book of Genesis. Chambers rejected the belief that the earth was about 6000 years old; he claimed that the death of living organisms existed long before the fall of Adam; and he disputed the sequence of events associated with the six biblical days of creation.

Throughout the 19th century, human variation was perceived as a topic for scientific investigation as well as a social problem. Speculative ideas from philosophy, science, religion, politics, and social commentary flowed back and forth among scholars. The result was a coexistent collection of contradictory views and few well-established facts or theories to sort out the meritorious from the superficial or biased treatments. One of the most influential participants in this ferment of debate was Herbert Spencer (1820–1903).[14]

HERBERT SPENCER AS SOCIAL PHILOSOPHER

Spencer was the eldest of seven children, and the only one of his siblings to survive infancy. His father was a private tutor for the well-to-do and a person of independent opinions; he left the Methodism of his youth for the Society of Friends (Quakers). Spencer was educated by his father until he was 13 and by his uncle, a reverend, for the next three years. He never attended college and taught himself by collecting specimens, hanging around his father's friends, and reading prodigiously on every topic he encountered. He studied engineering on his own, and at 17 became a civil engineer, devoting the next nine years of his life to the burgeoning railroad industry. During his spare time he read Lyell's work on geology and Lamarck's theory of evolution. He believed that if the universe is governed by laws, comparable laws must exist for society itself. He considered himself a radical liberal in politics and began to write articles for radical journals. He quit the railroads and tried his hand as a journalist and then became an editor of *The Economist,* where he began to develop his social theories based on natural law.

Spencer's first book, *Social Statics* (1850), established him as an original thinker, a philosopher with a profound grasp of contemporary issues, and a social critic whose ideas provoked discussion among all literate classes.[15] Through his writings he made the acquaintance of novelist Marian Evans (George Eliot), but despite her love for him, he never proposed to

SOCIAL STATICS,

ABRIDGED AND REVISED;

TOGETHER WITH

THE MAN *VERSUS* THE STATE.

BY
HERBERT SPENCER.

NEW YORK:
D. APPLETON AND COMPANY.
1893.

Herbert Spencer astounded his contemporaries with *Social Statics*, a radical interpretation of society based on scientific rather than religious principles. It led to Social Darwinism, a field Darwin never acknowledged. (*Left,* courtesy of CSHL Archives; *right,* courtesy of the New York Public Library.)

her although they remained life-long friends. She was the only woman he considered marrying and when that fell through, he remained a bachelor. In 1854 Spencer's uncle left him a modest bequest which made him independent for life. He chose to devote his life to writing a series of books that would explore all the major features of the universe and unite them in a "synthetic philosophy."

Spencer's output was prolific. He wrote *Education, First Principles, Principles of Biology, Sociology, The Man Versus the State,* and dozens of articles expressing his views. He lived frugally; he never bought property and preferred a reclusive life in his lodgings. He was eccentric in dress and habit, refusing all honors conferred upon him, declining the invitations of royalty, and preferring the work habits of a drudge. He introduced the term evolution in the 1850s to replace the development hypothesis which he derived from reading Lamarck and Chambers. He also coined the term "survival of the fittest" in 1852, six years before Darwin's publication of his theory of the origin of species by natural selection.[16] Spencer, like Chambers and Lamarck, derived his evolutionary views from philosophic principles rather than from scientific evidence.

SOCIAL STATICS AS AN OUTCOME OF NATURAL LAW

The idea of progress appealed to Spencer. It was part of Lamarck's principle of natural increases in complexity, and it was grafted by Chambers to biological development using von Baer's epigenetic unfolding of specific and predictable stages of organ formation. Spencer conceived of his *Social Statics* as a model for developing society from a morality based on natural law. Spencer rejected the idea that human beings are innately rigid in their behavior and favored a more positive image of human potential: "We must either affirm that the human being is wholly unalterable by the influences that are brought to bear upon him—his circumstances, as we call them; or that he perpetually tends to become more and more unfitted to those circumstances; or that he tends to become fitted to them. If the first is true, then all schemes of education and governance, of social reform—all instrumentalities by which it is proposed to act upon man—are utterly useless; seeing that he cannot be acted upon at all. If the second is true, then the way to make a man virtuous is to accustom him to vicious practices, and vice versa. Both of which propositions being absurd, we are compelled to admit the remaining one."[17]

In Spencer's scheme, evil arises from a failure to adapt to circumstances, "all imperfection is unfitness to the conditions of existence" and hence, in the long run, "all imperfection must disappear." Progress is not an accident that might fortunately happen in a society; it is a necessity. The basic unit for any society is the individual. The governing principle that leads to a society is freedom: "Every man has freedom to do all that he wills, provided he infringes not the equal freedom of any other man."[18]

To pursue this freedom, Spencer argued, the individual has a right to the use of the earth. Land might be leased, but it can never be owned. If individuals are free to reap the fruits of the earth, there should be no restriction on that pursuit. He criticized those who restrict women's equal right to pursue their freedom. "Equity knows no difference of sex," he claimed, and he believed that the alleged inferiority of women arose from inequality of education and opportunity.[19] Another implication from Spencer's opposition to the ownership of land is that colonialism and the attempt to acquire land through warfare are morally unjustified. "No invader ever raised standard but persuaded himself that he had a just cause. Sacrifices and prayers have preceded every military expedition, from one of Caesar's campaigns down to a border foray. 'God is on our side' is the universal cry. Each of two conflicting nations consecrates its flags; and whichever conquers sings a *Te Deum*. Attila conceived himself to have a 'divine claim to the dominion of the earth'; the Spaniards subdued the Indians under plea of converting them to Christianity, hanging thirteen refractory ones in honor of Jesus Christ and his apostles; and we English justify our colonial aggressions by saying that the Creator intends the Anglo-Saxon race to people the world."[20]

APPRAISAL OF SPENCER'S PHILOSOPHY

Spencer's views, seen more than a century later, are difficult to classify. Many of his views fit conservative philosophy, many more are pacifist, liberal, and radical. Some of his views would meet with almost universal rejection. He rejected government by need, restricting its role to the narrow function of protecting the right to be free. In his scheme both the rich and the poor had no claim on the government: "The ruling classes argue themselves into the belief that property should be represented rather than person—that the landed interest should preponderate. The pauper is thoroughly persuaded that he has a *right* to relief."[21] Spencer rejected the idea of public education because "all institutions have an instinct of self-preservation growing out of the selfishness of those connected with them." He feared that state education would be an indoctrination in public beliefs and policy. Instead, he argued, "education, properly so called, is closely associated with change, is its pioneer, is the never-sleeping agent of revolution, is always fitting men for higher things and unfitting them for things as they are."[22] Education should not be coercive, as it would be in a state school and as it was in the elite schools he rejected when he was a youth. "Do but gain a boy's trust," he argued, "convince him by your behaviour that you have his happiness at heart; let him discover that you are the wiser of the two; let him experience the benefit of following your advice, and the evils that arise from disregarding it; and fear not, you will readily enough guide him." Perhaps the most radical of all Spencer's views on the freedom of the individual was his claim that the individual has the right to ignore the state. He condemned the potential tyranny of a majority in a democracy that must have some checks to prevent its enacting murder, enslavement, or robbery.[23]

Spencer had sympathy for the poor and recognized the prejudices against them although he was firm in his belief that it was not the function of the government to feed them, house them, or find them employment. "It is a pity that those who speak disparagingly of the masses have not wisdom enough, or candor enough, to make due allowance for the unfavorable circumstances in which the masses are placed."[24] The poor behaved as they did, lived as squalidly as they did, and accomplished so little, not because of some innate failing, but because they were deprived by landowners of the right to use the earth. "Conceive yourself one of a despised class contemptuously termed 'the great unwashed'; stigmatized as brutish, stolid, vicious; suspected of harbouring wicked designs; excluded from the diginity of citizenship; and then say whether the desire to be respectable would be as practically operative on you as now."[25]

PERFECTION THROUGH THE ELIMINATION OF THE UNFIT

Although Spencer did hold sympathies for the problem classes of society and recognized how they arose, his response to their plight was limited. He opposed public charity because it would extinguish a sense of sympathy and justice in the public. Private charity, judiciously given, fostered that sympathy and restored the imbalance that fate sometimes deals to a suffering individual. But at the same time, Spencer accepted a higher value than sympathy—the public good served by nature, which eliminates the old, the weak, and the imperfect.[26] "The poverty of the incapable, the distresses that come upon the imprudent, the starvation of the idle, and those shoulderings aside of the weak by the strong, which leave so many 'in shallows and in miseries,' are the decrees of a large, far-seeing benevolence." Instead of an "invisible hand" guiding the economic fortunes of capitalists in a free market, Spencer saw a "purifying process" in the elimination of the socially unfit.[27] He condemned public charity for its checks on this process and berated those who "in their eagerness to prevent the really salutary suffering that surround us, these sigh-wise and groan-foolish people bequeath to posterity a continually increasing curse."[28]

Spencer believed that his age was in a state of transition to perfection. The progress that led to industrialization and the substitution of rational inquiry for superstitious or prejudicial belief would lead to more employment, better health, more democracy, a desire for self-education, a more tolerant and humane populace. Society would do so by the invisible guidance of the principles of complexity and progress when people are free to do what they want in accordance with the principle of the freedom of the individual and the withering away of the government's interference with that freedom. No process of improvement was without difficulty, however, and he recognized that those who did not adapt to the new directions of social evolution would be weeded out. "Every attempt at mitigation of this eventuates in exacerbation of it. All that a poor law or any kindred institution can do is partially to suspend the transition—to take off for a while from certain members of society the painful pressure which is effecting their transformation."[29]

Like Ambroise Paré three centuries earlier, Spencer denounced "organized begging; which has made skillful mendicancy more profitable than ordinary manual labor; which induces the simulation of palsy, epilepsy, cholera, and no end of diseases and deformities; which has called into existence warehouses for the sale and hire of impostor's dresses....The unthinking benevolence which has generated all this cannot but be disapproved by everyone."[30] If indiscriminate giving leads to bad results; judicious charity can be a corrective. "To that charity which may be described as helping men to help themselves it makes no objection—countenances it, rather. And in helping men to help themselves, there remains abundant scope for the exercise of a people's sympathies. Accidents will still supply victims on whom generosity may be legitimately expended....Even the prodigal, after severe hardship has branded his memory with the unbending conditions of social life to which he must submit, may properly have another trial afforded him."[31]

Along with the failings of beggary, Spencer similarly rejected private or state efforts to prevent crime. "Crime is incurable, save by that gradual process of adaptation to the social state which humanity is undergoing. Crime is the continual breaking out of the old unadapted nature—the index of a character unfitted to its conditions—and only as fast as the unfitness diminishes can crime diminish."[32]

Extensions of Spencer's Views

Many of Spencer's readers must have felt contradictory emotions as they read on and saw their own prejudices upheld in one chapter and condemned in another. A Victorian businessman might have applauded Spencer's views on the Poor Laws, beggary, and crime. But he would have shuddered to read that colonization is thievery on a large scale; that it is costly, does not permit fair trade; prevents self determination; and brutalizes the populace and makes them hostile.[33] Similarly, his ideas on the government's role in regulating business were contradictory for those who followed his reasoning. He felt it proper that the government should block anyone who "unnecessarily vitiates the elements and renders them detrimental to health, or disagreeable to the senses." But the government cannot take on positive or regulative actions, such as establishing public health policy, setting tariffs to protect domestic trade, imposing sales taxes, restricting industries (e.g., the sale of alcoholic beverages), or licensing professions. Spencer believed a person had a right to practice medicine, law, pharmacy, engineering, or any other trade without a state-mandated procedure for certification. Only those individuals practicing their competence or lack of competence would in the long run succeed or fail to meet the standards of private groups or to attract customers on their own.

Spencer saw in civilization "a progress toward that constitution of man and society required for the complete manifestation of everyone's individuality." As the state made the transition from the dominance of the state over the individual to the dominance of the individual over the state, the citizen "must become impressed with the salutary truth that no one can be perfectly free till all are free; no one can be perfectly moral till all are moral; no one can be perfectly happy till all are happy."[34]

SPENCER'S PHILOSOPHY TAKEN OVER AS SOCIAL DARWINISM

What was later to be called social Darwinism thus begins before Darwin's evolutionary views were published, with the *Social Statics* and its curious ethics. What is called "social Darwinism" should really be called "social Spencerism." There was no evolutionary implication, just a perfecting or stabilizing of the best features of the human species that was proposed by Spencer in *Social Statics*. Spencer is quite clear on this point when he states "Mark how the diseased are dealt with. Consumptive patients, with lungs incompetent to perform the duties of lungs, people with assimilative organs that will not take up enough nutriment, people with defective hearts that break down under excitement of the circulation, people with any constitutional flaw preventing the due fulfillment of the conditions of life are continually dying out and leaving behind those fit for the climate, food, and habits to which they are born....And thus is the race kept free from vitiation." It is vitiation, not the ascent to a new species that concerned Spencer. The term "social Darwinism" augments the basic Spencerian lack of sympathy for the unfit by adding the rationalization of progress in evolution to a higher type.[35] Spencer offered it as a nonconformist, a pacifist, a liberal, and a radical opposed to racism, sexism, the ownership of land, colonization, and the established Church of England. But he also opposed public education, public health, public welfare, and the regulation of business, trade, and the professions. Thirty years later, Spencer's evolutionary ethics became the underpinning for what he bitterly opposed—rampant imperialism, laissez faire exploitation of the poor, and state-imposed restrictions on the dangerous classes, the paupers, the feeble-minded, and the potentially criminal. He lamented the support of his old foes: "Oddly enough I am patted on the back by the Conservatives, which is a new experience for me."[36]

Although Spencer applied his ethics and principle of development or evolution to the emergence of a more perfect society, he did not conceive it then in the same evolutionary way that Lamarck and later Darwin were to do. He missed the implication of the weeding out of the unfit for species formation and saw it then as a way to maintain or perfect, rather than transform, the human species. After Darwin introduced the idea of evolution through natural selection, Spencer accepted that view as a logical extension of his own views. Spencer differed from Darwin in accepting Lamarck's major mechanism for evolution, the inheritance of acquired characteristics through the gradual accumulation of the effects of use and disuse.

Spencer used the biological metaphor of the organism, in an inverse way, for his model of society. The individual was the basic unit in society just as the cell, in the newly developed cell theory, became the unit composing the organism. In higher animals the cells are subservient in their tissue functions to the organism, whose dictates are founded in the brain. Society did not have such narrowly committed specialization, and all the units contributed to social policy. Furthermore, Spencer firmly believed in the supremacy of the individual over the state. For both the organism and society, development determines the stages of complexity and permits it to grow and differentiate.

While *Social Statics* created a stir and gave Spencer the recognition he wanted as a trenchant thinker of the times, he remained an outsider; no one political party or established institution was pleased with his overall assessment. His greatest appeal was to the new middle class who had fewer ties of loyalty to established classes, such as the nobility, that offered them no mobility or opportunity. Spencer's greatest success was in the United States where his works were widely published and revered through the 1890s. In Great Britain, Spencer's social views were rejected and then ignored as the policies he most denounced were put into action—a massive growth of the British Empire, protective tariffs, public education, and regulation of all institutions, businesses, and professions. Spencer became more pessimistic as this transformation took place, and he doubted the linearity of progress and accepted regression as well as a halting and jerking movement toward his ideals.

Social Statics is difficult to classify. It is neither economics, sociology, philosophy, biology, nor political science. It is a prescription for a nonfiction Utopia, based on projections from a principle of individual freedom and its proper function.

Footnotes

[1] About 5–7% of all newborns have an abnormal condition requiring medical attention. Prior to the 20th century, such children usually died within days or months of their birth, and hence most of the abnormalities were collectively lumped under categories like "failure to thrive," "neonatal mortality," or "multiple birth defects." After the introduction of antibiotics, such children would survive a year or more,

the more striking cases being referred to as "FLK" (funny looking kid). Since the era of human genetics as a medical science (after 1957), an attempt was usually made to analyze each case of a birth defect.

2 A.G. Roper, *Ancient Eugenics* (Oxford University Press, Oxford, 1913), p. 9.

3 Ibid. p. 12.

4 Environmental factors can alter the uterine environment and cause birth defects. Ergot poisoning could lead to thalidomide-type abnormalities. German measles (rubella) can cause blindness, deafness, mental retardation, or cardiac problems. There is no evidence, however, that maternal experiences of a behavioral sort, such as listening to music, reading, sudden frights, or witnessing crimes, while pregnant will cause talents or imperfections of a corresponding type in their children.

5 In fact the root *monstra*, in the word "demonstrate" is the same as in the word "monster"—to signify.

6 Ambroise Paré, *On Monsters and Marvels,* translated from the 1573 edition by Janis Pallister. University of Chicago Press, Chicago, 1982, pp. 3–4.

7 Stephen Jay Gould, "An Universal Freckle" *Natural History* 96(1987): 14–20, p. 18.

8 Two fine resources on Lamarck's life and work may be found in L.J. Burlingame, "Jean Baptiste Lamarck" *Dictionary of Scientific Biography* 7: 584–594 and L.J. Jordanova, *Lamarck* (Oxford University Press, Oxford, 1984).

9 The scientific rejection of Lamarck's theory of the inheritance of acquired characters is based in part on the absence of any chemical or physical model by which the environment can organize hereditary material, directly or indirectly, to fit the organism's adaptive needs. In addition, all experimental tests of alleged inheritance of acquired characteristics have failed. Despite the repeated history of fraud and error in the work of those proposing this popular theory of heredity, it recurs every generation. See Arthur Koestler's *The Case of the Midwife Toad* for the attempt to rehabilitate Paul Kammerer, one of the most effective advocates of this theory. Few scientists since 1948 have considered the theory of acquired characteristics as having much merit because that theory was championed with political support by Trofim D. Lysenko, who seriously damaged Soviet life sciences by his campaigns against genetics.

10 Jordanova, *Lamarck*, p. 101.

11 Milton Millhauser, *Just Before Darwin: Robert Chambers and Vestiges* (Wesleyan University Press, Middletown, Connecticut, 1959).

12 In plants they are often called "bud sports." Darwin described them in detail in his *Variation of Animals and Plants Under Domestication*. He felt they were too monstrous in their novelty to survive under natural conditions and that only gradual changes had an opportunity for survival. Navel oranges (and other naturally arising seedless fruit) owe their origins to bud sports.

13 "Karl Ernst Von Baer," *Encyclopedia Britannica* 2: 920. The idea of a "common form" was also part of a romantic movement in science (known as natural philosophy) particularly championed and developed by the poet Johann Wolfgang Goethe (1749–1832). This theory believed in a Platonic ideal type with variations emanat-

ing from it, such as an ideal leaf transformed into buds, petals, and other structures in a plant. It was not until the mid-19th century that the comparative anatomy of plants and animals sought an evolutionary basis (rather than a Platonic ideal) for the relation of similar structures (called homologous structures by anatomists). See Eric Nordenskiold, *The History of Biology* (Tudor Publishing Co., New York [reprint of 1928 Alfred Knopf, Inc. edition], Chapter XIII, pp. 268–285).

[14] J.D.Y. Peel, *Herbert Spencer: The Evolution of a Sociologist* (Basic Books, Inc., New York, 1971).

[15] Herbert Spencer, *Social Statics.* Reprint of the 1850 edition. (Robert Schalkenbach Foundation, New York, 1954).

[16] It should be kept in mind that Spencer thought of "the survival of the fittest" not as a mechanism for forming new species, but as a mechanism to weed out the unfit or degenerate members of a species. It was to preserve the species, not to create new ones, that his aphorism applied. Darwin liked the simplicity of the phrase and accepted it as a suitable consequence for his theory of natural selection leading to the origin of new species.

[17] Spencer, *Social Statics,* p. 57.

[18] Ibid. p. 95.

[19] Ibid. p. 138.

[20] Ibid. p. 142.

[21] Ibid.

[22] Ibid. p. 305.

[23] Ibid. p. 188.

[24] Ibid. p. 202.

[25] Ibid. p. 204.

[26] Ibid. p. 288.

[27] Ibid. p. 289.

[28] Ibid. p. 290.

[29] Ibid.

[30] Ibid. p. 291.

[31] Ibid. p. 292.

[32] Ibid. p. 314.

[33] Ibid. p. 322.

[34] Ibid. p. 409.

[35] I suspect that even without Darwinism the Spencerian philosophy of abandoning the unfit would have been popular in the latter half of the 19th century. For a detailed account of that philosophy in its "Darwinian" form, see Richard Hofstadter, *Social Darwinism in American Thought* (George Braziller, New York, 1959).

[36] Peel, *Herbert Spencer,* p. 229.

Hereditary Units and the Pessimism
of the Germ Plasm

F OR MANY CENTURIES PEOPLE, particularly farmers, were aware that
domestic plants and animals had to be culled of their weakest or
imperfect members, and the breeding stock or seed for the next generation
should be selected from among the best of the crop or herd. Biblical tradition, although not specific on matters of inheritance, implied that whatever mechanism might account for the origin of variations, such traits,
once produced, were generally fixed.[1] An Old Testament statement on
heredity appears in Genesis, where the theory of the inheritance of
acquired characteristics is used to describe Jacob's efforts at animal breeding. Jacob was frequently deceived by Laban, his father-in-law. In one incident, Jacob asked to be paid in variegated sheep, black lambs, and variegated goats from among the herds. Laban secreted the variegated goats
and black lambs from his flocks. Jacob then avenged the wrong done to
him by using rods derived from the partially peeled trunks of saplings
from several varieties of trees and planted these by the watering troughs of
the sheep and goats. The animals mated where they drank and the effect
of the variegated rods they saw was to produce numerous variegated offspring. Those that were not variegated he gave to Laban; those that were
he kept separately for his own family. "Whenever the stronger of the flock
were breeding, Jacob laid the rods in the runnels before the eyes of the
flock, that they might breed among the rods, but for the feebler of the

flock he did not lay them there; so the feebler were Laban's, and the stronger Jacob's" (Genesis 30:31–43).

Although Jacob's approach involved an environmental agent to modify directly the heredity of the sheep and goats, his technique is clearly not one of use and disuse. Jacob's approach might be considered the inheritance of sudden or temporary impressions, the rods being deliberately made variegated to convey that symbol to the mating animals. It also differs from later examples of maternal impressions and analogous phenomena of acquired characteristics because most (or all) of the offspring turned out as Jacob desired.[2]

In the early 19th century, Lamarck recognized that the problem of evolution was bound with the problem of heredity. There can be no evolution to more complex forms unless there are variations that produce them. Lamarck did not speculate on the nature of the material being transformed, but he believed it to be a chemical or physiological process brought about by the exercise of organs. Similarly, Chambers drew the parallel of embryonic development from a minute cell into a newborn organism and the appearance, over longer periods of time, of new complex forms from immediately preceding simpler ones. In Chamber's embryological model, "sporting variants" (what might be called mutations today) played a major role.

HEREDITY AS A PRODUCT OF PHYSIOLOGICAL UNITS

The first attempts to identify some underlying reality for hereditary units appeared in Herbert Spencer's *Principles of Biology* (1863) and were based on philosophic inference. The second appeared in 1865 in a Czecho-slovakian journal that was neither widely read nor appreciated among those who did read it, although some 150 copies of the journal were sent to major libraries around the world. The author of that theory of heredity was Johann Gregor Mendel (1822–1884), and he based his theory on the results of breeding peas. The third attempt to understand heredity at a more fundamental level was proposed by Charles Darwin (1809–1882). His ideas appeared in 1868 in *Variations of Animals and Plants under Domestication*. He made his inferences based on a review of the known published works on heredity and variation.[3]

Spencer, Mendel, and Darwin independently conceived that heredity depended on the existence of special units. Spencer called his "physiological units." Mendel's term for the rudiments that gave rise to "differentiating characters" might best be translated as "elements"; in 1902 Bateson translated them as "unit characters." Darwin referred to his units as "gemmules." All three differ from the term "gene" as we would understand a unit of inheritance today, but they share with that term the belief that heredity in some fundamental way is particulate rather than diffuse or holistic.

SPENCER'S THEORY OF HEREDITARY UNITS

Spencer's *Principles of Biology* was the second in his multivolumed work on synthetic philosophy; it followed *First Principles* and tried to relate evolution as a universal principle to the world of life. He claimed that "organic bodies, which exhibit the phenomenona of Evolution in so high a degree, are mainly composed of ultimate units having extreme mobility."[4] In contrast to "chemical units," such as atoms and molecules, which were too small; and morphological units, such as cells, which were too large and relatively immobile in the organism, Spencer predicted a third class of fundamental units: "...We must conceive it as possessed by certain intermediate units, which we may term *physiological*....In each organism the physiological units produced by this further compounding of highly compound atoms, have more or less distinctive character. We must conclude that in each case, some slight difference of composition in these units, leading to some slight difference in their mutual play of forces, produces differences in the form which the aggregate of them assumes."[5]

Spencer's physiological units, by undergoing new arrangement, led to "spontaneous" variations. Since each parent contributed a reproductive or germinal cell that was filled with physiological units, the offspring would blend most of the parent's traits. These units had the capacity to grow in numbers as the cells multiplied. Spencer used the idea to explain "genesis, heredity, and variation." Throughout the 19th century, the terms heredity and variation were considered separate phenomena caused by different mechanisms. Heredity was the like-to-like transmission of traits, especially those that mattered, such as the fundamental traits of a species. Variations were thought to be disturbances or alterations in traits, some of which might be inherited, but most of which were of significance only to the individual.[6]

MENDEL'S UNITS OF HEREDITY

Johann Gregor Mendel's contributions to an understanding of heredity had little effect on 19th-century thought, but once rediscovered, they totally eclipsed the ideas of Darwin and Spencer for both scientific and social investigations of heredity in the 20th century. For 35 years, Mendel's work, although published, remained ignored or rejected. Unlike almost all 19th-century scientists, Mendel arose from the peasant class and would have remained illiterate had it not been that his village of Heinzendorf, in what is now the western part of the Czech Republic (also known as Sudetenland), opened a public school. When his talents as a student were recognized by his teachers, his younger sister sacrificed part of her dowry so he could attend the equivalent of a community college where he studied physics. Overwork and malnutrition twice broke his health, and he finally obtained, in 1843, the only subsidized education available by joining an Augustinian seminary to study for the priesthood.[7]

Mendel adopted the name Gregor after being ordained in 1847. He turned out to be unsuitable for the priesthood, becoming ill at the sight of the sick and the dying. He then tried his hand as a substitute teacher of science, but by 1850 he had not learned enough on his own to pass a licensing examination. His superiors sent him to the University of Vienna to study physics and natural sciences. There he took a variety of courses in the sciences, including physics with Christian Doppler, combinatorial mathematics with Andreas von Ettinghausen, and biology with Felix Unger, a student of the founder of cell theory, Matthew Schleiden. While there, he joined a science club and presented two papers, both on agricultural insect pests. He had been given three years of support to finish a degree but failed to do so and attempted to pass the licensing examination on the basis of his coursework. Once again he failed to pass and returned defeated to his monastery in 1853, his head wrapped in bandages although he had no physical defect. For the rest of his teaching career he remained a substitute or supply teacher.[8]

Gregor Johann Mendel introduced breeding analysis of hereditary traits and worked out two major laws that are the basis of classical genetics. (Reprinted from H. Iltis [1924] *Gregor Johann Mendel: Leben, Werk und Wirkung,* Verlag von Julius Springer, Berlin [courtesy of CSHL Archives].)

Mendel began his breeding experiments shortly after his return; he used the monastery's garden and selected peas to test out a theory he developed to explain the sharply contrasting traits, such as flower color, that existed in such profusion among ornamental and domesticated plants. He wrote to seed supply companies and after a few years of testing some 34 varieties for the fidelity of their traits, he selected seven traits for serious study.

The atomic theory, then almost a half-century old, gave physicists and chemists a particulate view of the universe; cell theory, first introduced in 1838, gave a corresponding view of animal and plant life made up of fundamental morphological units, large enough to be seen by the crude microscopes of the mid-19th century. Mendel, like Spencer, assumed that there might be a particulate basis for heredity, because he was struck by the consistency of these traits as he followed their passage in and out of the hybrids he produced in his garden. The cell theory was formulated independently by Matthew Schleiden and Theodor Schwann in 1838. Schleiden had shifted from law to medicine and finally to botany. Although cells were named by Robert Hooke in 1665, they were thought by him to be empty boxes that provided bouyancy to cork. Schleiden realized that the cell's contents were significant in his studies of plant tissues and he proposed a universal argument that plants are communities of cells. He met Schwann by accident at a train station, and Schwann told Schleiden that he was led to the same conclusion studying animal tissues. They became friends and mutual supporters of the cell theory. Both Schleiden and Schwann erred in assuming that cells crystallized out of a liquid and that the nucleus seen in some cells was an example of such condensation. This early model was called the free formation of cells, a concept killed a generation later by Rudolf Virchow, who argued that cells arose from preexisting cells. The work of Schleiden and Schwann opened up a new field of medicine (microscopic anatomy) and a new field of biology (cell biology and cytology) as the contents of cells were made visible by advances in handling, preserving, and staining tissues, cells, and their components. These events of 1840–1870 occurred in parallel with the emerging fields of evolution, biology, applied breeding, and comparative anatomy. As the 20th century began, these fields were united by the findings in genetics, collectively called classical genetics.

MENDEL'S EXPERIMENTS AND THE LAWS OF HEREDITY

From his nine years of experiments, Mendel drew four major conclusions. He had established the constancy of the traits he extracted from hybrids. He discovered that the hybrids usually showed only one of the two parental traits he used. If he dusted pollen from a plant with green peas onto the emasculated flowers of a plant with yellow pea color, the offspring peas were uniformly and indistinguishably yellow. From such hybrids he obtained, by self-fertilization, offspring that produced the originally green or yellow pea color. Since Mendel kept a record of each pair of flowers he crossed and followed the fate of all their progeny, he applied his combinatorial mathematics and found that the ratios were consistent with his unit hypothesis. Each parent strain contributed a hereditary unit, one for the determination of the green color and the other for the determination of the yellow color. The resulting hybrid only expressed one of these traits. Mendel called the expressed trait the dominant form (in this case, yellow peas), and the latent or nonexpressed trait (the green peas) he called the recessive form. From the self-fertilization of such hybrids, he obtained   the extracted recessive and he obtained two types of the dominant or expressed trait. Two-thirds of the dominant forms obtained from these hybrids were hybrid like their parent. One-third were as pure and stable as the original packet of seeds he received from the supplier. Also, the extracted recessive or green pea plants remained true-breeding although they had resided for a generation in a hidden form within the hybrid yellow plants. When Mendel studied two or more pairs of traits at the same time, as in plants that had green or yellow peas residing in dented or inflated pods, the results were consistent with the independent associations of these traits and the ratios fell into a simple combinatorial algebraic ratio.

Mendel presented his work to the local science society at Brunn. His work was well received, but no one there thought it of great significance. Mendel also sent copies of the article to several prominent biologists who were publishing articles and books on heredity. He began a correspondence with the person Unger had recommended to him, Carl Nageli, and even sent packets of his hybrid seeds with predicted ratios for Nageli to confirm. Unfortunately, Nageli was breeding hawkweeds (*Hieracium*) and did not find Mendel's laws among them.[9] He sent hawkweed seeds to Mendel, and for several years Mendel tested them and could not confirm his own results. Neither Mendel nor Nageli knew that in hawkweeds most of the pollen dusted on the flowers does not enter the embryo sac after moving down the female apparatus. As a result, the flower produces offspring resembling the maternal plant and not the pollen donor, the process resembling parthenogenesis or "virgin birth." Additionally, the hawkweeds used by Nageli and Mendel were a mixture of separate species and varieties rather than simple varieties of a single species. In Mendel's work, all the varieties were of the same species because garden peas were domesticated from one species.

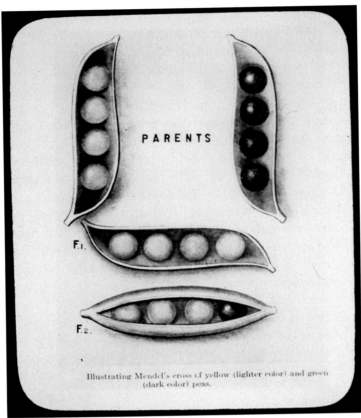

PARENTS

F.1.

F.2.

Illustrating Mendel's cross of yellow (lighter color) and green (dark color) peas.

A teaching slide used by the Eugenics Record Office to illustrate yellow × green peas in the parental generation produce yellow peas in the F_1 generation. When the F_1 peas are self-fertilized, a 3 to 1 ratio of yellow to green peas emerges in the F_2 generation. (Reprinted, with permission, from Harry H. Laughlin Archives, Truman State University, courtesy of CSHL Eugenics Web site.)

MENDEL'S FRUSTRATION WITH HAWKWEEDS

Mendel's second paper to the Brunn science society, which he also published, was on his hawkweed studies, and it is a melancholy admission that his pea laws were not universal. Mendel had a weak ego and did not pursue heredity in plants. He took up beekeeping, studied meteorology, wrote a weekly chess column for the local paper, served on the board of a local bank, and enjoyed his elevation to the status of prelate of the monastery. He later became involved in a bitter squabble with the emperor's new tax laws and refused to pay an annual tax on his monastery. He developed kidney disease and died, much to the relief of the bishop who had tried to get Mendel to compromise and restore the goodwill of the state. To prevent embarrassment, the monastery had all of Mendel's papers, including his scientific notes, burned.

Mendel's work was not appreciated by Nageli, then the foremost authority on heredity in Europe, because it was not universal. Many biologists who did look at it thought it suspiciously like numerology; nature was so complex and messy, it just did not seem right that a few mathematical laws would be of any significance to so fearsome a problem as heredity. Also, most of his contemporaries were convinced that the environment in some subtle way was involved in the hereditary process, and Mendel's paper made no mention of such an influence.

THE EDUCATION OF CHARLES DARWIN

While Mendel was working in relative obscurity on his pea crosses, Charles Darwin was bearing the brunt of international fame and notoriety. Darwin came from a well-to-do and famous family, his grandfather Erasmus being a poet and philosopher and his father Robert a successful physician.[10] They lived near the estate of their cousins, the Wedgwoods, whose porcelain and stone wares were shipped to all major ports of the world. Charles was not a good pupil for Latin and classics and preferred collecting bugs and hunting. His father thought he might like medicine and sent him to Edinburgh, where the young Darwin was more interested in studying tide pool animals and shells or going on geological walks. He ran out of the operating room as he witnessed a young boy undergoing surgery without anesthesia. After dropping out of medical school he returned home to rest and hunt, to the despair of his father. The second opportunity for a wealthy young man lacking direction was the ministry. Most naturalists were ministers who believed that studying plants and animals was the equivalent of reading "the Bible of nature." Darwin went to Cambridge for his seminary training and developed friendships with the naturalists on the faculty, especially J.S. Henslow, who appreciated his gifts

Charles Darwin at his prime (1854), a few years before publication of his master-piece, *The Origin of Species*. (Reprinted from G. West [1938] *Charles Darwin: A Portrait*, Yale University Press, New Haven, Connecticut [courtesy of CSHL Archives].)

for curiosity, skilled observation, attention to detail, and careful dissection of small invertebrates. It was Henslow who recommended Darwin for an appointment as naturalist on a voyage that would explore South America and the South Pacific.

From 1831 through 1836, Darwin enjoyed the adventure of his life-time on board the *H.M.S. Beagle*. He left with the hope that the biology and geology he would learn would provide a wealth of information for the Admiralty, for scholars eager to learn the works of the Creator, and for his future career as a curate naturalist. He was convinced of the correctness of his creed and hoped that the naturalist's supplies and modest library he took with him would be well used.

Darwin's gifts of observation opened new worlds for him, and his letters to Henslow were so impressive that Henslow had them privately printed, without Darwin's knowledge, for circulation among naturalists. He noted the relations of animals and plants to the geography and climate on the islands he explored in the Atlantic and Pacific oceans. He compared these observations with the land and organisms that inhabited the continents closest to the islands. He unearthed fossils and compared them to the forms of life living in those countries. He noted the differences among the land animals north and south of the Amazon River. He was surprised at the abundant fossil sea shells found high and inland among the Andes Mountains.

DARWIN'S EVOLUTIONARY VIEWS

Darwin returned with doubts. He no longer believed that species were created suddenly and thenceforth remained unchanged. He accepted the claim of geologist Charles Lyell (1797–1875), whose *Principles of Geology* (1830) he took with him on the *Beagle,* that the earth was considerably older than the traditional 7,000 to 10,000 years assigned to it by theologians. He was convinced that there was a descent of contemporary animals in South America from similar but no longer extant species he had unearthed as fossils. He knew that the organisms living on the Galapagos Islands off Ecuador resembled similar species on the mainland of Ecuador rather than the very different species living on the Canary Islands in the Atlantic. Yet the tropical location, the climate, and the volcanic composition of these islands were remarkably similar.

After he settled in London, he wrote his journal of travel and observations on the *Beagle* and established a reputation as a superb writer and naturalist. He married his cousin, Emma Wedgwood, in 1839 and moved to the outskirts of London, in Down, where he remained the rest of his life, never again to travel outside Great Britain. He also became chronically ill; some say from a tropical infection, some claim he was a hypochondriac. The onset was soon after he read Malthus's essay on population for relaxation. The thought overwhelmed him. If nature produces more offspring than the food and space available for all of them, most will perish, of course, but who are the survivors? Here Darwin realized that those best adapted to their environment would win out over those less adapted. If variations were inheritable, then those variations associated with better adapted survivors would be transmitted to the next generation. The idea for natural selection, as he later called this process, would explain evolution without the recourse of a guiding Creator. It would contradict the beliefs of his own church. It would no doubt be pilloried, as Chambers's *Vestiges* were throughout Chambers's life, and Chambers dared not reveal his authorship while he was alive. Darwin chose to keep his realization well guarded. He would not publish until he had proven his theory and tested it against every conceivable objection he could imagine from his fellow scientists who uniformly felt, as he himself did in 1831, that the life on earth was placed there, in essentially its present form, by a benevolent Creator, during the six days of creation described in Genesis.[11]

Darwin spent eight years working on the anatomy and distribution of barnacles. He kept up an immense correspondence while doing his barnacle studies and amassed the evidence for his theory of natural selection. He confided his secret to a few of his closest scientific friends, including botanist Joseph Hooker and geologist Charles Lyell. Hooker urged him to publish before he was scooped. Darwin felt confident that no one had as much evidence and no one would be so foolish as to invite public outrage without that evidence. He was wrong. In 1858 Alfred Russell Wallace (1823–1913), a naturalist in Malaysia, sent Darwin an essay describing a theory for the origin of species that was so similar to his own he swore that he himself could have written it.[12] Hooker and Lyell worked out an honorable solution and both the short account of Darwin's theory sent to his correspondents and Wallace's manuscript to Darwin were read before the Linnean Society in 1858 and published in their proceedings. Darwin promised his friends that in return for salvaging his priority, he would publish an "abstract" of his theory. In 1859, that abstract of over 350 pages appeared as *The Origin of Species* in a first edition of 1250 copies, all of which were sold on the first day of issue.

RECEPTION OF DARWIN'S VIEWS

Darwin's readers were greeted by a clear prose that avoided any mention of theological belief. "How is it that varieties, which I have called incipient species, become ultimately converted into good and distinct species which in most cases obviously differ from each other far more than do the varieties of the same species? How do these groups of species, which constitute what are called distinct genera, and which differ from each more than do the species of the same genus, arise? All these results ... follow from the struggle for life. Owing to this struggle, variations, however slight and from whatever cause proceeding, if they be in any degree profitable to the individuals of a species in their infinitely complex relations to other organic beings and to their physical conditions of life, will tend to the preservation of such individuals, and will generally be inherited by the offspring. The offspring, also, will thus have a better chance of surviving, for, of the many individuals of any species which are periodically born, but a small number can survive. I have called this principle, by which each slight variation, if useful, is preserved, by the term Natural Selection, in order to mark its relation to man's power of selection."[13]

Darwin acknowledged that his natural selection "is the doctrine of Malthus applied with manifold force to the whole animal and vegetable kingdoms; for in this case there can be no artificial increase of food, and no prudential restraint from marriage."[14] Darwin marshaled his evidence in this tightly packed abstract of some 300 pages. He used the evidence of slight changes observed in domestic variation, he used the geological record to show that circumstances had changed; he discussed the rapid colonization of land following disasters and the successions of animals and plants that took place; he surveyed geographical distribution and the role of isolation in making each separate continent have its unique collections of species; he used comparative anatomy to show the relations of organisms; he used embryology to show the similarity of the stages of development among the mammals. But as to the cause of these variations on which natural selection worked, he had no original insight. "Our ignorance of the laws of variation is profound."[15] At the very end of his book he made a token statement to placate those readers who would have noticed what was missing throughout their reading. "Thus, from the war of nature, from famine and death, the most exalted object which we are capable of conceiving, namely, the production of the higher animals, directly follows. There is grandeur in this view of life, with its several powers, having been originally breathed by the Creator into a few forms or into one; and that, whilst this planet has gone cycling on according to the fixed law of gravity, from so simple a beginning endless forms most beautiful and most wonderful have been, and are being evolved."[16]

Darwin did not discuss the evolution of man in his *Origin of Species*, wisely recognizing that this would require a separate book-length justification. Whereas Darwin avoided the issue in the Linnean Society announcement of the theory of natural selection, Wallace put a disclaimer in his essay, claiming that evolution did not apply to man, whose existence required the Creator's assistance. Wallace, a lapsed Christian, retained a belief in the separate creation of man. He also took an interest in spiritualism, publishing on that topic in 1875, much to Darwin's consternation.

DARWIN'S THEORY OF HUMAN EVOLUTION

Darwin published *The Descent of Man* in 1871, using comparative anatomy, comparative animal behavior (ethology, as it is called today), sexual selection, and race formation as the evidence to support the origin of the human species from ape-like ancestors. "Man is descended from some less highly organized form," he claimed. Anyone who seriously studies zoology, "will be forced to admit that the close resemblance of the embryo of man to that, for instance of a dog—the construction of his skull, limbs, and whole frame on the same plan with that of other mammals...all point in the plainest manner to the conclusion that man is the codescendant with other mammals of a common progenitor."[17] Although humans have since diverged into races or what Darwin was willing to call sub-species, he pointed out that "the races agree in so many unimportant details of structure and in so many mental peculiarities that these can be accounted for only by inheritance from a common progenitor; and a progenitor thus characterized would probably serve to rank as man."[18] At the time of Darwin's study of human evolution, no fossil record of man's past had yet been discovered.

The evolution by natural selection of species or of humans faced torrents of criticism from the general public. It was denounced by almost all the clergy, and deeply divided naturalists, many of whom were troubled by the implications this theory had for their own very strong religious beliefs. Darwin felt he could weather this storm, but he was troubled by lack of a satisfactory theory of heredity to account for the origin and transmission of variations. While writing the greatly expanded version of the first chapter of his *Origin of Species*, Darwin inferred a particulate basis for inheritance. He described his idea as a "provisional hypothesis of pangenesis," and in 1868 made it public in *Variation of Animals and Plants under Domestication.*[19]

DARWIN'S THEORY OF HEREDITARY UNITS

Darwin's views were strikingly similar to Spencer's. "It is universally admitted that the cells or units of the body increase by self-division or proliferation, retaining the same nature, and that they ultimately become converted into the various tissues and substances of the body. But besides this means of increase I assume that the units throw off minute granules which are dispersed throughout the whole system; that these, when supplied with proper nutriment, multiplied by self-division, and are ultimately developed into units like those from which they were originally derived. These granules may be called gemmules. They are collected from all parts of the system to constitute the sexual elements, and their development in the next generation forms a new being; but they are likewise capable of transmission in a dormant state to future generations and may then be developed."[20] The gemmules are emitted throughout development and have a tendency to accumulate in plant organs such as buds or in the animal reproductive organs. Darwin considered them extremely minute, similar to the infectious agent found in smallpox or rinderpest, where even a pin dipped into the infectious matter can bring down an adult organism within a few days. He believed they are dispersed throughout the body, and that this is why a piece of a leaf of a begonia plant could be used to generate an entire plant.

The theory of pangenesis was not helpful in predicting why some offspring showed reversion to ancestral traits, or atavism, as it was then called; nor did it explain the origin of sporting variation; it gave no clue to which traits would be expressed and which hidden; and it did not explain how gemmules became altered within the cell to reflect the present status of that cell. Although it received a cool reception from biologists, it did suggest an experimental test to Darwin's cousin, Francis Galton (1822–1911), who shared with Darwin a common grandfather in Erasmus Darwin.[21] Galton was a precocious child tutored by his older sister, and he grew up treated as a child prodigy, showing off before guests how well he learned his lessons from his sister. His family were Quakers, although his

Francis Galton is best known for his efforts to measure human abilities and for founding the field of eugenics. (Reprinted from Galton F. [1908] *Memories of My Life*, Methuen & Co., London.)

father did not mind being both a banker and a gunsmith. At Cambridge, Galton tried his hand at mathematics, but he was only an ordinary student and did not live up to his early promise. He chose a more practical skill and studied medicine but did not complete his medical studies. Instead, he inherited a fortune when his father died and he decided to use it as a "gentleman scientist." He arranged with the Geographical Society to fund an expedition to the Middle East and traveled south along the Nile into areas of Africa that were only vaguely explored. His reports from southwest Africa, in particular, were well received by the Geographical Society and he had established himself as a scholar. Like Darwin, he never held an academic post and worked with the security of a family fortune to care for his personal and scholarly needs. Galton was an eccentric individual with a penchant for measuring quantitative phenomena and social customs. He also made some significant, but not brilliant, contributions to many fields.

FINGER PRINTS

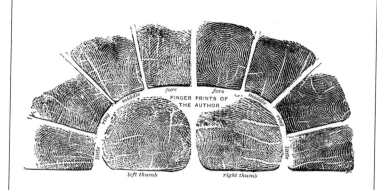

BY

FRANCIS GALTON, F.R.S., ETC.

London

MACMILLAN AND CO.

AND NEW YORK

1892

Galton's own fingerprints illustrate his success in classifying their patterns and proving their value for uniquely identifying individuals. (Reprinted from Galton F. [1892] *Fingerprints*, Macmillan and Co., London.)

He named and described the anti-cyclone for meteorology; he applied statistics to heredity; he devised the correlation coefficient to seek causal relations; he campaigned successfully to use fingerprints as a means of identification; and he named and founded the field of eugenics. He shocked the public by studying the efficacy of public prayer, citing the myriads of prayers offered up to the royal family every week and showing that they fared no better in health and mortality than wealthy families for whom such public prayer was lacking.

GALTON'S EXPERIMENTAL TEST OF PANGENESIS

Galton assumed that gemmules, if they are shed constantly by the cells, probably pass through the blood in huge numbers as they make their way to the reproductive organs. He secured the help of two surgeons to help him transfuse rabbits. If the gemmules did circulate, then the gemmules from one strain could be introduced into the other and by appropriate breeding of these strains their reproductive organs might yield the evidence of their new gemmules. He used several approaches. He transfused blood by syringe from a bowl of blood obtained from silver-grey strains into several different strains. He also tied two rabbits together so that their carotid arteries could be surgically crossed, each pouring blood of the other strain into their circulations. No matter what strain transfused its blood into the silver-greys, their progeny were uniformly silver-grey, and none of the characteristics or fur color appeared from their blood donor types, including black and white, yellow, Himalaya, Angora, and pure white strains.[22] Galton submitted the manuscript to the *Proceedings of the Royal Society* and it was published in 1871. His final remarks were unambiguous: "The conclusion from this large series of experiments is not to be avoided, that the doctrine of Pangenesis, pure and simple, as I have interpreted it, is incorrect."[23]

(*See facing page.*) Francis Galton and Charles Darwin are cousins. Charles Darwin married his cousin Emma Wedgwood. The Wedgwoods were wealthy manufacturers of china. The Galtons made their fortune in armaments. Charles Darwin's immediate paternal ancestors were wealthy physicians. (Reprinted, with permission, from Harry H. Laughlin Archives, Truman State University, courtesy CSHL Eugenics Web site.)

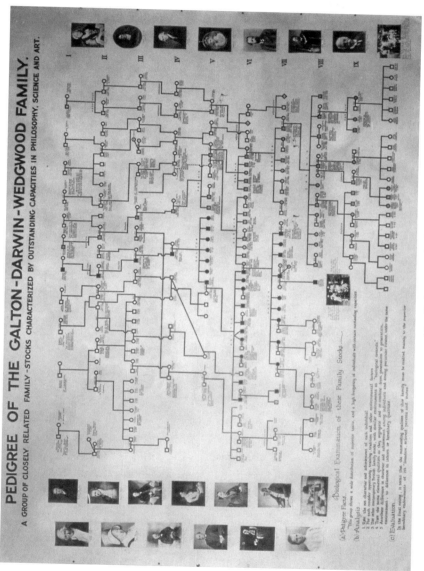

(See facing page for legend.)

Darwin was appalled. His own cousin, too! In a letter to *Nature* he defended himself. He never said blood was the means of transmission; after all, protozoa and plants do not have blood but they do have gemmules. Darwin lamely explained that transmission could be cell-to-cell and totally bypass the circulatory system. The wound was nevertheless felt, "As it is, I think every one will admit that his experiments are extremely curious, that he deserves the highest credit for his ingenuity and perseverance. But it does not appear to me that Pangenesis has, as yet, received its death blow; though, from presenting so many vulnerable points, its life is always in jeopardy; and this is my excuse for having said a few words in its defence."[24]

Galton caved in. Out of deference to his cousin's fame or out of family loyalty, he apologized, noting how easy it was in his mind to confuse "diffusion" of gemmules with "circulation." In a graceful way he told a story to save both their reputations and subtly to let the reader know that he believed in his heart that he was right. In the story he imagined himself and Darwin to be early ancestors of man. Galton "heard his trusted leader utter a cry, not particularly well articulated, but to my ears more like that of a hyena than any other animal."[25] The animal cry was a warning that Galton interpreted as a hyena in the bush. He carefully searched his surroundings but found no such hyena but there was a threatening leopard instead. "I am given to understand for the first time that my leader's cry had no reference to a hyena down in the plain, but to a leopard somewhere up in the trees; his throat had been a little out of order—that was all. Well, my labour has not been in vain; it is something to have established the fact that there are no hyenas in the plain, and I think I see my way to a good position for a look out for leopards among the branches of the trees. In the meantime, *Vive Pangenesis.*"[26]

AN ALTERNATIVE TO CIRCULATING HEREDITARY UNITS

Despite Galton's apology, pangenesis did not catch on and most naturalists considered it dead. But throughout the 1880s, doubts grew about the theory of acquired characteristics, and no effective mechanism seemed workable to bring the effects of the environment present in the individual organs to the reproductive material itself. By the 1880s the mechanism of fertilization was clear. There were two components, the sperm and the egg. Mature sperm or spermatozoa were formerly thought to be parasites in the male's ejaculates from the time of their discovery by Anton von Leeuwenhoek in the 17th century to Karl von Baer himself. But improved microscopy made it possible to observe the actual fertilization process in many organisms. Furthermore, all cells were shown to arise from preexisting cells, a point elevated to the status of the cell doctrine by pathologist Rudolf Virchow. None was demonstrated to arise as a condensation from poorly formed living matter, as Schleiden and Schwann had originally proposed in their model of free formation of cells. Erwin Haeckel interpreted the difference in size between the minuscule sperm and the immense egg to mean that the hereditary material must be only a minute part of the egg because male and female contributed equally to the heredity of their offspring. A more influential idea came to August Weismann (1834–1914).[27]

Weismann attended medical school in Gottingen, receiving his medical degree in 1856. After his graduation he studied the development of insect eggs and noted that at a very early stage, the egg partitioned off a group of cells that later formed the germinal or reproductive cells. He was appointed to the faculty at Freiburg where he had hoped to make his career as a microscopic anatomist. By 1864, however, his vision had deteriorated and he remained partially blind the rest of his life. He became a theoretical biologist, using the published works of his colleagues as a basis for interpreting the fundamental problems of life.

Weismann accepted Darwin's theory of evolution by natural selection and incorporated his ideas into the less explored relation of heredity to cellularity and the nature of variation. In 1883 he published an essay, *On Heredity*.[28] In it he cast doubt on the Lamarckian mechanisms of acquired characteristics. It was not exercise that set the limits for organ function, but heredity which set the limits on what exercise could accomplish. "We cannot by excessive feeding make a giant out of the germ destined to form a dwarf; we cannot, by means of exercise, transform the muscles of an individual destined to be feeble into that of a Hercules; or the brain of a predestined fool into that of a Leibnitz or a Kant, by means of much thinking. With the same amount of exercise the organ which is destined to be strong, will attain a higher degree of functional activity than one that is destined to be weak. Hence natural selection, in destroying the least fitted individuals, destroys those which from the germ were feebly disposed."[29]

The theory of use and disuse in transforming heredity Weismann questioned also, using an example drawn out of the human experience.

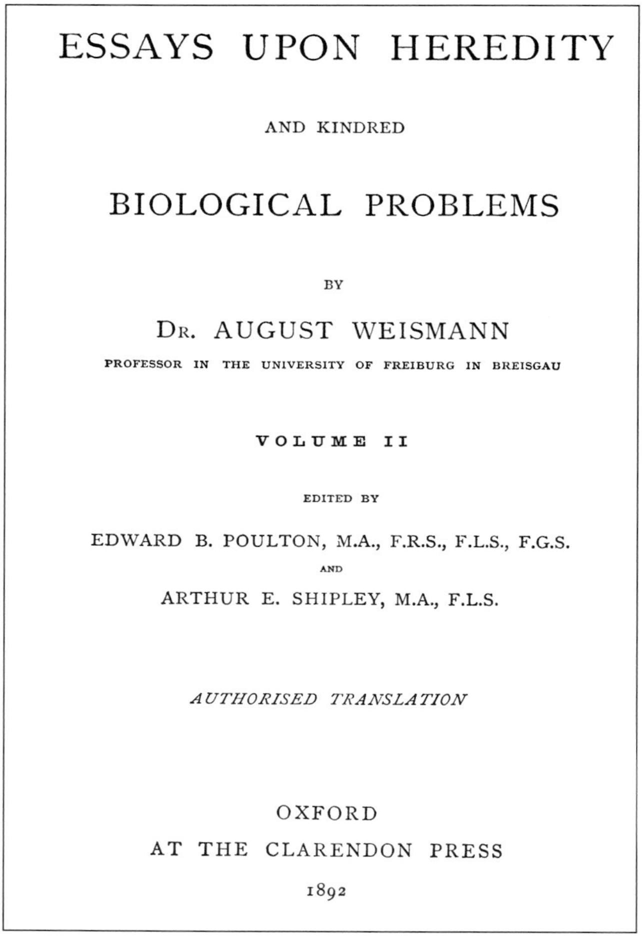

ESSAYS UPON HEREDITY

AND KINDRED

BIOLOGICAL PROBLEMS

BY

Dr. AUGUST WEISMANN

PROFESSOR IN THE UNIVERSITY OF FREIBURG IN BREISGAU

VOLUME II

EDITED BY

EDWARD B. POULTON, M.A., F.R.S., F.L.S., F.G.S.

AND

ARTHUR E. SHIPLEY, M.A., F.L.S.

AUTHORISED TRANSLATION

OXFORD

AT THE CLARENDON PRESS

1892

August Weismann accepted Darwin's theory of evolution by natural selection and incorporated those ideas in his theories about the relationship of heredity to cellularity and the nature of variation. His essays on these topics are collected into *Essays Upon Heredity and Kindred Biological Problems*. (Reprinted from Weismann A. [1891] *Essays Upon Heredity and Kindred Biological Problems*. Clarendon Press, Oxford, United Kingdom.)

"The Bach family shows that musical talent, and the Bernouilli family that mathematical power, can be transmitted from generation to generation, but this teaches us nothing as to the origin of such talents. In both families the high-water mark of talent lies, not at the end of the series of generations, as it should do if the results of practice are transmitted, but in the middle."[30] Many talents, in fact, did not seem to have a biological basis for their presence: "How many poets arose in Germany during the period of sentiment which marked the close of the last century; and how completely all poetic gifts seem to have disappeared during the Thirty Year's War."[31]

WEISMANN'S THEORY OF THE GERM PLASM

Although most of Weismann's writings are strictly biological and do not draw on human examples, this popular university lecture was influential through the remaining years of the 19th century. Whatever his intent in presenting it, portions of it found their way to the thinking of European and American intellectuals struggling to solve social problems that refused to go away. Weismann asserted that "human intelligence in general is the chief means and the chief weapon which has served and still serves the human species in the struggle for existence. Even in the present state of civilization—distorted as it is by numerous artificial encroachments and unnatural conditions—the degree of intelligence possessed by the individual chiefly decides between destruction and life; and in a natural state, or still better in a state of low civilization, this result is even more striking."[32]

Weismann had an explanation for the failure of Lamarck's theory of acquired characteristics. He distinguished between the body tissues and their cells, collectively called the soma, and the cells of the reproductive tissue, which he called the germ plasm. "If, as I believe, the substance of the germ cells, the germ-plasm has remained in perpetual continuity from the first origin of life, and if the germ-plasm and the substance of the body, the somatoplasm, have always occupied different spheres, and if changes in the latter only arise when they have been preceded by corresponding changes in the former, then we can, up to a certain point, understand the principle of heredity."[33]

Two years later Weismann developed this idea in his most provocative and lasting essay, *The Continuity of the Germ-plasm as the Foundation of a Theory of Heredity*. New observations by microscopic anatomists called cytologists revealed that within the cell a nucleus resides which often shows fine threads or "chromatic loops" (or chromosomes, as we would call them today). Even a sperm cell has its origin from a cell with a well-defined nucleus. Weismann speculated freely on the possibilities: "The physical causes of all apparently unimportant hereditary habits or structures, of hereditary talents, and other mental peculiarities, must all be contained in the minute quantity of germ-plasm which is possessed by the nucleus of a germ cell."[34] In 1864, while still at the start of his career, he had noted the separation of germinal and somatic cells in the embryos of flies (Diptera); now he could assign importance to that event: "The germ cells are separated from the somatic cells during embryonic development, sometimes even at its very commencement." He could also refine the event to a level beyond that of the cell itself: "Only the nuclear substance passes uninterruptedly from one generation to another."[35]

WEISMANN'S EXPERIMENTAL TEST OF LAMARCKISM

Even if the germinal tissue were set aside early in the embryo, what was to prevent a Darwin-like flow of fundamental units into it? Weismann assailed such a Lamarckian mechanism as unproven and unlikely. He focused on one class of Lamarckian changes that were the only instances cited in the literature or widely believed in human anecdotal experience. In 1888 he discussed *The Supposed Transmission of Mutilations*. He admitted that he, like Darwin, had to make use of Lamarck's use and disuse model to account for variation, but "in the course of further investigation I gradually gained a more decided conviction that such transmission has no existence in fact."[36] Use and disuse, of course, could not be put to experimental test, but the theory of transmitted mutilations could. He cited numerous examples of cats brought to him or put on display by naturalists who claimed that the tailless form arose from a mother whose tail had been cut off by a wagon or cart. Weismann used to assure such persons that their cat was a Manx cat, descended from the Isle of Man where such a breed abounds. The original progenitor, he believed, had no Lamarckian origin but was instead a monstrosity arising "from unknown changes in the germ."[37] A similar situation developed in Japan where tailless cats were considered better mousers. Selection for shorter tails, including occasional monstrosities that do arise, would lead to a short-tailed strain.

Weismann thought he would put the theory of mutilations to a test. He started to cut off tails, beginning in October 1887, of male and female mice; from these 12 original parents he obtained 18 progeny, all with tails.

He cut these tails off and bred 12 of the brothers and sisters to obtain 333 progeny, all with tails. He chose 15 of these for amputation and obtained 237 in the third generation, again with tails. A sample of 14 mutilated tail-less mice were bred to yield 152 fully tailed offspring. Twice more he repeated his efforts obtaining an additional 138 and 41 fully tailed off-spring. Altogether, some 900 young were obtained from six generations of mutilation. The original parents had tails measuring 10.5 to 12 mm when newborn. Their six generations of progeny showed not a single offspring with a tail size below 10.5 mm.[38]

Weismann recognized that he had not disproved long-range effects of Lamarckism, but he claimed he had discredited those who offered anecdo-tal experiences favoring phenomena such as mutilations and maternal impressions. Much longer attempts at mutilation in many cultures and nations also failed to yield change in the affected organ. "Such hereditary effects have been produced neither by circumcision, nor the removal of the front teeth, nor the boring of holes in the lips or nose, nor the extraordi-nary artificial crushing and crippling of the feet of Chinese women. No child among any of the nations referred to possesses the slightest trace of these mutilations when born: they have to be acquired anew in every gen-eration."[39]

Weismann, like Darwin, did not know the source of variations on which natural selection acted. Weismann, however, severely damaged the belief in the inheritance of any form of acquired characteristics. The shift among scientists was dramatic; before the 1870s almost all scientists believed Lamarckism, in some form, must be correct. By the mid-1880s serious doubts were expressed that this could be so, and after Weismann's analysis, adherents of Lamarckism dwindled year by year as the 19th cen-tury entered its waning years. Natural selection worked, with or without a Lamarckian mechanism for the origin of variations. The variations them-selves were real. A parallel debate on the social significance of heredity was quite different in its outcome. Lamarckism implied the potential reversibil-ity of undesirable traits; Weismann's theory of the germ plasm with the refutation of Lamarckism created a pessimism in which those saddled with defective heredities were doomed to live less adaptive lives and to be stig-matized as possessing defective germ plasm. Neither Lamarckism nor Weissman's germ plasm applied to most social traits, but virtually no one who took part in the debates was aware of that assessment.

FOOTNOTES

[1] Jesus used this prevailing view in his Sermon on the Mount when he warned about false prophets: "You will know them by their fruit. Are grapes gathered from thorns, or figs from thistles? So, every sound tree bears good fruit, but the bad tree bears evil fruit. A sound tree cannot bear evil fruit, nor can a bad tree bear good fruit. Every tree that does not bear good fruit is cut down and thrown into the fire. Thus you will know them by their fruit" (Matthew 7: 16–20).

An equally fixed or determinate view of heredity stems from the Old Testament when Moses went before God on Mount Sinai to receive the tablets for the Ten Commandments and was confronted by His words: "The Lord, the Lord, a God merciful and gracious, slow to anger, and abounding in steadfast love and faithfulness, keeping steadfast love for thousands, forgiving iniquity and transgressions and sin, but who will by no means clear the guilty, visiting the iniquity of the fathers upon the children and the children's children, to the third and the fourth generation" (Exodus 34: 6–7). Unlike the simile used in Matthew, the episode in Exodus is less clear about heredity itself. The children might be punished in a legal sense, like a sentence passed by court, and the calamities awaiting the children may be external, such as a life of beggary or bad luck dogging every effort to succeed.

This was discussed in Chapter 1 in characterizing what is meant by the term "unfit persons." It is, from a geneticist's perspective, not very helpful. But to the stigmatized family a century of what looks like degenerate behavior may look to them and their community as a hereditary pathology. Even families today who have two or three generations of alcoholics perceive their family problem as a hereditary one.

[2] There is no counterpart in the history of genetics for a theory based on visual impressions during intercourse (in contrast to pregnant women receiving shocks to the embryo from the visual experience). The literal acceptance of this episode requires miraculous (divine) assistance, and this story may not really signify a "folk genetic" belief in Jacob's era.

[3] A more technical account of the idea of hereditary units may be found in my book, *The Gene: A Critical History,* Chapter 3, "The need for hereditary units" (Wm. B. Saunders Co., Philadelphia, 1966).

[4] Herbert Spencer, *The Principles of Biology,* Volume 1 (Appleton, New York, 1875 [the first edition appeared in 1864]), p. 1.

[5] Ibid. p. 183.

[6] William Bateson in 1906 was the first to recognize that these were related phenomena when he introduced the word "genetics" to replace the phrase "heredity and variation."

[7] The best biographies of Mendel are by Hugo Iltis (*Life of Mendel* translated by Eden and Cedar Paul, George Allen & Unwin, London, 1922) and by Vitezslav Orel (*Gregor Mendel: The First Geneticist,* Oxford University Press, Oxford 1996). Details of Mendel's life are sketchy and sparse, but the Mendel Museum in Brno (formerly Brunn), Czechoslovakia, publishes an annual collection of essays on Mendel and his times (*Folio Mendeliana*).

⁸ Many scientists show consistent excellence from youth to old age. Others, like Darwin and Mendel, were not distinguished as students at a college age. Mendel's personality reveals chronic insecurity which, with his vows to the church, may have limited his opportunities to apply or publicize his findings.

⁹ The Mendel–Nageli correspondence was translated into English and published as a supplementary pamphlet, *The Birth of Genetics*, referenced as *Genetics* volume 35, number 5, part 2, September 1950, 47 pp. The hawkweeds are unusual in their mode of inheritance. Whether Nageli would have independently discovered Mendel's laws had he chosen a more typical plant we do not know, but even if he had not, he, like Mendel's rediscoverers in 1900, would have quickly confirmed Mendel's laws. Technically, Mendel's two hereditary laws are called the "law of segregation" (learned by students as a 3:1 ratio of dominant to recessive traits from the hybrid parents) and the "law of independent assortment" (learned by students as a 9:3:3:1 ratio of the progeny of dihybrids showing, respectively, both dominants, either of the two dominants, or both recessives as the expressed traits).

¹⁰ Many fine biographies of Darwin have appeared. These include the fictionalized popularization by Irving Stone, *The Origin*, and the comprehensive biography *Charles Darwin: A Man of Enlarged Curiosity* by Peter Brent (Harper and Row, New York, 1981). For Darwin's early years, see *Charles Darwin: Voyaging* by Janet Browne (Alfred A. Knopf, New York, 1995).

¹¹ Few people today realize how difficult it would have been for Darwin, had he been a university professor, to publish his work on the origin of species by natural selection. Darwin may have rejected all offers for a university appointment prior to his work on evolution to protect his freedom, his family, and his reputation. Evolution as Darwin presented it did not need a Creator. Human beings did not need divine help to evolve. His world of life would have come into being without a God. Few scientists of his day were willing to abandon the almost universal belief that evolution, if it did occur, had to be set in motion by God, led to an ultimate destination by God, or had to unfold from God's plan. The shock of Darwin's theory was its absence of crediting God for the design of natural selection. God's role, if any, is the creation of the laws of science, or in Darwin's compromise, the first microscopic life form to appear on earth. Although there are more atheists and agnostics among scientists in the century since Darwin offered his theory, the general public has never come around to accepting Darwinian evolution without some form of divine assist or design.

¹² See Brent, *Charles Darwin*, pp. 412–416 for an account of the scramble Darwin made to have his 20 years' unpublished priority preserved.

¹³ Charles Darwin, *The Origin of Species* (John Murray, London, 1859), p. 32.

¹⁴ Ibid. p. 33.

¹⁵ Ibid. p. 78.

¹⁶ Ibid. p. 243.

¹⁷ Charles Darwin, *The Descent of Man* (John Murray, London, 1871), p. 590. This was extended and popularized by the German evolutionist Ernest Haeckel, who coined

the phrase "ontogeny recapitulates phylogeny." Although much criticized, that phrase in its broad sense is true. Each new generation of evolving organisms uses what it has received from the past and modifies by natural selection to meet new needs.

[18] Ibid. p. 591. The human genome analysis of 2001 confirms the universal kinship of all humanity as Darwin had predicted.

[19] Charles Darwin, *Variation of Plants and Animals Under Domestication* (John Murray, London, 1868 [1905 popular edition, two vols.]).

[20] Ibid. vol. 2, p. 456.

[21] N.T. Gridgeman, "Francis Galton (1822–1911)." *Dictionary of Scientific Biography* 5: 265–266. Also, for more details, see Galton's autobiography, *Memories of My Life,* (Methuen & Co., London, 1908). There is some dispute about how much of a prodigy and genius Galton was. Lewis Terman, in his *Genetic Studies of Genius* (Stanford University Press, Stanford), gives him one of the highest IQ scores possible (about 190–200) based on childhood achievements. But Galton's lackluster showing in college and his failure to probe any one field in depth or come up with any great discovery have made his alleged genius less spectacular than Terman believed.

[22] Francis Galton, "Experiments in pangenesis, by breeding from rabbits of a pure variety, into whose circulation blood taken from other varieties had previously been largely transfused." *Proceedings of the Royal Society (Biology)* 19(1871): 393–404.

[23] Ibid. p. 404.

[24] Charles Darwin, "Letters to the editor." *Nature,* April 27, 1871, p. 503.

[25] Francis Galton, "Letters to the editor." *Nature,* May 4, 1871, p. 5.

[26] Ibid. p. 6.

[27] "August Weismann (1834–1914)." *Encyclopedia Britannica* 23: 491–492.

[28] August Weismann, *On Heredity* (1883) translated into English and included, pp. 67–106, in *Essays Upon Heredity and Kindred Biological Problems,* two volumes, edited by E.B. Poulton, S. Schonland, and A.E. Shipley (Oxford University Press, Oxford, 1891).

[29] Ibid. p. 85.

[30] Ibid. p. 97.

[31] Ibid. p. 98.

[32] Ibid. p. 99.

[33] Ibid. p. 105.

[34] August Weismann, *The Continuity of the Germ-plasm as the Foundation of a Theory of Heredity,* 1885, in Poulton et al., *Essays upon Heredity,* p.194.

[35] Weismann, *Germ-plasm,* p. 209.

[36] August Weismann, *The Supposed Transmission of Mutilations,* 1888, in Poulton et al., *Essays on Heredity,* p. 434.

[37] Weismann, *Mutilations,* p. 439. Weismann may not have been aware that in stating that the sporadic mutation to taillessness was from "unknown changes in the germ" he was describing mutation theory as it would be interpreted by H.J. Muller in the 1920s. Such mutations can be induced as random errors in the genes by X-rays and by a variety of potent chemicals. Most human sporadic mutations probably arise from chemical products of our own metabolism rather than from outside radiation or from ingested or inhaled chemicals.

[38] Ibid. p. 445.

[39] Ibid. p. 447. Weismann's thesis still stands today. There is no evidence for the inheritance of mutilations or any known environmental agent that directs mutation in a Lamarckian way. There is also no evidence that changes in the soma of animals can cause similar changes in the heredity of the affected individual. Of course, the fact that a hereditary trait occurs in an individual does not mean that it cannot be treated or ameliorated. Many traits do show such a response to altered environments, but there are many physical and behavioral genetic defects that cannot be altered by medical or social means.

Part II: Eugenics Takes the Spotlight

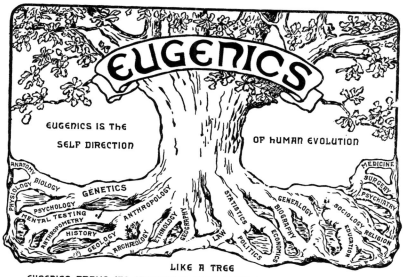

EUGENICS

EUGENICS IS THE SELF DIRECTION OF HUMAN EVOLUTION

ANATOMY BIOLOGY PHYSIOLOGY GENETICS PSYCHOLOGY MENTAL TESTING ANTHROPOMETRY HISTORY ANTHROPOLOGY GEOLOGY ARCHÆOLOGY ETHNOLOGY GEOGRAPHY LAW POLITICS ECONOMICS STATISTICS BIOGRAPHY GENEALOGY SOCIOLOGY EDUCATION RELIGION PSYCHIATRY SURGERY MEDICINE

LIKE A TREE
EUGENICS DRAWS ITS MATERIALS FROM MANY SOURCES AND ORGANIZES
THEM INTO AN HARMONIOUS ENTITY.

10

The Jukes and the Tribe of Ishmael

U LSTER COUNTY IN NEW YORK INCLUDES some of the most scenic vistas of the Hudson River valley and includes the forests and the granitic outcroppings of the Catskill Mountains. Magnificent castle-like mansions, like the Mohonk Mountain House near New Paltz, sprang up at the end of the 19th century as places for the wealthy and middle-class professionals to enjoy their meetings and vacations, with paddleboating in mountain lakes, bridle paths for relaxed rides in the woods, and dozens of gazebos for summer privacy and conversations in the open air.

More than a century before this region became a retreat for the privileged, occasional immigrants who did not fare well in the urban life in New York City sought the isolation and opportunity to fish and hunt rather than to endure the permanence of life on a farm or to work as hired laborers. As the river boat trade along the Hudson River increased, so did the opportunities for businesses and services tied to it. Those who established permanent homes and jobs looked down on those early inhabitants who seemed indolent, indifferent to opportunity, and shabby in their living habits. As the valley settlements grew, the abundance of wildlife for hunting and fishing diminished. The purchase, development, and settlement of property in Ulster County also diminished the territory in which unregulated hunting and fishing could take place. Those who established legal claims to property and enjoyed regular habits of employment soon treated those who shunned their life-style as social outcasts and failures. With no place to go and an unwillingness to take on the habits of

161

View of the Hudson River Valley from the Catskill Mountain House, Catskill Mountains, New York. (Courtesy of the Library of Congress.)

respectability expected of them, the outcasts were gradually enveloped in the vices of the demi-monde. Crime, prostitution, and alcoholism became the identifying hallmarks of this rejected population.

ELISHA HARRIS DISCOVERS THE JUKES

The first study of this Hudson Valley community of social failures was carried out by Dr. Elisha Harris (1824–1884) in the early 1870s.[1] Harris received his M.D. from the New York College of Physicians and Surgeons in 1849. He was interested in infectious diseases and became the superintendent of the Quarantine Hospital on Staten Island. His prominence in this field made him an active member and founder of the U.S. Sanitary Commission for the Federal Armies during the Civil War. After the war, Harris became Registrar of Vital Statistics for the Board of Health in New York City. He identified many appalling morbidity and mortality differences between the poor and the middle class in New York City. In 1867, while he was a member of the Department of Sanitation in New York City, he successfully launched a campaign to get the city to put 40,000 windows and 2,000 rooftop ventilators into the bleak, hastily built tenement buildings and railroad flats in which the poor managed their daily survival in crowded unhealthy environments. He established the first free vaccination service, and between 1869 and 1876, 600,000 inhabitants were immunized against smallpox. He initiated house-to-house visits by public health workers to identify inadequate waste disposal, illegal dumping, and other health hazards that threatened the lives of the inhabitants and their neighbors. Among Harris's contributions to medicine was the railroad ambulance, a means of delivering medical services to the needy and evacuating the wounded in times of war; it was first put to use during the Franco-Prussian War.

While in New York City, Harris took an interest in the conditions that led not only to ill health but to crime, alcoholism, and dependency on public charity. He met regularly with a small group of citizens, amateurs as well as professionals, who discussed these problems. The club included the publisher, George Haven Putnam; the economist and lawyer, Edward Morse Shepard; and the manufacturer and sculptor, Richard Louis Dugdale, at whose home they met. Their sociology club served as a forum for scholarship on the problems of the day. Later they gave it a formal name, the Society for Political Education, enabling it to publish occasional pamphlets to educate voters.[2] Harris became the corresponding secretary of the New York Prison Association, and using his skills as a statistician working with vital statistics, he identified some interesting recurrences of family names in the county prisons, especially in Ulster County. Harris spoke to the prison officials there and traced the lineage of the family back some six generations to a woman he dubbed "Margaret, mother of criminals." He presented his preliminary results in 1874 to the New York Prison Association, and these were in turn given wider publicity in the January 25, 1875 *Boston Medical and Surgical Journal.*[3] Harris had discussed his preliminary findings with the members of the sociology club, and in 1874, Richard Dugdale, who was also a member of the New York Prison Association, and who had more time to devote to the project, began a more thorough investigation of both the county and state records on this lineage of criminals.

The first account of Harris's work was reported in an article on *Pauperism* by Charles L. Brace in the *North American Review* for 1875.[4] Brace described the work with interest: "An extraordinary instance of inherited

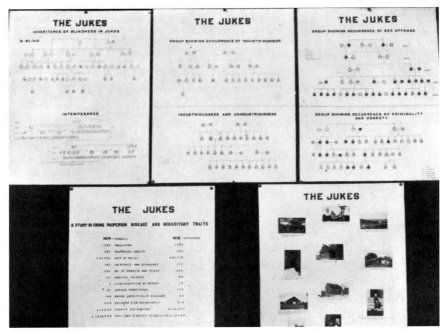

Pedigrees of members of the Jukes family and photographs of their dwellings in Ulster County, New York. Exhibit from the Second International Congress of Eugenics, 1921. (Reprinted from H. Laughlin [1923] *Exhibits Book: Second International Exhibit of Eugenics*, Williams & Wilkins, Baltimore [courtesy of CSHL Archives].)

pauperism was given recently at a meeting of the State Charities Aid Society, in New York, by Dr. Elisha E. Harris, registrar of the Board of Health. A pauper child, named Margaret, was suffered to grow up neglected in a village of Ulster County, New York, some eighty-five years since. She and two neglected sisters have begotten six generations of criminals and paupers. The total number of descendants now known, mainly of this pauper child Margaret, both living and dead, convicts, paupers, criminals, beggars, and vagrants, is six hundred and twenty-three. In a single generation there were seventeen children. Of these only three died before maturity. Of the fourteen surviving, nine served an aggregate term of fifty years in the state's prisons for high crimes and the other five were frequently in jails and almshouses. This 'mother of criminals' cost the county hundreds of thousands of dollars."[5]

Estabrook reexamined and updated the fate of the Jukes. In this letter (from Kingston, New York, in Ulster County), he sends Laughlin pedigrees of recent members of the kindred. (Reprinted, with permission from American Philosophical Society, courtesy of CSHL Eugenics Web site.)

OLIVER WENDELL HOLMES PUBLICIZES THE JUKES

Oliver Wendell Holmes (1809–1894), a physician, poet, much admired essayist, and self-styled "Autocrat of the Breakfast Table," read the *Boston Medical and Surgical Journal* account of Harris's work and briefly discussed it that same year in a more encompassing essay review for the *Atlantic Monthly,* a journal of opinion to which he had contributed since its founding.[6] Holmes was unhappy with the views of Prosper Despines, whose *Psychologie Naturelle* had appeared in three volumes in 1868. Holmes had studied medicine in Paris for three years as a young scholar and retained his fluent knowledge of French and his interest in French culture. Despines looked upon criminality as a disease and believed that prisons should be eliminated and replaced by "moral hospitals" in which scholarly study of the pathology of criminals could be undertaken with attempts to treat or cure criminals developed by alienists and other specialists in psychology.

Holmes acknowledged that human behavior might have a stronger hereditary component than he had previously thought and cited the work of Dr. Harris. "Finding crime and poverty out of proportion prevalent in a certain county on the upper Hudson, he looked up the genealogy of the families whose names were oftenest on the criminal records. He found that a young girl called Margaret was left adrift about seventy years ago in a village of the county. Nine hundred descendants can be traced to this girl, including six generations. Two hundred of these are recorded as criminals, and a large number of the others, idiots, imbeciles, drunkards, and of otherwise degraded character. If genius and talent are inherited, as Mr. Galton has so conclusively shown; if honesty and virtue are heirlooms in certain families; if Falstaff could make King Henry know his son by a villainous trick of the eye and a foolish hanging of the nether lip,—and who that has seen two or three generations has not observed a thousand transmitted traits, villainous or other, in those around him?—why should not deeprooted moral defects and obliquities show themselves, as well as other qualities, in the descendants of moral monsters?"[7]

Holmes readily acknowledged the hereditary nature of physical traits such as supernumerary fingers, "bleeders," or those with deep-dimpled chins, single strands of prematurely white hair, and other "trivial peculiarities." He generalized readily from the physical to the behavioral, asserting that "we cannot add one cubit to our stature, and there is no more reason for believing that a person born without any moral sense can acquire it, than there is that a person born stone-deaf can become a musician."[8] In contrast to Despines, who opposed capital punishment as being unjust to moral idiots, immoral as revenge, useless for intimidation, and dangerous to society because it cheapens the values of life, Holmes believed the vast number of criminals are responsible for their acts and should be punished, including by execution. Despines shared the Lamarckian view that environments shape the heredity of both degenerative and positive behavior. Unlike Holmes, Despines sought to suppress the causes of moral degeneration by attacking poverty, luxury, "popular excitements," drunkenness, and bad passions. He wanted journalists to refrain from publicizing sensational criminal trials and he hoped debased literature would be suppressed.

Holmes was skeptical of this soft-headed account of criminality and believed society had no interest in such approaches. Society, he claimed, wants "the cheapest and surest protection against the effects of crime." He ridiculed "simple-hearted reformers who look forward to the time when ginger will not be hot in the mouth," or the day

> When the roughs as we call them, grown loving and
>   dutiful,
> Shall worship the true, and the pure and the
>   beautiful,
> And, preying no longer as tiger and vulture do,
> All read *The Atlantic* as persons of culture do.[9]

DUGDALE'S EXTENSIVE STUDY OF THE JUKES

Richard Dugdale (1841–1883) was born in Paris of English parents who were wealthy manufacturers.[10] After the uprisings of 1848, the Dugdales lost much of their wealth and returned to England, where young Dugdale continued his education in London, showing an aptitude for art. In 1851 the family moved to New York City, and Dugdale took a special liking to sculpture. He developed rheumatic fever when he was 14, and the family bought a farm in Indiana, hoping the country life would be more agreeable to their son. At the age of 19, Dugdale returned with his parents to New York City and chose a career in business, educating himself at Cooper Union. He inherited a modest sum of money from his father and used the bequest to reduce his working hours and pursue intellectual activities. He admired social reformers and the attempts to use statistics and scientific methods to study the problems of society. His home in Greenwich Village became the meeting place for the sociology club he helped establish. Through his association with Dr. Harris, he became a member of the Executive Committee of the New York Prison Association in 1868. In 1874 Harris appointed him to investigate the jails in Ulster County and several other county jails to pursue the history of "Margaret, mother of criminals." The following year, Dugdale took on the more encompassing task of studying the New York State jails for their registers of family-related prisoners. The initial reports of Dugdale to the Prison Association were edited for publication by G.P. Putnam's Sons, and in 1877 appeared with the title *The Jukes: A Study in Crime, Pauperism, Disease, and Heredity.*[11]

Elisha Harris obligingly wrote the foreword for the 1877 edition of *The Jukes*. He pointed out that Dugdale was the first to be "investigating the natural history of crime and pauperism."[12] He applauded this effort because "in the progress of medical science the close study of healthful as well as morbid conditions has resulted in defining the rules of hygiene, which treats of the prevention and extinction of the causes of disease."[13]

Unlike Holmes and Harris, who looked upon the study as a documentation of fixed hereditary social behavior, Dugdale approached the problem in a more complex way. He said that the pseudonym, Jukes, derived from the slang term, "to juke," described the habits of 42 families that traced themselves to a single lineage. The term "to juke" described the erratic behavior of chickens who kept no permanent nests and deposited their eggs wherever a convenient spot was to be had.[14] People who "juked" did not like to be tied down. Dugdale explored the question raised by the Italian Inspector of Prisons, M. Beltrani-Scalia, "What is crime and who commits it?" Dugdale conducted interviews, constructed genealogies, discussed the family members with employers and physicians, consulted prison wardens, and investigated the records of the town poorhouses, county clerk's office, sheriff's arrest logs, and prison registers.[15]

DUGDALE'S ENVIRONMENTALIST INTERPRETATIONS

From this thorough collection of notes and recorded data, Dugdale found a relatedness of criminal families extending across six generations. With a few exceptions, Dugdale claimed that what was inherited was a bad environment rather than a bad physiology. He condemned the prison system as without merit in preventing crime. "Indeed, so conspicuous is the failure of the entire machinery of the punitive and reformatory institutions of our state, that we cannot call these establishments the results of the wisdom of our generation, but rather the cumulative accidents of popular negligence, indifference and incapacity."[16] He pointed out that the "tendency of heredity is to produce an environment which perpetuates that heredity" and that "the correction is change of environment."[17] In this outlook, Dugdale shared the beliefs of Morel, Despines, and French physicians whose outlook was strongly shaped by Lamarck's doctrine of the inheritance of acquired characteristics. Although bad environments did cause degeneracy, good environments, Dugdale believed, could reverse the damage, with the possible exceptions of those with congenital syphilis and infants born to alcoholic mothers. He rejected the idea of an innate goodness or evil: "Men do not become moral by intuition, but by patient organization and training."[18]

Dugdale's environmental reforms included decent housing, because "love of home and pride in it are the most powerful motives in checking vagrancy, and in organizing the environment that can perpetuate these essential domestic sentiments."[19] He did not limit educational reform to the provision of vocational training, a suggestion of growing popularity in the larger cities and with penal reformers; instead, he argued, "when the term 'industrial training' is used, much more is meant than formal instruction in a trade. It is contemplated that, in a properly ordered scheme of reformation, something like a general training of the faculties must be provided for. Our reformatories must reform and develop the senses of touch, hearing, sight, smell and taste, so that the mind shall be filled with the knowledge of things, instead of being left vacant of everything except a memorization of words."[20]

Dugdale's environmental optimism pervaded his study of the Jukes. He believed the control of crime and pauperism "becomes possible, within limits, if the necessary retraining can be made to reach over two or three generations."[21] Even a single generation of intense social reform would produce dramatic results, "Such an energetic, judicious and thorough training of the children of our criminal population would, in fifteen years, show itself by the great decrease in the number of commitments."[22] Dugdale's proposals included the introduction of the German kindergarten program as a head start for the children of the poor and the criminal classes. He advocated early marriage to establish a family bond; and for those failures leading to illegitimate children, placement of the children at infancy in adoptive homes because "illegitimates who are placed in a favorable environment may succeed in life better than legitimate children in the same environment."[23] In contrast to Holmes's pessimism about hereditary behavior and moral imbecility, Dugdale asserted that "in knowledge of moral obligation (excluding insanity and idiocy), the environment has more influence than heredity."[24]

The first of the six generations of the Jukes family was Max, of Dutch ethnicity, who left New York City and headed up the Hudson, arriving in Ulster County, where he lived as a trapper and hunter. He was reputed to be of a jovial character whose major fault was drinking too much and whose failing vision in old age made him dependent on others. One of Max's sons married Effie, one of six sisters who are more properly the progenitors of the Jukes line. A second son may also have married one of the Jukes sisters. One of the sisters left New York and could not be traced by Dugdale. Ada (the original Margaret, the mother of criminals as Harris called her) married an unspecified laborer, and had both legitimate and illegitimate children.

The Conversion of the Jukes to a Hereditary Class

Although Dugdale's account of the Jukes family is full of hope for the future, the study became converted over the years to one of hopelessness, with the hereditary relation of the criminals and paupers eclipsing all other facets. Dugdale himself could not defend his environmental position against this hereditarian pessimism because he died of heart failure in 1883, his last few years being spent as an invalid.[25] Dugdale's last publication was a three-part series on "The origin of crime in society" that appeared in the *Atlantic Monthly* in 1881.[26] Dugdale stressed the role of environment in increasing crime after "disturbances of the social order" such as epidemics, as in the London bubonic plague of 1666 and the yellow fever epidemic in Memphis, Tennessee in 1879. He also claimed that embezzlement became epidemic during the Civil War when the temptation of monies obtained through war profiteering proved too much for bankers entrusted with the management of such wealth. He also cited the draft riots in New York City in July 1863. Hordes of vagrants followed in their wake and pillaged stores during the breakdown of municipal order following the rioting of immigrants who resisted military service.

Dugdale acknowledged that the statistics of crime made it appear that criminal tendencies are inherited. About 44% of New York City's inhabitants were of Irish origin or descent, but they constituted two-thirds of the prison population. "This strikingly establishes the force of hereditary tendencies in the formation of the criminal character,"[27] he claimed, but actual hereditary criminals were few in number, most increases of crime being associated with economic changes. People will turn away from crime if "wages are worth more than gain from theft,"[28] if the "instruments and methods of culture" are used to prepare the criminal class for "legitimate occupation," and if more gradations of occupation and wealth existed in society so mobility among classes becomes easier. Dugdale pointed out that many prominent wealthy families made their fortunes by cheating other investors or the public, but once they attained their wealth, they sought respectability for themselves and their families.[29]

Although Dugdale's writings reflect his environmentalist positions, not all his associates in the Sociology Club carried away that impression. In his obituary for Dugdale, the economist and lawyer Edward Morse Shepard, in 1884, described Dugdale's efforts to better conditions of prisoners, and he contrasted Dugdale's views on paupers and criminals. Dugdale, he said, "concluded that there is hope in the physical and mental vigor of criminals; that there is hopelessness in the absence of that vigor in paupers; that the misdirected energy of the former may, by proper discipline, be diverted to useful work; but that the sooner death comes the better, to the 'undervitalization and consequent untrainableness' of pauperism; that it is a crime to maintain paupers that may breed another and perhaps more numerous generation of their own kind; that licentiousness is the concomitant and a chief cause both of pauperism and crime; and that the open abandonment of virtue in women is the dreadful analogue of pauperism and crime in men."[30]

Sociologist Edward S. Morse in 1892 took the side of the hereditarians. He quoted with approval Robert Fletcher, President of the Anthropological Society of Washington, who condemned the criminal: "He breeds criminals; the taint is in the blood, and there is no royal touch which can expel it." He proclaimed "vagabonds, like criminals, spring largely from a degenerating stock." And he cited August Weismann's theory of the germ plasm as evidence that such traits are inherited. He exhorted his readers to "quarantine the evil classes as you would the plague and plant on good ground the deserving poor."[31]

Not all of Dugdale's contemporaries or later commentators drew this negative image of Dugdale's views of the Jukes. Sociologist Franklin H. Giddings, who edited the fourth edition, wrote a new introduction in 1901 and noted that "an impression quite generally prevails that "The Jukes" is a thorough-going demonstration of 'hereditary criminality,' 'hereditary pauperism,' 'hereditary degeneracy' and so on. It is nothing of the kind, and its author never made such claim for it."[32]

INFLUENCE OF THE JUKES STUDY ON OSCAR McCULLOCH

It was the unhappy fate of *The Jukes* to be largely misinterpreted as the history of a condemned, unredeemable kindred. In this portrayal it soon stimulated other investigations, in the United States and in Europe, that stressed the role of heredity in social failure. In the early 20th century this included groups like the Nams, the Hill Folk, the Jackson Whites, and especially the Kallikaks. The Kallikaks were a New Jersey family of degenerates descended from the dalliance of a Revolutionary War soldier. Allegedly his later marriage to a respectable woman led to a line of outstanding and successful people. Henry H. Goddard, a psychologist and the author of *The Kallikak Family*, naively believed his Kallikak study was controlled science.[33]

The first of these confirmations of the hereditarian view of crime and pauperism was compiled by the minister of the Plymouth Congregational Church in Indianapolis, Oscar Carleton McCulloch (1843–1891).[34] McCulloch was born July 2, 1843, in Fremont, Ohio; he was the eldest of five children of a pharmacist. When he was seven years old, the family moved, first to Wisconsin and then to Illinois. Oscar grew up in a religious home, his father being a devout Presbyterian. Oscar worked in his father's store, an arrangement he appreciated because he was in frail health, having been born with a clotting disorder that resulted in episodes of bleeding and severe migraine headaches. His ill health exempted him from military service during the Civil War, but he volunteered to visit sick and wounded soldiers sent to Chicago to recuperate. About this time, McCulloch had completed business college in Poughkeepsie, New York, and became a salesman for a drug company. He was successful in this activity and covered the far West, still a largely unsettled area. On one of his return trips to Chicago, he visited a mission for the poor in a slum and decided to abandon business for the ministry after securing his father's permission.

McCulloch received his divinity degree from the Chicago Theological Seminary in 1870. The seminary represented a consortium of faculties from Congregationalists, Unitarians, and the Disciples of Christ. McCulloch was drawn to the Congregationalists because he liked their democratic traditions and he still retained much of the Protestant faith of his youth. He was called to the Sheboygan, Wisconsin, Congregational Church, where his sermons and pastoral skills made him a popular minister. In 1870 he married a teacher and pianist who nurtured him in his bouts with ill health, but she soon contracted tuberculosis, dying in 1874 and leaving him two sons to raise. McCulloch read widely in the sciences and humanities and started a monthly reading club for the church and community. His growing interests in Darwinism and his secular tastes in periodical literature found their way into his sermons and eventually created a schism in the church, some believing him too liberal to meet the creedal standards of their Congregational Church. In 1877 McCulloch began keeping a diary which he maintained until his death some 14 years later. In it, he kept clippings of newspaper accounts of his sermons and he added reflective comments on his personal life and on his extensive readings. His religious views evolved as he read the writings of the transcendentalists and other reform-minded ministers. After reading a biography of Theodore Parker, one of the founders of the modern Unitarian church, McCulloch noted in his diary: "...only better one such man than a thousand of the average ministers. O God, my Father, touch my heart with just such fire and love for men."[35]

McCulloch was recommended by the Presbyterian minister, Myron W. Reed, to compete for the Plymouth Congregational Church in Indianapolis, which sought a new minister.[36] Reed knew McCulloch when they were at the Chicago Theological Seminary. The Plymouth Church was in a difficult financial condition with poor attendance and a foreclosed mortgage. McCulloch left Sheboygan reluctantly after he had survived a vote for his removal and felt that he could not endure a divided church. In 1877, he accepted the offer from the Plymouth Church and began to put into practice a reform ministry based on an "Open Door" policy that followed "more of Christ and less of Calvin." His sermons were well received, attracting much favorable newspaper comment, and he used his financial skills to make the church solvent and to raise funds for its relocation in a new, larger building designed to seat 1,000 members and guests at Sunday services.[37] McCulloch remarried in 1878, taking as his wife the former Sunday School teacher of the Sheboygan Church, who helped him raise the two boys and soon added three daughters to the family.

McCulloch as a Social Reformer

McCulloch attracted the intellectuals of Indianapolis. David Starr Jordan, then a young professor of zoology at Butler College, enjoyed the sermons and discussions so much he asked to be a member and McCulloch even waived the baptism most Congregational churches required for membership. Jordan said it was the "only religious organization I ever formally joined."[38] What attracted the well-read and liberal-minded to McCulloch's church was his enthusiasm for social reform. He created the Plymouth Institute, a secular vehicle to launch new social programs. He started a free library so that working people could read periodicals and books they could not afford to purchase. He began a dime savings bank to teach thrift to the laboring class. He championed the right of the laboring class to organize and was an enthusiastic supporter of the Knights of Labor. He revamped the existing, often neglected or ineffective, charity organizations in the city and established the Charity Organization Society. Very soon thereafter, he organized a Children's Aid Society, the Friendly Inn for transients seeking temporary shelter, a training school for nurses, free baths to promote public health, a visiting district nursing association for the chronically ill, and a summer program to get the sick and the impoverished children into camps or farms in the countryside. His persuasive efforts won over his critics and he attracted national notice, becoming president of the National Conference of Charities and Corrections in 1891 when it held its annual meeting in Indianapolis.

Shortly after his arrival in Indianapolis, McCulloch visited some destitute families and gave a sermon on the plight of the poor in Indianapolis. His earliest description of one of these families appears in his diary entry for January 18, 1878. He visited a "family composed of a man, half-blind, a woman, two children, the woman's sister & child, the man's mother, blind, all in one room ten feet square. One bed, a stove, no other furniture. When found they had no coal, no food. Dirty, filthy because of no fire, no soap, no towels. It was the most abject poverty I ever saw."[39] About this same time he and Myron Reed, his friend and fellow minister, were browsing in a bookstore. "I remember," Reed later noted in his eulogy for McCulloch, "we bought that wonderful book 'Margaret, the Mother of Criminals,' the same week; and we read it, and we preached about it the same Sunday night."[40]

For ten years McCulloch followed the families of the paupers and criminals in Indianapolis and asked J. Frank Wright, an employee of the Marion County Commission, to help him collate the interviews and follow up on the families. At the Fifteenth National Conference of Charities and Correction held at Buffalo, New York, in July 1888, McCulloch presented *The Tribe of Ishmael: A Study in Social Degradation*. McCulloch began his address with a biological model derived from the zoological account of a crustacean parasite, Sacculina, that was "not only interesting to the student of physical science, but suggestive to the student of social science."[41] The host for Sacculina is the hermit crab. The free-swimming larval form of the parasite attaches itself to the crab and "loses the characteristics of the higher class, and becomes degraded in form and function." McCulloch attributed this to a hereditary tendency "because some remote ancestor left its independent, self-helpful life, and began a parasitic, or pauper life." The adult parasite is amorphous, "with only the stomach and organs of reproduction left."[42]

Having established the biology of the parasite, McCulloch drew the moral that "the Sacculina stands in nature as a type of degradation through parasitism or pauperism."[43] Human counterparts to such hereditary parasitism existed, in McCulloch's view, and he informed his audience that he studied one such family in Indiana. "It resembles the study of Dr. Dugdale into the Jukes, and was suggested by that." He claimed that the name, "the tribe of Ishmael," is not fictitious; it is "the name of the central, the oldest, and the most widely ramified family."[44]

THE TRIBE OF ISHMAEL AS SEEN BY MCCULLOCH

McCulloch's portrayal is bleak and stresses the wretchedness of the families and the burdens they have imposed on society. "In this family history are murderers, a large number of illegitimacies and of prostitutes. They are generally diseased. The children die young. They live by petty stealing, begging, ash-gathering. In summer they 'gypsy,' or travel in wagons east or west. We hear of them in Illinois about Decatur, and in Ohio about Columbus. In the fall they return. They have been known to live in hollow trees on the river-bottoms or in empty houses. Strangely enough, they are not intemperate to excess."[45] McCulloch identified a "wandering blood," a "licentiousness" with a "diseased and weakened condition" and "mental weakness" as the characteristics of the tribe. He pointed out that they had been recipients of public aid since the 1840s and that such public charity encouraged them "to be idle." In his summing up, he claimed that "the individuals already traced are over five thousand, interwoven by descent and marriage. They underrun society like devil-grass. Pick up one and the whole five thousand will be drawn up."[46]

McCulloch shared the belief of many 19th-century clergymen that public relief was a failure and needed to be replaced. Public relief became a premium "paid for idleness and wandering." The solution was "not to give alms but to give counsel, time, and patience to rescue such as these."[47] He admitted that this was not easily done and his own success rate was low: "I have tried again and again to lift them, but they sink back." Because he perceived them as a "decaying stock," he believed the only hope was to "get hold of the children" at an early age and try to reform them before their families and their environments shaped them beyond salvation.[48]

ORIGINS OF THE TRIBE OF ISHMAEL

The tribe of Ishmael was retraced in 1977 by Hugo P. Leaming.[49] Much of the information on the tribe, its history and customs, comes from Wright's unpublished but extensive manuscript notes in the Indianapolis State Library. According to Leaming, the tribe began to form during the American Revolutionary War with a three-way miscegenation and coalition of escaped and freed blacks, nomadic Indians, and poor whites, including indentured servants and convicts deported from England. They moved about Maryland, Virginia, the Carolinas, and Tennessee, entering Kentucky about 1785. In the 1790s they had crossed the Ohio into Cincinnati and remained under the leadership of the founder of the tribe, Ben Ishmael and his wife Jennie. After their disappearance about 1810, Ben's

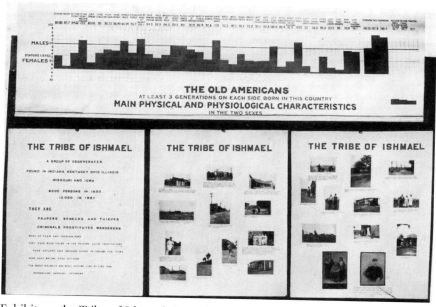

Exhibit on the Tribe of Ishmael, Second International Congress of Eugenics, 1921. Photographs of dwellings are shown, mostly in Indianapolis. The individual homes for members of the Tribe of Ishmael were usually temporary. They would spend their winters in shacks and sheds in Indianapolis and then go on the road for the rest of the year. (Reprinted from H. Laughlin [1923] *Exhibits Book: Second International Exhibit of Eugenics*, Williams & Wilkins, Baltimore [courtesy of CSHL Archives].)

son John led the tribe to the Indian territories of the old Northwest. The tribe was accepted, when they reached the White River, the future site of Indianapolis, as one more of the Eastern Indian tribes displaced from their homelands by European colonists.

The nomadic tradition persisted throughout the 19th century. The tribe, at one time numbering 10,000 persons, followed a triangular route; they would leave Indianapolis in the spring, migrate Northwest to the Urbana-Champaign, Illinois, region, move Southwest in the summer to Decatur, Illinois, and return in the fall to over-winter in Indianapolis. They may have been composed initially of Shawnee Indians, blacks of the nomadic Islamic Fulani tradition, and a Celtic gypsy-like population called the Tinkers.[50] Because they shunned full-time steady jobs, farming, or the purchase of property, they were associated with the least desirable activities in Indianapolis. They tended horses, repaired umbrellas, cut ice, did

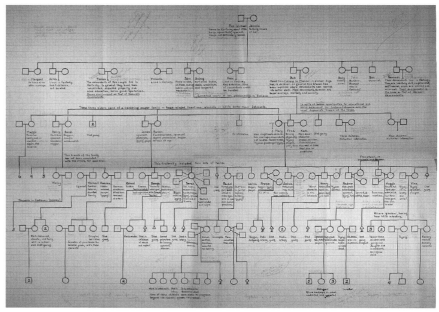

Pedigree of some members of the Tribe of Ishmael. Eugenicists hoped to establish a hereditary basis for social failure (pauperism, feeblemindedness, vagrancy) in this kindred through construction of elaborate pedigrees. (Reprinted, with permission from American Philosophical Society, courtesy of CSHL Eugenics Web site.)

home laundry, hauled rubbish and ashes, scavenged, engaged in prostitution, begged by feigning blindness, and carried out petty thefts. They lived in shanty towns or built homes that were pulled on skids along the mud of the banks of the White River.

The tribe was described in literature by James Fenimore Cooper in *The Prairie*, by Edward Eggleston in his *Hoosier Schoolmaster*, and by Booth Tarkington in the *Conquest of Canaan*. The ethnicity of the tribe is usually described as white with some admixtures of the other two cultures. Their children were often described as "tow-headed" and their complexion was tan or olive with "mixed, indeterminate features."[51] According to Leaming, the tribe dispersed in the early 20th century, with those members who retained a black identity moving to Chicago, Detroit, and Philadelphia. Since there were thirteen clans in the original tribe with about 250 family names, the scattered descendants have long since lost their identity as members of the tribe of Ishmael.[52]

FOOTNOTES

[1] "Elisha Harris," *Appleton's Cyclopedia of American Biography* 3: 91; see also Harris's letters to Henry I. Bowditch in Bowditch's *Public Hygiene in America* (Little Brown & Co., Boston, 1877), pp 191–192. Harris established his reputation after editing the *Report of the Council of Hygiene and Public Health of the Citizen's Association of New York Upon the Sanitary Condition of the City* (1865). The city was divided into 29 districts, and physicians did a thorough survey of the health problems of the inhabitants. Harris's contribution is discussed in George Rosen's *A History of Public Health* (MD Publications, Inc., New York, 1958), pp. 244–245.

[2] The club is discussed by one of its members, publisher G.H. Putnam, *Memories of a Publisher* (G.P. Putnam's Sons, New York, 1915), pp.171–172.

[3] This was the source used by Oliver Wendell Holmes and Charles Loring Brace in their popularizations of "Margaret the mother of criminals." Richard Dugdale, who began his studies in 1874 was, of course, not well known at the time, whereas Harris was quite prominent; thus, both Holmes and Brace attributed the work exclusively to Harris, who did initiate it.

[4] Charles L. Brace, "Pauperism," *The North American Review* 120(1875): 316–334. Brace (1826–1890) was a pioneer social worker and theologian who devoted his life to the poor although he was born to wealth. He was a founder of the Children's Aid Society in New York City.

[5] Ibid. p. 321. Margaret, the mother of one line of the Jukes family, had produced enough descendants by the early 19th century to give the region a notoriety that attracted the attention of Washington Irving. In *The Legend of Sleepy Hollow*, a typical description of the Jukes family habits are described for Rip van Winkle and his clan. They are portrayed as shiftless, rowdy, inebriate paupers. The father of the Jukes, Max, was of Dutch origin, coming to the Colonies about 1720 (there was a catastrophic crop failure in Europe about 1716–1718 that sent many of the Rhineland peasants, from Germany to Holland at the North Sea, to settle the New World). Names like van Dunk and van Dyke were commonly found among the Jukes.

[6] Oliver Wendell Holmes was also prominent in the public health movement. Independently of Ignaz Semmelweis he deduced, in 1843, that puerperal ("childbed") fever was caused by infection from unwashed hands. Fortunately, Holmes's advice was eventually respected in the United States, and physicians adopted the habit of scrubbing their hands thoroughly before examining patients. Semmelweis's colleagues derided him and Semmelweis died unappreciated. Holmes's essay "Crime and automatism: with a notice of M. Prosper Despine's *Psychologie Naturelle*" appeared in the *Atlantic Monthly* 35(1875): 466–481. Holmes named *The Atlantic Monthly* after being invited by James Russell Lowell to help him found and edit a new literary journal. It was Holmes's essays, under the title "Autocrat of the Breakfast Table," that established both the magazine's and Holmes's fame.

[7] Holmes, *Crimes and Automatism*, p. 475.

[8] Ibid.

[9]   Ibid. p. 477.

[10] Details of Dugdale's life are scanty. His obituary by Edward Morse Shepard (a member of the sociology club) was privately printed as *The Work of a Social Teacher Being a Memorial of Richard L. Dugdale* (The Society for Political Education, Economic Tracts No. XII, New York, 1884). Appleton's *Cyclopedia of American Biography* 2: 250 provides a brief sketch of his education and activities in liberal causes. A later account, by Arthur Estabrook, was published as a preface to *The Jukes in 1915* (Eugenics Record Office, Cold Spring Harbor, New York).

[11] Richard L. Dugdale, *The Jukes: A Study in Crime, Pauperism, Disease, and Heredity* (G.P. Putnam's Sons, New York, 1877). The book was based on the "Report of special visits to county jails for the year 1874," a pamphlet of 66 pages printed by the state at Albany, New York in 1875. It was reprinted in the "30th Annual Report of the Prison Association of New York" and entered book form as the third edition. Elisha Harris wrote the introduction, and the book had grown to 115 pages. In 1884 the fourth edition appeared with an introduction by M.F. Round. The fifth edition appeared in 1891 and the seventh in 1902. A reissue of the fourth edition appeared in 1910 with an introduction by Franklin H. Giddings.

[12] Dugdale, *The Jukes,* 4th edition, p. 4. Harris may have felt like a graduate student's major professor and wanted his student to share the full benefit of the book's authorship. Dugdale was well read and he was familiar with the work of Spencer, Morel, and Maudsley, all of whom were heavily committed to the idea that behavioral traits were associated with hereditary tendencies but all of whom shared a Lamarckian view of that inheritance.

[13] Ibid. p. 3.

[14] A.E. Winship, *Jukes-Edwards: A Study in Education and Heredity* (R.L. Myers & Co., Harrisburg, Pennsylvania, 1900). Winship claims: "It is not the real name of any family, but a general term applied to forty-two different names borne by those in whose veins flows the blood of one man. The word 'jukes' means to roost. It refers to the habit of fowls to have no home, no nest, no coop, preferring to fly into the trees and roost away from the places where they belong. The word has also come to mean people who are too indolent and lazy to stand up or sit up, but sprawl out anywhere." (pp. 8–9).

[15] Dugdale, *The Jukes,* 4th edition, p. 2.

[16] Ibid. p. 62.

[17] Ibid. p. 64.

[18] Ibid. p. 56.

[19] Ibid. p. 61.

[20] Ibid. pp. 61–62.

[21] Ibid. p. 63.

[22] Ibid. p. 57.

[23] Ibid. p. 28.

[24] Ibid. p. 65.

[25] Elof Axel Carlson, "R.L. Dugdale and the Jukes Family: A Historical Injustice Corrected," *BioScience* 30(1980): 535–539.

[26] Richard L. Dugdale, "Origin of crime in society," *The Atlantic Monthly* 48(1881): 452–462; 735–746.

[27] Ibid. p. 458.

[28] Ibid. p. 462.

[29] Ibid.

[30] Shepard, *Memorial of Richard L. Dugdale,* p. 8.

[31] Edward S. Morse, "Natural selection and crime," *Popular Science Monthly* 41(1892): 433–446.

[32] Dugdale, *The Jukes,* 4th edition, pp. iii–iv.

[33] Henry H. Goddard, *The Kallikak Family* (MacMillan, New York, 1912). See also a critique of the Kallikak family by J. David Smith, *Minds Made Feeble: The Myth and Legacy of the Kallikaks* (Aspen Systems Corp., Rockville, Maryland, 1985).

[34] For an excellent biography of McCulloch, see Genevieve C. Weeks, *Oscar Carleton McCulloch 1843–1891 Preacher and Practitioner of Applied Christianity* (Indianapolis Historical Society, Indianapolis, Indiana, 1976).

[35] Ibid. p. 43.

[36] Ibid. p. 49.

[37] Junius B. Roberts, "Plymouth Church, Indianapolis," *Indiana Magazine of History* VII(1911): 52–60.

[38] David Starr Jordan, *Days of a Man,* volume I, p. 132.

[39] Weeks, *Oscar Carleton McCulloch,* p. 169.

[40] "Tributes to Oscar C. McCulloch," *Proceedings of Charities and Correction at the Nineteenth Annual Session Held in Denver, Colorado* (George Ellis, Boston, 1892). Essay of Myron Reed, pp. 247–250.

[41] Oscar C. McCulloch, "The tribe of Ishmael: a study in social degradation," reprinted from *Proceedings of the Fifteenth National Conference of Charities and Correction, Held at Buffalo, July 1888* (Charity Organization Society, Plymouth Church Building, Indianapolis, Indiana), p. 5.

[42] Ibid.

[43] Ibid. p. 1.

[44] Ibid.

[45] Ibid. p. 5.

[46] Ibid. p. 4.

[47] Ibid. p. 6.

[48] Ibid. p. 7.

[49] Hugo P. Leaming, "The Ben Ishmael tribe: A fugitive nation of the old Northwest," pp. 97–142 in *The Ethnic Frontier,* edited by Melvin G. Holli and Peter d'A. Jones (Wm. B. Eerdmans Publishing Company, Grand Rapids, Michigan, 1977).

[50] The Tinkers are a wandering population found in Ireland and parts of Great Britain. They are like gypsies in traveling in wagons, living as tinsmiths, horse-traders, and trash scavengers. Like the gypsies, they have a private language they use among themselves and do not reveal to outsiders. The Tinkers enjoy wearing brightly colored outfits. In his "Old time slums of Indianapolis" (*Indiana Magazine of History* VII: 170–173), George Cottman describes the Ishmaelites; they had a social life of their own and wore red, blue, yellow, striped trousers, "red shirts crossed with bright hued suspenders," gaudy neckerchiefs, cowhide boots, and a broad-brimmed brown hat (p. 173).

[51] Leaming, *The Ben Ishmael tribe,* p. 122. In the three-way miscegenation, two of the groups, the Tinkers and the Fulanis, were already miscegenated, the Tinkers with gypsies and the Fulanis with whites. The Fulanis are a white–black miscegenated Islamic people who extended from the Nile to Nigeria. They were nomads and cultivated cattle in the sub-Saharan regions. The low alcoholism noted by McCulloch may have stemmed from this Islamic tradition; so too, the progenitor, Ben Ishmael may have been an altered version of Ibn Ishmael, people of Ishmael, Ishmael being the founder of Arabic people. The Fulanis were particularly populous in Northern Nigeria, French Guinea, and Senegal, areas well within the region raided for the American slave trade. The Shawnees were an Algonquin tribe near Tennessee and Kentucky who extended to Pennsylvania in the early 18th century and then migrated to Ohio and the mid-West in the late 18th century. They were described as "restless and inclined to wander" (*Encyclopedia Britannica* 20: 472).

[52] Leaming (p. 123) believes the Ishmaelites underwent a diaspora about 1910, moving to Chicago, Detroit, and Philadelphia, and their descendants may have pioneered the Marcus Garvey and Black Muslim movements. They were last described as a degenerate tribe by Arthur Estabrook in "The tribe of Ishmael", pp. 398–404 in *Eugenics Genetics and the Family, vol. 1,* The 2nd International Congress of Eugenics, September 22–28, 1921 (Williams and Wilkins, Baltimore, Maryland, 1923). Estabrook claimed that in 1921 there were still some 10,000 members in the mid-West (p. 400).

11

A Minor Prophet of Democracy

FEW MOVEMENTS ARE CONCEIVED AND ORGANIZED through the efforts of one person. Ideas are often scattered like burning leaves and twigs in a forest fire and ignite when they land in favorable areas. The original ideas change, pull in other ideas, and emerge years later in forms and intensity that would surprise those who were first to bring forth tentative suggestions. The eugenics movement in the United States had a different history from that of its parallel and earlier movement in England. Francis Galton's aristocratic views and wealth, and the elite of British culture in whose company he belonged, were not the model nor the ideals for American democracy. Social class in America was fluid. Mobility for most Americans, and even its near-penniless immigrants, was largely a matter of merit rather than connections and position in society.

One of the innovators of the American eugenics movement was David Starr Jordan (1851–1931). His adult life straddled the last quarter of the 19th century and the first quarter of the 20th century. He witnessed and contributed to the shifting views of a nation initially buoyed with an optimism that it could solve all its problems with harm to no one which then found itself transformed into a nation that despaired of its problems, seeking draconian measures to solve them.

JORDAN'S EDUCATION AND EARLY WORK

Jordan was a naturalist, an explorer, an educator, and an essayist. He defined himself as "a minor prophet of democracy."[1] Jordan, born in New York state near Albany, attended Cornell University, where he took a liking to natural history. He was pleased to be invited in 1873 by Louis Agassiz to the newly launched marine biology station on Penikese Island off the Southwest corner of Cape Cod.[2] This forerunner of the Woods Hole Marine Biological Station was short-lived. The seed money for the station, provided by tobacco merchant John Anderson, rapidly ran out, and Agassiz's death the following year left it without effective leadership. Agassiz was one of the world's most celebrated naturalists. He had made his name known to scholars throughout the world for his daring explorations of the flowing waters that existed under glaciers. He was a paleontologist who unearthed fossils and tried to make sense out of their classification, claiming that there were numerous episodes of worldwide catastrophes which caused large numbers of species to become extinct.

Agassiz did not accept Darwin's theory of natural selection, nor did he believe that the fossil record supported the idea that life had evolved on earth. He was a popular writer, and his hostility to evolution made him a favorite with American intellectuals who were still comfortable with the creationist views of most Protestant churches. Long after most naturalists in North America had accepted the basic ideas of Darwinism, Agassiz held out in class and in his writings for the biblical view of a divine creation of life. At Penikese Island, Jordan noted that "he seldom spoke without a piece of chalk in hand."[3] He revered the living world and "in each natural object he saw a thought of God."[4] Despite Agassiz's intense opposition to Darwinism "he believed in the absolute freedom of science," and all his students became evolutionists.[5] Jordan himself found the transition from childhood faith to evolution a difficult one: "I went over to the evolutionists with the grace of a cat the boy 'leads' by its tail across the carpet."[6]

Agassiz's death meant that Jordan was on his own and without financial support. He taught in Appleton, Wisconsin, for a year, but the school folded. In 1874 he went west to Indianapolis and taught high school for a year. He joined the staff of Butler University (then called the Northwest Christian College) and in his spare time got an M.D. at the Medical College of Indianapolis. Jordan was not proud of his M.D.; he described the educational standards of American medical schools in those days as of "the medieval period." While he was in Indianapolis, he read about the sermons of Oscar McCulloch at the Plymouth Congregational Church. Jordan shared with McCulloch an enthusiasm for evolutionary biology and its social implications. They corresponded and shared reprints of their writings until McCulloch's death in 1891.

In 1879, Indiana University appointed Jordan to its faculty as professor of natural history. Jordan was making a name for himself as an ichthyologist, and he studied the fishes in the lakes and rivers of the midwest. As his opportunities for travel and support for research increased, he extended his studies of fish to other parts of the United States and to South America. Indiana University was founded in 1821 as a seminary school provided for in the charter of statehood in 1816. It was assigned a location in Bloomington by President James Madison. By 1838 it had become Indiana University and served the state by educating its future leaders. Jordan enjoyed teaching and got along well with his colleagues as a productive scholar with energy, ambition, and vision. In 1885 he was named president of the university and initiated a reform of its liberal arts degree. He introduced the idea of giving students electives from the numerous courses available in the university. He assigned to the first two years the teaching of general courses among several disciplines. He argued successfully that in the last two years students should elect a major and be assigned to faculty supervisors who would guide them in their scholarly habits.[7] The prototype of the modern undergraduate liberal arts program made Jordan widely known among educators, and he was often invited to speak on his ideas. Jordan devoted his energies to building Indiana University, but he never abandoned his teaching. In his remaining 33 years as a university president he taught an undergraduate course in the philosophy of biology, giving it the name "bionics," including in it his emerging views on evolution, eugenics, hygiene, ethics, and the role of the informed citizen.

JORDAN'S MOVE TO CALIFORNIA

Jordan left Indiana University in 1891 and accepted the invitation from a committee of scholars who were charged by the will of industrialist Leland Stanford to select a president for the newly created Stanford

University. Jordan took on the challenge of building a university from scratch and sought outstanding scholars to chair the departments he set up. He collected his speeches and published them, mostly through the American Unitarian Association. They reflected the views of a social Darwinist whose political outlook was liberal and idealistic.

While he was still president of Indiana University, Jordan used the annual commencement address as an opportunity to reach the young and to say something more than a pleasant farewell. In 1886 he stressed ethics and the value of a state university education to the progress of democracy. In 1888 he discussed "The Ethics of the Dust."[8] He pointed out the near identity of the early embryos of dogs and humans and how they diverge through development, which somehow unfolds in response to an "inherent power, the subtle production of heredity." When the embryos are normal, both dogs and men carry out their normal activities. Abnormalities of development, whatever their causes, may abort development altogether or lead to "perpetual immaturity." He claimed that "there are dwarfs in body and in mind—those reach the age of manhood, while retaining the stature or the intellect of childhood." To Jordan, "badness is the evidence of distorted development." Although most individuals may be described as average in form and capacities, Jordan asserted that this status was belied by "constant rebellion" that led a few individuals in every generation to excel in one aspect or another. At the same time he pointed out that among the average there is "an individuality we did not see before." He assured his students that they would serve society well as average members of it and each would have a unique personality and life to lead, but among them would be a few who would aspire to leadership and creativity, giving to the new generation higher standards than it had before. Each generation would likewise share in this small number of excelling individuals, adding to the progress of civilization and, he believed, reflecting a divine direction.

JORDAN'S VIEWS ON HEREDITY

Jordan's views on heredity were shaped through discussions with McCulloch on the significance of the Jukes and the Tribe of Ishmael. He shared McCulloch's appraisal that these were degenerate stocks, and they discussed the similarities of parasitism in nature with the role of almsgiving in society.[9] Jordan summed up the beliefs of his generation that there were three kinds of poor: the Lord's poor who are temporary victims of misfortune and who deserve charity; the Devil's poor who deserve the wretchedness they brought on themselves by their vice; and paupers, who inherited feeble minds and feeble wills. The paupers, Jordan claimed, "of necessity fall to the bottom, being destitute of initiative and self respect." Jordan believed that the Tribe of Ishmael were "poor whites" who were descendants of debtors sent from England to Jamestown, Virginia, where they went on "to become ancestors of a forlorn group of ne'er-do-wells scattered through the middle West."[10]

A second source of Jordan's emerging views on the characteristics of the unfit came from his trips abroad. He visited Switzerland in 1881, 1883, 1890, and 1910. On his first three trips, he observed in the village of Aosta numerous cretins who roamed the streets, often begging and living miserable lives because of their limited capacities to care for themselves. Cretins arise from defects of the thyroid glands.[11] These can be caused by a diet low in iodine, by an autoimmune reaction during pregnancy that destroys the fetal thyroid, or by various genetic defects in thyroid hormone production or the response to the hormone. Cretins are mentally retarded, stunted in growth, and show characteristic physical deformities, such as a goiter or enlargement of the malfunctioning thyroid gland. Jordan's views were harsh and pessimistic. "The idiot has received generous support, while the poor farmer or laborer with brains and no goiter has had the severest of struggles. In the competitions of life a premium has thus been placed on imbecility and disease. Cretin has married cretin, goiter with goiter, and charity and religion have presided over the union. The result is that idiocy is multiplied and intensified. ...True charity would give them not less helpful care, but guarantee that each individual cretin should be the last of his generation." Jordan referred to the cretins unsympathetically, describing them as "feeble little people with uncanny voices, silly faces, and sickening smiles, incapable of taking care of themselves."[12]

When Jordan returned to Aosta in 1910, the cretins had virtually disappeared. He was referred to the Mother Superior of a convent, Sister Lucia, who responded to his query about their absence. "Il n'y en a plus" (There are no more), she told him.[13] Starting in 1890 the cretins were segregated in the Asylum for the Aged Poor and the Church no longer sanctioned their marriages as they had in the past. Jordan noted that "only one cretin still survived—an old woman four feet high, with the manners of an affectionate lap dog, even licking my hands like a dog."[14] Instead of medical sterilization, the Swiss provided a "social castration" through isolation in an asylum.

JORDAN'S SOCIAL PHILOSOPHY OF PARASITIC DEGENERACY

In 1898 Jordan assembled many of the topics he covered in his bionics course and extension lectures in a collection he called *Footnotes to Evolution*.[15] He gave prominence to McCulloch's analysis of the Tribe of Ishmael, preparing the reader for its full impact by assigning a preceding chapter to the parasitism of Sacculina and the biological and evolutionary significance of degeneracy of organs in the parasitic state.[16] Jordan's biological theory assumed that quiescent or sessile animals, such as tunicates, "descended from fish-like ancestors" and were "reduced to motionless sacs, buried in the sand or anchored to rocks or wharves." Even more degenerate, he believed, were parasitic forms, such as fleas or scale insects which lose their wings or become confined to a host's body. He referred to this degenerate state as "animal pauperism."

In his following chapter on human parasitism,[17] Jordan did not hesitate to draw an analogy. "Recent studies, as those of Dugdale, McCulloch, and others, have shown that parasitism is hereditary in human species as in the Sacculina."[18] He limited his sympathies for the unfit, claiming that "the pauper is the victim of heredity, but neither Nature nor Society recognizes that as an excuse for his existence."[19]

Jordan pointed out that "no community was ever built up of thieves and imbeciles."[20] He condemned universal suffrage, and urged society not to "extend the right to vote to venal, cowardly, or ignorant voters." Although he had opposed slavery as immoral, he felt uncomfortable with giving freed blacks the right to vote; it was the "least of the evils, no doubt, but an evil nevertheless."[21] Coupled with his concern about degenerate classes in America was his belief that Europe was no longer sending its best, but rather its worst, citizens to America. He advocated restrictive immigration. "Every family of Jukes and Ishmaels which enters at Castle Garden carries with it the germs of pauperism and crime."[22] His objection was not to the ethnic origins of the immigrants but to their abilities to function as good citizens. "No race is so perfect that judicious weeding could not improve it."[23]

Jordan was particularly interested in social degeneracy, and he had read widely the various theories and typologies of these alleged pathological forms of humans. He preferred the term "the higher foolishness," coined by Israel Zangwill that included Nordau's degenerates, Hirsch's "monkey geniuses," Maudsley's "borderland dwellers," Magnan's "dégénérés supérieurs," and Lombroso's "mattoids." These were variant forms of "inspired idiots and educated fools of all ages and climes."[24] Most of them, Jordan argued, were probably not the result of a degenerate heredity but of a defective culture. They differed in his mind from the true parasitic humans, such as paupers.

THE DEVELOPMENT OF A NEGATIVE EUGENIC OUTLOOK

Throughout the 1890s, Jordan's interest in degeneracy became intense. Accompanying the medical and religious debates on the causes of degeneracy were social and artistic debates. Millet's painting, *The Man with the Hoe,* portraying a tired peasant in the fields whose countenance and dress seemed unchanged since the middle ages, inspired those who saw it to write poems or essays on the causes of this forlorn appearance. Edwin Markham's poem attributed the arrested development to the evils of the industrial revolution. For Jordan, *The Man with the Hoe* was a symbol of the unfit, the Jukes and the Ishmaels, and a warning to civilization. This crude figure "was not brought to his own estate by centuries of industrial repression. He was rather primitive and aboriginal, persisting in a competitive world mainly because wars had destroyed generations of self-extricating, freedom-loving peasantry. He represents 'the man who is left.'"[25]

Jordan turned to the study of war in the 1890s, believing that this practice was a major cause of degeneracy. "During this period, ...I first began to study seriously the effect of war on the human breed, the constant elimination of the strong and the brave, as well as of the bully and the soldier of fortune."[26] He published his views in a series of short books, *The Blood of a Nation* in 1902, *The Human Harvest* in 1907, *Unseen Empire* in 1912, *The Heredity of Richard Roe* in 1913, and *War and the Breed* in 1915.

THE

HUMAN HARVEST

*A STUDY OF THE DECAY OF RACES THROUGH
THE SURVIVAL OF THE UNFIT*

BY

DAVID STARR JORDAN

President of
Leland Stanford Junior University

"La guerre a produit de tout temps une selection à rebours."
(NOVICOW)

IN·LUCE·
VERITATIS

Boston
AMERICAN UNITARIAN ASSOCIATION
1907

Title page and table of contents from *The Human Harvest* (1907) by David Starr Jordan reveal the influences of degeneracy theory on Jordan's eugenic philosophy. (Courtesy of CSHL Archives.)

Although Jordan was aware, by 1901, when he first presented *The Blood of a Nation* in *Popular Science Monthly*, that the term blood, as a synonym for heredity, was biologically inaccurate, he retained it for its metaphorical value and because no other term was so immediately recognized by the general reader as a symbol for heredity.[27] For Jordan, the blood of a nation determined its history, and conversely, the history of a nation was determined by its blood. He clearly believed there were inborn cultural behaviors. "Wherever an Englishman goes, he carries with him the elements of English history.... Thus, too, a Jew is a Jew in all ages and climes, and his deeds everywhere bear the stamp of Jewish individuality. A Greek is a Greek; a Chinaman remains a Chinaman."[28] He acknowledged that "education may intensify their powers or mellow their prejudices; oppression may make them servile or dominion make them overbearing, but these traits and their resultants, so far as science knows, do not 'run in the blood.' They are not 'bred in the bone.' ... It is always 'blood which tells.'"[29]

Expanding on his theme of *The Man with the Hoe* as the "Man who is left," Jordan asserted that it is this less desirable survivor who becomes the father of the next generation. "By the sacrifice of their best, or the emigration of the best, and by such influences alone, have races fallen from first-rate to second-rate in the march of history."[30] As wars waste the nation's young, he claimed, "the weak, the vicious, the unthrifty will propagate, and in default of better, will have the land to themselves."[31] Articulating a Spencerian outlook on human evolution and human society, Jordan asserted that "the survival of the fittest in the struggle for existence is the primal cause of race progress and race changes." In a complementary way, "the survival of the unfittest is the primal cause of the downfall of nations."[32]

Jordan's views of heredity were based on the prevailing view in the 1890s that traits were permanently fixed and not modified by the environment as Lamarck had supposed, but that selection could readily sift out any desired traits: "In selective breeding with any domesticated animal or plant, it is possible, with a little attention, to produce wonderful changes for the better. Almost anything may be accomplished with time and patience."[33] By applying selective forces in society, Jordan believed he could interpret the genetic history of a nation. He argued that primogeniture was the reason the American stock was so good. The inheritance of property is a problem for many families, especially the inheritance of land. "In England the eldest son is chosen for this purpose, a good arrangement according to Samuel Johnson 'because it ensures only one fool in the family.'"[34] This left the other children without property and forced them to consider other careers and opportunities. "The evil of primogeniture has furnished its antidote. It has begotten democracy." The unpropertied sons "manned the *Mayflower*."[35]

Jordan studied the effects of the Civil War on heredity. "North and South, it was the same. 'Send forth the best ye breed' was the call on both sides alike, and to this call both sides alike responded." He claimed that the Civil War destroyed the plantation aristocracy with its cousin marriages, diminished its First Families of Virginia marriage practices, and created a leftover society less able than its predecessors.[36] In *The Human Harvest*, Jordan identified Benjamin Franklin as the source for his ideas on war being dysgenic; he quoted from one of his essays arguing why the new nation should disband its army and not have a permanent military establishment: "A standing army not only diminishes the population of a country, but even the size and breed of the human species."[37] Altogether, Jordan attributed seven factors that cause the decline of nations. These were primogeniture, repression or intolerance, monasticism, abuse of charity, alcoholism, the migration to cities, and war. All were important, but war was the major focus of his attention because in modern times war had become an immensely expensive and formidably destructive enterprise.

FOUNDING THE AMERICAN EUGENICS MOVEMENT

Just before the outbreak of the first World War, as the industrial nations built up their flotillas of battleships and amassed standing armies, Jordan campaigned for peace. He became friends with Jane Addams in her efforts to combat war. He enjoyed Theodore Roosevelt's friendship and praised him for his efforts to bring about peace among nations but condemned him for his imperialist aspirations. In 1912 he explored the costs of war and the nongenetic or "euthenic" effects, as he called it, on society. He published his ideas in *Unseen Empire: A Study of the Plight of Nations that do not Pay their Debts.*[38] The unseen empire is one of finance that has created an "empire of debt" among warring nations and nations that prepare for war. It included the brokers of wars, such as the Rothchilds and other major bankers, the munitions makers, propagandists for the profits of war who created war scares, and politicians who practiced secret diplomacy that led to wars. All of these components of the "unseen empire" circumvented democracy and prevented the free citizens of nations from determining their own destinies or setting their own priorities. He pointed out that the cost of one U.S. battleship exceeded the entire endowment of Harvard University and the yearly maintenance of the ship exceeded the annual budget of the university. He opposed war because it consumed "the fruits of progress" and it denied rational solutions to international problems. He favored more international courts, with the hope that humanity would move "from the rule of force to that of law."[39]

In a desperate effort to bring the views of pacifism to a world preparing for war, he left for Europe and spoke on "The Eugenics of War" to the newly founded London Society of Eugenics chaired by Major Leonard Darwin, Charles Darwin's son and a loyal supporter of the British Army and its role in establishing the British Empire. Also in attendance were the

Duchess of Marlborough, at whose home the meeting took place, and other stalwarts of the Empire. As Jordan talked of the dysgenic effects of war on the European nations that fought in the Napoleonic wars and that savaged the United States during the Civil War, his audience murmured its opposition, and, at the close of his address, Major Darwin had to discard politeness and inform Jordan that he disagreed with him. Darwin defended the view that the battlefield permitted the bravest and ablest to survive. From London, Jordan went to Berlin and spoke to the German eugenic society, which included many of the generals of the imperial staff. It was with great difficulty that Jordan spoke; the silence was oppressive, and midway through his talk one general broke the tradition of respect for the guest speaker and shouted "Genug!" (Enough).[40]

The American eugenics movement was the product of many years of effort by clergymen, sociologists, biologists, physicians, and popular commentators on American life. Jordan's essays and books, his innumerable speeches, his prominence as a scientist and educator, and his standing as an intellect made him the reasonable person to be asked to chair a Committee on Eugenics established by the American Breeder's Association at its December 9, 1909 meeting. The American Breeders Association was founded December 29–31, 1903 at St. Louis, Missouri, by a committee of the American Association of Agricultural Colleges and Experimental Stations. Originally its focus was on animal and plant breeding, fields greatly stimulated by the rediscovery of Mendel's laws. As interest in eugenics increased, a recommendation for a third section, on eugenics, was made, and the Committee chaired by Jordan presided while awaiting the outcome of a mail vote on the proposed changes to the constitution of the American Breeders Association. The membership was enthusiastic in support of this change, voting 499 to 5 in its favor.[41]

Liberty Hyde Bailey was president of the association and with Jordan selected Charles Benedict Davenport as secretary; for members they invited Alexander Graham Bell, Vernon Kellogg, Luther Burbank, William Ernest Castle, Adolf Meyer, H.J. Webber, and Frederick A. Woods.[42] The committee assigned to the section on eugenics the functions of "investigation, education, and legislation." A number of subcommittees were initiated. The subcommittee to investigate feeblemindedness was chaired by A.C. Rogers; it was asked to look into the question "Do two imbecile parents ever beget normal children?" A subcommittee on insanity was chaired by

Adolf Meyer of the New York State Commission on Lunacy. Within a few months, new subcommittees were to be added from a list that included inherited medical disorders, criminality and pauperism, mongrelization of races, consanguineous marriages, differences in human physiology and anatomy, and intelligence. Davenport, reporting the committee's activities, appealed for funds.

EUGENICS AS A SUBJECT FOR RESEARCH

The committee supervised the founding of the Eugenics Record Office (discussed in more detail in Chapter 15) which was to be located in Cold Spring Harbor, New York, with Harry H. Laughlin as superintendent.[43] The money for its support was provided by Mrs. E.H. Harriman, the mother of Averell Harriman, later to become governor of New York state.[44] Its close association with Davenport's Long Island Biological Society permitted Davenport to coordinate the activities of its basic genetic programs and maintain separate staff and facilities for the greatly expanded research activities in human heredity.[45] In Davenport's report of the deliberations of Jordan's committee, he noted the following agreements. For research: "...We have become so used to crime, disease and degeneracy that we take them as necessary evils. That they were, in the world's ignorance, is granted. That they must remain so, is denied."[46] For education: "As precise knowledge is acquired it must be set forth in popular magazine articles, in public lectures, in addresses to workers in social fields, in circular letters to physicians, teachers, the clergy and legislators. The nature and the dangers of unfit matings, the way to secure sound progeny, must ever be set forth."[47] For legislation: "Society must protect itself; as it claims the right to deprive the murderer of his life so also it may annihilate the hideous serpent of hopelessly vicious protoplasm. Here is where appropriate legislation will aid in eugenics and in creating a healthier, saner society in the future."[48]

The American Breeders Association was "the first body of scientists to give recognition officially to the new science of eugenics and the first to set to work a large number of scientists for specific research in eugenics."[49] Galton's death in 1911 was the basis for the second formal eugenics research program. In his will, Galton left his estate to the University of London to endow "The Galton Professorship of Eugenics" with an accompanying laboratory, office, and library.[50] In 1912 the American Breeders Association recommended changing its name to the American Genetics Association and the following year it suggested a new name for its journal; the *American Breeders Magazine* became the *Journal of Heredity* beginning with its fifth volume in 1914. Its editor was Paul Popenoe, a student of Jordan and an enthusiast for eugenical sterilization.

FOOTNOTES

[1] David Starr Jordan, *The Days of a Man, Volume 1 1851–1899* (World Book Co., Yonkers-on-Hudson, New York, 1922), p. vii. Jordan was one of the first prominent scientists to speak out on public issues. Most scientists did not, especially after the Ph.D. was granted as the scholar's degree, beginning about 1875 in the United States. Jordan missed out on the new graduate tradition and belonged to an older school of naturalists rather than experimentalists. His work was both descriptive and theoretical, the former for his studies of the fish in North America and the latter for his applied evolutionary ideas. Jordan followed a European tradition of popularizing science through essays and trade books.

[2] Ibid. p. 115. Agassiz was one of the first American scientists to shift from the study of museum specimens to live specimens. At the Marine Biology Laboratory at Woods Hole, his aphorism, "Study Nature, not books" still greets visitors. He also told his students, "A laboratory is a sanctuary which nothing profane should enter." The experimental biology station had its start in Italy when Thomas Huxley and Anton Dohrn founded the Naples Marine Biology Station. Until Woods Hole copied this format, it was part of the earned doctorate's professional training to leave the United States for a year abroad with part of that time spent at the Naples Station.

[3] Ibid. p. 110.

[4] Ibid. p. 111.

[5] Ibid. p. 113.

[6] Ibid. p.114.

[7] Jordan set a tradition of academic excellence and enthusiastic teaching as the hallmarks for college education. His model of the scholar–teacher is still an ideal for professors. Indiana University recognized his contributions by naming its life science building Jordan Hall. He also set a tradition for academic freedom and resisted community and state pressure against the teaching of evolution.

[8] David Starr Jordan, *The Ethics of the Dust*, Commencement Address, 1888, Indiana University (Library call number: LD2524 J8 1888).

[9] Both McCulloch and Jordan, who may have communicated his biological ideas to McCulloch, owe their theory of social degeneracy to the ideas of Sir Edwin Ray

Lankester (1847–1929), a zoologist who was a professor at University College, subsequently a professor at Oxford, and finally Director of the Natural Science Division of the British Museum. Lankester's book, *Degeneration, A Chapter in Darwinism* (MacMillan, London, 1880), brought the message to the general public that parasitism was a form of degeneracy. Lankester later accepted the theory of feeble-minded people as hereditary degenerates and endorsed the eugenics movement. See his essay, "The feeble-minded," pp. 271–282 in his *Science From an Easy Chair* (Methuen & Co., London, 1910).

[10] Jordan, *Days of a Man*, p. 113.

[11] The condition called cretinism is rare today because most people use iodized salt, and iodine-bearing foods are readily transported inland to communities that once had limited access to this mineral. The goiter can arise from the dietary lack of iodine or from genetic causes, including autoimmunity. Infants born without thyroids are usually diagnosed shortly after birth and given the hormone their body cannot make. Cretins showed a dwarf stature in addition to mental retardation and other physical impairments.

[12] Jordan, *Days of a Man*, p. 314.

[13] Ibid. p. 315.

[14] Ibid.

[15] David Starr Jordan, *Footnotes to Evolution: A Series of Popular Addresses on the Evolution of Life* (Appleton & Co., New York, 1898).

[16] Ibid. Chapter 11, "Degeneration."

[17] Ibid. Chapter 12, "Hereditary inefficiency."

[18] Ibid. p. 303.

[19] Ibid.

[20] Ibid. p. 308.

[21] Ibid. p. 309.

[22] Ibid. p. 310.

[23] Ibid. p. 311.

[24] Ibid. p. 290.

[25] Ibid. p. 456.

[26] Ibid. p. 618.

[27] David Starr Jordan, "The blood of a nation," *Popular Science Monthly* 59(1901): 90–100; 129–140. The essay was later made into a book. Many of Jordan's books were published by the American Unitarian Association, Boston (later Beacon Press).

[28] Ibid. p. 91. Jordan shared a fallacy of his age, the identification of culturally transmitted traits with biologically transmitted traits. That fallacy is still part of folk genetic belief, and ethnic identity (cultural heritage) is often assigned a blood (genetic) status.

29 Ibid.

30 Ibid.

31 Ibid. p. 94.

32 Ibid. p. 95.

33 Ibid. p. 92.

34 Ibid. p. 96.

35 Ibid.

36 Jordan, *Days of a Man,* vol. 2, p. 423.

37 David Starr Jordan, *The Human Harvest* (American Unitarian Association, Boston, 1907), p. 27.

38 David Starr Jordan, *Unseen Empire: A Study of the Plight of Nations that do not Pay their Debts* (American Unitarian Association, Boston, 1912).

39 Ibid. p.179.

40 Jordan, *Days of a Man*, vol. 2, p. 460.

41 *American Breeders Magazine* 1(1913): 153.

42 Most people recognize Bell as the inventor of the telephone, but he was well known among scholars of his generation as a student of heredity. He wrote articles on deaf mutism and supernumerary breasts. Burbank was a Lamarckist but believed in the power of selection in his botanical studies and did not hesitate to destroy thousands of plants to cultivate the one good specimen he thought worthy of propagating. Castle was one of the first American geneticists to promote the rediscovery of Mendelism. Unlike Bateson (whose lectures converted Castle to Mendelism), Castle succeeded both among the academic scientists and the practical scientists at America's fine network of agricultural field stations supported by the Department of Agriculture. Bateson found enormous resistance to the acceptance of Mendelism because British Darwinists erroneously thought it negated Darwinism.

43 Cold Spring Harbor, on the north shore of Long Island, has also housed the Carnegie Institution of Washington's biological research station for most of the 20th century. The basic research in genetics carried out by summer investigators and some of the full-time staff is world renowned. Its three most prominent directors have been Charles Davenport, Milislav Demerec, and James Watson. Under Davenport, classical genetics flourished, especially the close ties to fruit fly and corn genetics. Demerec led the transition to molecular genetics by shifting from fruit flies to bacteria and encouraging the formation of the bacteriophage school. Watson shifted the laboratory to molecular and developmental cell biology in both normal and tumor cells. Cold Spring Harbor, to 20th century geneticists, is what the Naples Station was to biologists around the world in the late 19th century. Even in Davenport's days, the research work was separately funded from the eugenic work.

44 Jordan, *Days of a man*, vol. 2, p. 298, discusses how Mrs. Harriman's sister, who studied biology at Columbia University, influenced her to endow Cold Spring Harbor with the funds needed to study eugenics.

[45] There were usually three separate programs: the Eugenics Station, the Carnegie Institution, and the Long Island Biological Association. The latter was often used to fund summer institutes and education programs for school children. Over the years these waxed and waned, although the research function (heavily subsidized by the Carnegie Institution) usually predominated. After Davenport's retirement the eugenic function was allowed to die. In 1963, the Carnegie Institution and LIBA programs formed the Cold Spring Harbor Laboratory of Quantitative Biology, whose name was shortened to Cold Spring Harbor Laboratory in 1970.

[46] *American Breeder's Magazine* 1: 128.

[47] Ibid.

[48] Ibid. 1: 129.

[49] Ibid. 2: 62.

[50] Unlike the Eugenics Record Office, which played a significant role in shaping American political philosophy and legislation, the Galton laboratory did not go beyond occasional popular articles in promoting eugenics. This restraint reflected a major difference in the approaches of the two eugenics movements. Galton sought more intelligent people through encouragement of the brightest to consider it a duty to have larger families. The American eugenics movement was more concerned about preventing its stock from deteriorating or becoming contaminated by undesirable immigration.

12

Isolating the Unfit through Compulsory Sterilization

T HE HOPED-FOR REFORMS THAT DUGDALE envisioned as a means to lift the Jukes from their degenerate state to one of normalcy never took place. A pessimism of irreversible innate degeneracy was replacing the optimism of social reform as the United States built up its heavy industries and emerged as a world power.[1] Immigration was increasing every year, and the population shift from the farm to the city had become a national trend. Loring Moody, a Boston merchant who made modest but not substantial wealth was convinced that charitable reforms would not check the evil of degeneracy in American society. He founded, in 1880, a short-lived Institute of Heredity for "improving our race by the laws of physiology."[2] He rejected Dugdale's optimism and claimed that the reforms of the past failed because "the causes are congenital. People who are born with theft and murder in the blood will steal and kill. The jailor and hangman neither cure them, nor check their tendencies, nor thin their ranks. For as fast as we imprison and hang criminals others are born to take their places; so that all our conflicts with evil result in long-drawn battle."[3]

REPRODUCTIVE ISOLATION AS A WAY OUT

Moody hoped to raise money for the institute from private philanthropy and wrote hundreds of letters to potential benefactors explaining his physiological theory of degeneracy, citing Galton as an exemplar of the new view of heredity that assigned differences in human faculties primarily to physiological differences at birth. One of his correspondents, Elizabeth Thompson, a wealthy heiress, wanted more evidence supporting his views and specific proposals for the program his institute would initiate. The correspondence of Moody and Thompson resulted in a book, appearing in 1882, entitled *Heredity: Its Relation to Human Development*. Moody's major idea was to contain the unfit.[4] He suggested to Thompson that "as a means of eliminating the inherited effects of disorders from posterity, I would have the government establish and maintain good, comfortable, attractive hospital homes for the care, treatment and life residence of all habitual drunkards, confirmed criminals, idiots and incurable lunatics, who should be treated as people suffering from dangerous congenital diseases liable to propagate through heredity; and so they should be strictly guarded from having any offspring, as far as possible by moral, and the remainder by legal, restraint."[5] Despite Moody's attempts to raise an endowment, the undertaking fizzled and the institute remained an aborted dream.

REPRODUCTION AND MENTAL ILLNESS

While social reformers sought a medical basis for defining the unfit as diseased and thus proper patients for treatment and preventive medicine, physicians were exploring the relation of reproductive organs to mental and degenerative diseases.[6] In Germany the condition identified as hysteria, thought since antiquity to be associated with the uterus, was now being looked at as a problem of the ovaries. In 1892, Dr. Bernhard Heilbrun reported the case of a hysterical woman who had been bedridden for seven years. She suffered severe muscular cramps and had difficulty retaining her food. Heilbrun removed her ovaries, proclaiming one of them was diseased, small in size, and tuberculated. The patient left her bed after 12 days and went home on the 20th. Ten months later she was able to walk to her physician's office for her checkups.[7]

Heilbrun's alleged success was followed by a report the following year from Dr. Wilhelm Tauffer, a Budapest physician, who had performed 12 castrations or removals of both ovaries in hysterical patients, including what he classified as hystero-epileptic patients. Hysteria, he believed, was traceable to ovarian disease. American physicians, commenting on the work in the *Journal of the American Medical Association*, were uncertain whether the treatment could be extended to women suffering from "psychoses" (insanity). More work was needed to show the relation of ovarian pathology to mental disorders.[8]

A symposium to discuss this growing interest was held in October 1886.[9] Sir Spencer Wells argued that castration or oophorectomy of the female was not suited for "nervous excitement and madness." The surgery should never be performed on a sane patient without her consent; and the procedure was, he believed, unjustifed for nymphomania. Opposing Wells's view was Dr. Robert Battey, who asserted that in cases of females with mental and nervous disorders, he only removed abnormal ovaries and that all his removals were confirmed by pathological examination of the removed tissue. He described 36 cases and substituted for the uterine term (hystero) the ovarian term (oophoro), renaming the hysterias as oophoro-mania, oophoro-epilepsy, and oophoralgia. Most of his 36 cases he claimed were cured of their nervous disorders. He rejected Wells's claim that the surgery was ineffective for nymphomania and cited the success of Dr. W.H. Bifford for such cases. Dr. Alfred Hegar provided the more moderate view taken by the AMA—it's a good operation; don't abuse it.[10]

In 1887, Dr. E.W. Cushing discussed his surgical treatment of a 33-year-old French Canadian woman who had been troubled, since she was 15, by dysmenorrhoea, melancholia, and masturbation. She felt she was damned for her sins and begged for a surgical treatment that would relieve her. Cushing obliged by removing her ovaries, and the patient remarkably improved, claiming that "a window has been opened in heaven" for her.[11] Cushing acknowledged that some of his colleagues scoffed at her recovery. Cushing had his supporters at the medical meeting where he presented his results. Dr. H.I. Bowditch claimed that "a physician who would not do this operation or would not permit it to be done would be wanting in humanity."[12] He said "that there are many cases in our insane asylums, similar to the one described, which might be cured by removal of the ovaries. Dr. L.F. Warner asked Bowditch if he would perform a similar operation in a male. Bowditch assured him that he would. Warner, unimpressed, discounted the patient's recovery, and opposed the operation because the "favorable results reported came from the powerful moral influence of the operation."[13]

The debate regarding castration for mental illness carried over into the 1890s. Dr. Walter Lindley, president of the State Society of Medicine in California gave his annual address in 1890 with the admonition: "Knowing as all surgeons do today, that castration and spaying are simple operations that can be performed with about as little danger as the ancient rite of circumcision, I do not hesitate to advise that the following classes be required by law to submit to this procedure: idiots, those who commit or attempt to commit rape, wife-beaters, murderers, and some classes of the insane."[14] The editor of the *Medical Record* who cited Dr. Lindley's views scoffed at the suggestion. "Fortunately," he claimed, "it is a physiological law that the degenerate classes have little procreative power and tend to die out of themselves. What society needs most, therefore, is the adoption of such ethical and social measures as will prevent the development of new vicious and criminal families."[15]

The editor of the *Medical Record* was wary of the new reproductive strategies for mental defects. In 1892, he reviewed a case, similar to that of Cushing, of "Castration for melancholia" performed at the Eastern Michigan Asylum on a 57-year-old male with a severe "sickening neuralgia."[16] Both testes were removed and the medical superintendent of the asylum reported a complete cure. The removal of testes and ovaries was becoming widespread and merited some comment. The editor identified the trend to perform oophorectomies in America as having come from the southern states and "thence diffused its genial and unsexualizing influence over the East and North; but testectomy, if we may coin a word or so on so great an occasion, comes from the West."[17] The editor was concerned that "neuralgia is very common and so is depression of the spirits." He feared widespread abuse and imagined seeing reports flooding his desk with such titles as "My second series of one thousand castrations, with hints on technique."[18]

Some physicians toyed with the idea that castration would be a suitable alternative to capital punishment or imprisonment for certain crimes. Dr. W.A. Hammond presented a paper in 1892 to the New York Society of Medical Jurisprudence. He favored castration because "punishment would be continuous, and not momentary and intense, and in the continuance of a punitive award will be found its greatest effect."[19] He thought the castrated prisoner would be identified in society with a condition worse than a mark upon Cain because of the elevated voice, changed facial appearance, and "effeminate and cowardly" disposition. Those who were reformed by the punishment he believed would be welcomed in church choirs. Hammond suggested the experiment begin with pimps: "If it worked well with them, it might be carried upwards into higher walks of vice."[20]

Whereas Hammond thought of castration as a purely punitive operation to terrify criminals and thus prevent crime, Dr. Robert Boal, speaking to the Illinois State Medical Society, extended the value of the procedure by pointing out its therapeutic aspects. Boal argued that killing and imprisonment of prisoners was archaic and inhumane. "In my opinion all criminals who indicate constitutional depravity transmissible by heredity should be subjected to surgical unsexing enforced by law."[21] This would satisfy the demand for justice, deter criminals, and reform offenders. He believed that crime was associated with a defective sexual faculty and that criminal and defective classes of society are perpetuated by heredity. Unsexing would "limit the productive capabilities of these classes, thus aiding 'natural selection' and insuring if extensively applied the 'survival of the fittest.'" Medicine, he claimed, had a duty to protect society.[22]

CASTRATION AND OOPHORECTOMY FOR THE UNFIT

Suggesting castration for social evils was one thing, but carrying it out was another. Repugnant to the editor of the *Medical Record* was the report of Dr. J.J. Putnam, who castrated a male in 1890 and reported in 1893 on his status. The patient was a 41-year-old dipsomaniac with "excessive and openly erotic" behavior, who had been "unbalanced all his life, and especially so since a severe blow on the head in childhood." The castration "was done at his own urgent desire."[23] The patient, while still "unreliable and suspicious" abated in his desire for liquor and his erotic tendencies and disposition became quieter. The case prompted the editor to condemn castrations and oophorectomies for behavioral reasons as "founded upon equally irrational and absurd principles."[24]

With the exception of Battey's series of oophorectomies for hysteria, the cases were isolated, although numerous, as surgeons tried out new techniques based on the reported successes in the medical journals. The first large-scale, planned program for the treatment of insanity by oophorectomy was carried out at the Norristown Insane Asylum in Pennsylvania in 1893. The trustees of the asylum approved a plan by Dr. Joseph Price and his colleagues[25] to castrate "fifty patients selected as being cases likely to be benefited by the operation." The fifth patient to be operated on died during surgery and the remaining operations were halted. The Lunacy Committee of the State Board of Charities investigated the incident and condemned oophorectomy except for cases of gross physical disease associated with the ovaries. The lawyer on the committee assailed the operations as "illegal,... brutal and inhuman, and not excusable on any reasonable ground."[26] He warned that such surgery, unless intended to save life, puts the physician at risk "of a criminal prosecution." He summed up replies of the medical experts he consulted and argued that "it is regarded by the best medical authorities as a useless and improper expedient for the cure or relief of insanity, and the operation of oophorectomy in a public hospital upon indigent insane women must be regarded as largely experimental, and for that reason bound to reflect upon hospital authorities now boasting of modern humane methods."[27]

Most notorious in his campaign for the legal sterilization of the unfit was Dr. F.E. Daniel of Austin, Texas. In an address to the Medico-Legal Congress held in Chicago in 1893, Daniel outlined his case. "No fact is better established," he asserted, "than that drunkenness, insanity, and criminal traits of character, as well as syphilis, consumption, and scrofula may descend from parent to child."[28] He argued against containment as a policy. "The wealth of all the Czars would not be adequate to provide asylum and medical treatment for the progeny of these people in fifty years from now; for, while insane people do not marry, those do in whom disease exists undeveloped, and with the lower classes, particularly negroes, it is known that illicit intercourse is extremely common."[29] Among the various causes of hereditary degeneracy, Daniel singled out onanism. "No one at all acquainted with the subject will deny that masturbation (a perverted sexual sense) may, and frequently does, become a cause of mental alienation; and I cannot subscribe to the belief that it is always a symptom of mental disease already existing. There is no doubt that habitual masturbation is often a manifestation of mental disease; nor, on the other hand, that it will lead to, and become a cause of, insanity."[30]

RACISM AND DEGENERACY

Daniel thought of sexual criminals as diseased or insane individuals and thus not proper subjects for criminal prosecution. He was particularly struck by the real and alleged rapes of white children by black males. "In a case where a powerful man, especially a negro man, who, in the South, at least, should have little excuse for unsatisfied sexual desire, amongst a race whose ideas of morality are crude and sexual virtue is not a striking characteristic, attempts to effect sexual intercourse with a small child of a different race—a physical impossibility—and that, too, in knowledge that if caught he will surely meet with a speedy and horrible death, it would be, in my opinion, prima facie evidence of an unsound mind—insanity in some degree."[31]

Daniel argued that criminal prosecution in such a case was inappropriate. "The aim of jurisprudence should be, in addition to the repression of crime, a removal of the causes that lead to it, and reform, rather than the extermination of the victims."[32] To this end, Daniel suggested that "the offender should be rendered incapable of a repetition of the offense, and the propagation of his kind should be inhibited in the interests of civilization and the well-being of future generations. These ends are not fulfilled by hanging, electrocution, or burning at the stake."[33] Daniel denied that capital punishment deterred crime; in fact he felt some deranged individuals are excited by the dramatic way in which death would be visited on them by the state or by a mob.

Daniel proposed castration not as a punishment, but as a humane response to sexual criminals. "Castration has been advocated by numerous writers in various parts of the world as a punishment for crime," he acknowledged, "but that it is or has ever been practiced to any extent anywhere for the purpose of curing mental disorders, or with any intention or thought of arresting the hereditary transmission of either disease or vices of constitution; in short, for the purpose of prophylaxis as applied to race improvement and for the protection of society, I am not advised."[34]

Daniel approvingly cited W.A. Hammond and Frank Lydston, who discussed the problem of rape in the *Virginia Medical Monthly*. They "advise castration as a remedy for the evil; and there is much wisdom in the advice." Lydston "would castrate the rapist, thus rendering him incapable of a repetition of the offense, and of propagating his kind, and turn him loose, on the principle of the singed rat, to be a warning to others. Dr. Lydston says and very truly, that a hanging or even a burning, is soon forgotten; but a negro buck at large amongst the ewes of his flock, minus the elements of manhood, would be a standing terror to those of similar propensities."[35] Daniel did not want to go as far as Dr. Orpheus Everts, Superintendent of the Cincinnati Sanitorium, who, in 1888, urged the castration of all criminals to arrest the "descent of their respective vices of constitution." That would be risky because "we might cut the wrong man."[36]

Castration as a Treatment for Masturbation

Daniel's more limited vision was to "substitute castration as a penalty for all sexual crimes or misdemeanors, including confirmed masturbation."[37] Returning again to his dread of habitual masturbation as the cause of degeneracy, Daniel urged the legalization of castration; it would be "an advisable hygienic measure in habitual masturbation, whether the practice be cause or effect, by arresting the wasting of vital force by seminal losses, and consequent impairment of physical health."[38] Reflecting on the mood of his times, Daniel noted: "Is it not a remarkable civilization that will break a criminal's neck, but will respect his testicles?"[39]

Among the permissible reasons for castration, Daniel included "rape, sodomy, beastiality, pederasty, and habitual masturbation." Enactment of such a law was more than beneficial to the degenerate: "This we owe to ourselves, if we would not merit reproach; to posterity, if we would secure to future generations the full fruits of sanitation in the practice of the great science of preventive medicine."[40]

Despite his rhetorical fervor, Daniel did not persuade either his local Texas society or the national medical congress who heard his pleas. Although no law was enacted to carry out castrations for the unfit, the idea simmered. The failure at Norristown was attributed to the higher risks females experience with abdominal surgery. No major risk of death existed for castration of males. With that shift in mind, Dr. F. Hoyt Pilcher, of the Institution for Feebleminded Children at Winfield, Kansas, obtained the approval of the trustees in 1898 to castrate 58 boys, primarily for the eugenic consequences of the operation.[41] Despite some opposition and outcry from the medical journals and public press, Pilcher was backed by his trustees who issued a resolution upholding "his work in seeking in this manner to purge the race of certain defective strains."[42] As the 19th century drew to a close, there was less opposition to the legalization of sterilization than there was to its chief methods of surgery—oophorectomy and castration. For its strongest advocates there was something new to close out the century; a procedure was emerging, nonmutilating but remarkably effective. The time, the opportunity, the patient, and the physician would come together, and a little-noticed event in Jeffersonville, Indiana, would send increasing concentric ripples of effects to engulf the lives of millions in the next half-century.

Harry Sharp's Early Career

Dr. Harry Clay Sharp (1869–1940) received his M.D. in 1893 at the Louisville Medical Institute and served from 1895 to 1910 as the reformatory physician at Jeffersonville, Indiana.[43] Sharp was born in Charleston, Indiana, and attended Ohio State University before entering medical school. He was a scholarly physician, well-read in literature, interested in the issues of his day, and a staunch advocate of the health of his patients. In his progress report prepared for the first biennial report of the reformatory to the governor for the period ending 31 October 1898, he listed close to 1,000 illnesses, including 197 cases of malaria, 52 cases of tuberculosis, 8 of typhoid fever, 46 of tonsillitis, 36 of syphilis, 86 injuries, and numerous instances of boils, abscesses, burns, hemorrhoids, and flu.[44] He also performed two circumcisions. In his accompanying letter to the Governor, Dr. Sharp praised the state for its support of the prison hospital and the improved sanitation of the prison dormitories, which greatly reduced the infectious diseases that plagued the prisoners only a few years earlier.[45]

At the meetings of the Mississippi Valley Medical Association, held in Nashville, Tennessee, in October 1898, Sharp discussed his growing interest in abnormal behavior, stimulated by his association with the inmates at the reformatory and his eagerness to read as much as he could on the causes and treatment of mental defect. He presented a paper on *Neurasthenia and its treatment*.[46] The condition, he believed, was caused by the stress and pace of contemporary life. He worried about the burnout of over-achieving professionals and the neuroticism arising among harried middle-class housewives who were overcommitted in their social circles. He related the physical consequences of psychological stress, including ulcers, headaches, muscular pains, and other symptoms that could even lead to early death.[47] He suggested a treatment consisting of rest, reduced mental activity, a bland diet, and changed work habits. At this stage of Dr. Sharp's career, he revealed no public expression of a supplementary interest in the problems of heredity and degeneracy.

THE AMA PUBLICIZES AN ALTERNATIVE TO CASTRATION

In the spring of 1899, Sharp read two significant articles in the *Journal of the American Medical Association (JAMA)*. Dr. Albert John Ochsner (1858–1925), a Chicago surgeon, suggested using vasectomies on criminals as an alternative to imprisonment.[48] Also appearing, a few months later, in the June 10 issue of *JAMA* was the address of another Chicago physician, Daniel R. Brower, on "Medical Aspects of Crime."[49] Sharp was pleased when he noted Brower's observation that "the medical aspects of crime have not been sufficiently considered."[50] Brower believed physicians should do for the criminal what physicians since Pinel have done for the insane, "substituting patience and scientific treatment for brutality and chains."[51] Physicians such as Gall, Lombroso, Ellis, and many others "have established the fact that the habitual criminal is an abnormal man, the abnormality manifesting itself: 1, physically in the criminal physiognomy, in stigmata of degeneration, in anomalies of the muscular, sensory, respiratory, and circulatory systems, and 2, psychically, in moral insensibility, lack of forethought, low grade intelligence, vanity and egoism, and emotional instability."[52] Brower made use of Dugdale's and McCulloch's studies, supplemented with the Jurkes family studied by Dr. Pelman in Bonn, Germany, as well as Boies's portrayal of the unfit classes in *Prisoners and Paupers,* to characterize the unfit as degenerate hereditary stocks. "We should make it impossible for there to be more Jukes, Ben Ishmael, or Jurkes families, and to do this marriage should be regulated."[53] Brower, too, had read Ochsner's paper in *JAMA* early that spring, and endorsed vasectomies for confirmed criminals. Children under 7, of the criminal class, should be placed in suitable environments so that "the great bulk of them might be made useful citizens" through manual training programs.[54]

SHARP'S FIRST VASECTOMY ON A PRISONER

Sharp's next biennial report to the governor, covering the period to 31 October 1900, listed more than 2000 treatments. Malaria was still high, at 418 cases, tuberculosis was equally a problem with 416 cases. Typhoid was contained at 36, syphilis at 63, and tonsillitis was down to 26. Added to the list were 4 castrations, and 24 cases of spermatorrhoea. Circumcisions had risen to 30 cases.[55] The report omits an interesting case that Dr. Sharp encountered in 1899.

A young man named Clawson visited him and was troubled by his habit of compulsory masturbation.[56] He asked Sharp to help him. Sharp's first account of this case is described in a paper he submitted to the *New York Medical Journal* in 1902.[57] Clawson was "a boy nineteen years of age, operated upon October 11, 1899. He was born in Missouri, of criminal parentage; had masturbated since twelve years of age. He was very dull, unable to make progress in school, and mental concentration was impossible. He attributed his mental condition to his excessive masturbation, and applied to me for relief. Sixty days after the operation he had gained twenty-two pounds, felt well, asserted that erections were even more vigorous than they were prior to the operation, and believed that I had performed the operation simply for the purpose of deceiving him, and requested that I perform castration. One year following the operation his weight had increased thirty-eight pounds, his mental condition had greatly improved, and he had ceased to masturbate. Upon questioning him he stated, to use his own words, that 'the desire is as great as ever, but I have the will to resist.'"[58]

Sharp used Clawson's story several times in the next few years, not naming him until he was interviewed in 1937; by that time Sharp was no longer in Indiana but at a veteran's hospital in Lyon, New Jersey.[59] In 1908 Sharp published, at the reformatory print shop, a pamphlet entitled *Vasectomy: A Means of Preventing Defective Procreation.*[60] Here Sharp begins Clawson's story in a slightly different way: "A boy 19 years old came to me and asked that he be castrated, as he could not resist the desire to masturbate. I first had him put in a cell with a fellow inmate, thinking that perhaps he would be abashed and the sense of shame would prevent him. He came to me again still insisting on castration, saying it was as bad as ever. I did the operation, and two weeks afterward he came to me and said I was just fooling him, that I had not operated on him and he wanted the other operation. I asked him to wait two months and then, if he was no better I would perform castration."[61]

Why did Clawson come forward and ask to be castrated? This may be the same Harry Clawson who is reported to have died in 1902 from tuberculosis, and Dr. Sharp's initial discussion of weight gain might reflect the alleged beneficial effect of vasectomy on the consumptive process.[62] Clawson in 1899 may have listened to Dr. Sharp instruct the prisoners on proper health habits while at the Jeffersonville Reformatory. Sharp may have warned the inmates about masturbation, cautioning them that it could destroy their physical as well as their mental health. Clawson could then have associated his medical problems as having arisen from his masturbatory habit and sought help from Sharp. When the vasectomy initially failed to cure him of his masturbation he may have panicked and repeated his desire to be castrated, that being the only known treatment available, the outcome being preferable to continued degeneration or death.

SHARP CARRIES OUT 176 ILLEGAL VASECTOMIES

Sharp said that he continued the vasectomies over the following years: "From 1899 to 1907 this operation was done on 176 men in the Indiana Reformatory on request. The request was solely for the purpose of relief from the habit of masturbation."[63] In his report for 1901–2, Sharp lists 34 vasectomies, 60 circumcisions, and 3 castrations among the nearly 3000 visits by his patients at the reformatory.[64] The report for 1903–4 lists 78 circumcisions, 42 vasectomies, and no castrations among the genital procedures for that interim.[65] But in his letter to Governor Winfield T. Durbin summarizing the work for November 1, 1902 through October 31, 1904, he somehow adds an additional 100 vasectomies to the 76 he officially reported, and shifts his view of the operation from that of treatment to that of preventive medicine.[66] He also adds the new element of going directly to the top in enlisting the state to legalize the procedure for this new, nontherapeutic use of vasectomies. Sharp had preached the doctrine of compulsory sterilization to his fellow physicians in his 1902 paper; it was now time for him to reach out to other professionals. What follows is one of the least quoted and most historically significant documents that characterizes 20th-century social thought gone awry: "During the past five years I have severed the vas deferens of 176 boys and young men for the relief of excessive masturbation and spermatorrhoea. The result has been very satisfactory in that it brought relief from the trouble named and in addition both the physical and mental condition has improved. I therefore suggest that you endeavor to secure such legislation as will make it mandatory that this operation be performed on all convicted degenerates. It renders them powerless to reproduce their kind, and it is an undoubted fact that the progeny of degenerates becomes a charge upon the state."[67]

INDIANA PASSES THE FIRST COMPULSORY STERILIZATION LAW

Two years later, Sharp repeated, somewhat more forcefully, his plea for legislative action, this time to Governor J. Frank Hanly: "In the past two years I have performed thirty vasectomies making 206 in all. Two years have elapsed since my last operative observation. I have found no ill effect either immediate or remote, thus further proving the advisability of this operation. I therefore wish to urge you to insist upon the General Assembly passing such a law or laws as will provide this as a means of preventing procreation in the defective and degenerate classes."[68] The following year, in 1907, Indiana passed the first compulsory sterilization law for the unfit.

Sharp is not specific on the sources of his ideas. He implies that he carried out animal experiments and studied the medical literature before seeing Clawson to make sure that vasectomies did not change the masculinity or sexual habits of the males. "It was on account of these facts that I suggested that the vas deferens in the male and the oviduct in the female be severed as a means of preventing procreation in defectives."[69] If Sharp had read the medical literature on the procedure, he would have felt assured.[70] In 1823, Sir Astley Cooper vasectomized a dog on one side and cut the spermatic blood vessels on the other. The testis lacking the blood supply atrophied; and the dog, with its one functional testis, remained sexually active although sterile. Cooper concluded that bilateral vasectomy only causes sterility. Vasectomy was not used as a surgical procedure in humans until 1893 when it was used in England by R. Harrison and by the Swedish physician Lennander in an attempt to arrest the swelling of an enlarged prostate gland.[71]

OCHSNER'S REASONS FOR ADVOCATING VASECTOMIES

Sharp was certainly influenced by Ochsner's article that appeared in *JAMA* on April 22, 1899. Ochsner was born in Baraboo, Wisconsin, the son of Swiss pioneers who farmed near Baraboo. The Ochsners claimed a direct ancestry from the famed anatomist Andreas Vesalius.[72] After teaching in public school for several years, Ochsner attended the University of Wisconsin and then entered Rush Medical College in Chicago. After receiving his M.D. in 1886, he went to Vienna and Berlin, where he studied pathology with Rudolf Virchow and microbiology with Robert Koch. From 1891 until his death he was chief surgeon at Augustana Hospital in Chicago and a professor at Rush and the University of Illinois College of Medicine. Ochsner performed many vasectomies for treatment of enlarged prostates, mostly on older men past their sexual prime.[73] In 1897 he performed vasectomies on two males, one age 42 and the other 54, both of whom resumed active sexual activity after recovery from the surgery. In the *JAMA* article, Ochsner advocated its use on prisoners, rejecting the view among some physicians that castration be used to protect the next generation. Ochsner's justification for sterilizing prisoners by vasectomy instead was based on the prevailing views that he had read. "It has been demonstrated beyond a doubt that a very large proportion of all criminals, degenerates, and perverts have come from parents similarly afflicted. It has also been shown, especially by Lombroso, that there are certain inherited anatomic defects which characterize criminals, so that there are undoubtedly born criminals."[74] Ochsner added to this two additional ideas—that some criminals are made so by the corrupting influence of other criminals as in children growing up in the home of a parent who is a criminal; and that female criminals are usually sterile anyway because their sexual vices lead to infections of the uterus or oviducts. The conclusion followed that "... if it were possible to eliminate all habitual criminals from the possibility of having children, there would soon be a very marked decrease in this class, and naturally, also a consequent decrease in the number of criminals from contact."[75]

Ochsner's long-range goal was not simply the elimination of criminality through vasectomies. It protected society without mutilating the criminal or preventing his participation in society when reformed. The method would "do away with hereditary criminals from the father's side" and "the same treatment could reasonably be suggested for chronic inebriates, imbeciles, perverts, and paupers."[76]

Ochsner was a highly regarded surgeon who had published several books and dozens of professional articles up to this time. These works included books on the preparation of pathological specimens (1888), antiseptic surgery (1892), cleft palate (1894), general pathology (1894), abdominal surgery for cancer (1896), and hernia (1897).[77] He began using vasectomies for the treatment of prostate enlargement after reading the work of Harrison and Lennander. There were no scientific standards of controlled experiment for the procedure in those days, and eventually the technique was abandoned as ineffective.

In his distinguished career, Ochsner was a founder of the American College of Surgeons and served as its president in the last years of his life. He was known at meetings for his habit of wearing a white bow tie and, despite his halting manner of speech, his ideas and new surgical techniques commanded attention.

How Vasectomies Were Medically Misused

Sharp acknowledged the priority of Ochsner for the suggestion and, indeed, stated that he had used Ochsner's surgical technique for several years before switching to the more effective procedure described by the British physician, Robert Reid Rentoul. Rentoul in 1903 promoted the idea of legalizing vasectomies for the unfit in his book with the awkward title, *Race Culture; or, Race Suicide (A Plea for the Unborn)*, but his campaign met with no success in Great Britain.[78] Like Ochsner, he used the operation as a treatment for enlarged prostates, and his enthusiasm for vasectomies led many physicians to refer to the operation as Rentoul's procedure.

Sharp's ideas of heredity were shaped by social Darwinism and by the growing scientific preference for Weismann's theory of the germ plasm, although both Lamarckian and Weismannian views coexisted in his thinking. He strongly believed, as late as 1908, in the inheritance of maternal impressions, citing cases from his own practice and reminding his fellow practitioners that "many of you know or have seen physical deformities as a result of profound or mental impression during the period of gestation.

As the mental faculties are much more highly organized, they are a great deal more susceptible to such impressions."[79] He lamented that "there are many unfit who are allowed to go about at will,"[80] and he approvingly cited Herbert Spencer as having said that "to be a good animal is the first requisite in life, and to be a nation of good animals is the first condition to national prosperity."[81]

SHARP'S SOCIAL PHILOSOPHY

In 1907, Sharp stated that "heredity is but one of many causes of degeneracy, and while refined parents may beget degenerate children, I make the statement without fear of contradiction that no confirmed criminal or other degenerate ever begot a normal child...."[82] He extended his views the following year, asserting that "degeneracy is a defect, and that a defect differs from a disease in that it cannot be cured."[83] He shared the view of David Starr Jordan and the early founders of the American eugenics movement that "the degenerate class is increasing out of all proportion to the increase of the general population."[84] The net cast over these social failures was vast; Sharp claimed that "most of the insane, the epileptic, the imbecile, the idiotic, the sexual perverts; many of the confirmed inebriates, prostitutes, tramps, and criminals, as well as the habitual paupers found in our county poor asylums; also many of the children in our orphan asylums belong to the class known as degenerates."[85]

Sharp reinforced his preventive medicine outlook with the alleged therapeutic benefits to the patients on whom he operated. He described the procedure as simple, "I do it without administering an anaesthetic either general or local. It requires about three minutes' time to perform the operation and the subject returns to his work immediately, suffers no inconvenience, and is in no way impaired for his pursuit of life, liberty, and happiness, but is effectively sterilized."[86] Sharp, describing his work in a pamphlet prepared for the National Christian League for Promotion of Purity, states with evident pleasure that the "patient becomes of a more sunny disposition, brighter of intellect, ceases excessive masturbation, and advises his fellows to submit to the operation for their own good."

In Sharp's procedure, the cut vas deferens is ligated (tied) at the end leading to the penis, but left open in the piece leading to the testis. Sharp believed that the fluids bathing the testes from the open duct were beneficial for the patient; his "mind is strengthened and his nervous system benefitted from the reabsorption of sperm." [87] For females, Sharp proposed and carried out a similar operation. He describes severing the oviduct of an epileptic female child who was 11 years old. He used general anesthesia and the girl stayed about ten days in the hospital. She had her menses at age 14 with full bust development. He believed there would be no need for state institutions for feebleminded women if oviduct surgery were routinely performed. It was only their potential for getting

TABLE
OF STERILIZATIONS DONE IN STATE INSTITUTIONS UNDER STATE LAWS
UP TO AND INCLUDING THE YEAR 1940

| STATE | DIAGNOSIS AND SEX | | | | | | | | | SEX SUMMARY | | |
| | INSANE | | | FEEBLEMINDED | | | OTHERS | | | TOTALS | | |
	MALE	FEMALE	TOTAL	MALE	FEMALE	TOTAL	M.	F.	T.	MALE	FEMALE	TOTAL
ALA.	0	0	0	129	95	224	0	0	0	129	95	224
ARIZ.	10	10	20	0	0	0	0	0	0	10	10	20
CAL.	5,329	4,310	9,639	2,166	2,763	4,929	0	0	0	7,495	7,073	14,568
CONN.	19	337	356	6	56	62	0	0	0	25	393	418
DEL.	206	71	277	116	193	309	0	24	24	322	288	610
GA.	6	68	74	22	31	53	0	0	0	28	99	127
IDAHO	2	10	12	2	0	2	0	0	0	4	10	14
IND.	213	177	390	338	305	643	0	0	0	551	482	1,033
IOWA	83	91	174	25	121	146	4	12	16	112	224	336
KANS.	1,035	724	1,759	342	205	547	38	60	98	1,415	989	2,404
ME.	0	10	10	14	104	118	0	62	62	14	176	190
MICH.	71	234	305	417	1,324	1,741	25	74	99	513	1,632	2,145
MINN.	113	266	379	273	1,228	1,501	0	0	0	386	1,494	1,880
MISS.	135	320	455	14	42	56	0	12	12	149	374	523
MONT.	16	20	36	42	108	150	0	0	0	58	128	186
NEB.	53	90	143	101	144	245	0	0	0	154	234	388
N.H.	24	166	190	49	141	190	0	50	50	73	357 *	430
N.Y. *	0	41	41	0	0	0	1	0	1	1	41	42
N.C.	90	150	240	93	538	631	42	104	146	225	792	1,017
N.D.	123	174	297	51	158	209	11	17	28	185	349	534
OKLA.	70	232	302	27	141	168	0	0	0	97	373	470
ORE.	287	321	608	235	506	741	30	71	101	552	898	1,450
S.C.	0	0	0	1	34	35	0	0	0	1	34	35
S.D.	0	0	0	206	354	560	6	11	17	212	365	577
UTAH	44	43	87	99	66	165	0	0	0	143	109	252
VER.	1	12	13	56	118	174	9	16	25	66	146	212
VA.	976	1,365	2,341	660	923	1,583	0	0	0	1,636	2,288	3,924
WASH.	141	245	386	33	242	275	4	2	6	178	489	667
W.Va.	0	18	18	0	9	9	1	18	19	1	45	46
WIS.	0	0	0	165	991	1,156	0	0	0	165	991	1,156
TOTALS:	9,047	9,505	18,552	5,682	10,940	16,622	171	533	704	14,900	20,978	35,878

* New York's law was declared unconstitutional, repealed, and never re-enacted. The reason for the court's decision was the failure of the law to provide for a hearing for the patient before sterilization could be ordered.

Table of sterilizations done in state institutions under state laws. Note that California led the United States in sterilizations. By 1940, some 35,000 involuntary sterilizations had been carried out. (Reprinted, with permission, from Harry H. Laughlin Archives, Truman State University, courtesy of CSHL Eugenics Web site.)

pregnant that kept such women restrained by the state. As was true for the therapeutic effects of his surgery for males, Sharp believed females would improve from the secretions released by the untied oviduct at the ovarian end of the cut.[88]

SHARP'S ADVOCACY FOR MORE STATE STERILIZATION LAWS

Sharp felt at his peak as a social evangelist in 1909. He presented a paper to the public health and preventive medicine section of the American Medical Association at their annual meeting in Atlantic City, New Jersey on June 9 of that year.[89] He summed up his work and the positive results he believed he obtained with the happy conclusion that "we have a means of preventing procreation of the unfit, at the same time improving the condition of the unfortunate individual."[90] After the applause, Sharp responded to questions from his fellow practitioners. He had his share of supporters. Dr. F.C. Valentine, from New York, was not quite convinced, but if it were true, he noted, Sharp would go down in medical history as a pathfinder. Sharing Sharp's enthusiasm was another New York physician, Woods Hutchinson, who looked forward to the day when "the physician will be looked upon as the criminologist of the country."[91] Far less sanguine was Dr. W. Forrest Dutton, from Pennsylvania; he believed education to be the key to social change for the unfortunates of society. He pointed out that too many radical remedies have failed.

Sharp fared somewhat worse in the editorials of medical journals. The *Indianapolis Medical Journal* opposed compulsory sterilization as a punishment of criminals on the grounds that it violated the constitutional rights of the individual.[92] The editor, Dr. Samuel E. Earp, urged caution with this experiment in social medicine, questioning "the real as compared with the alleged influence of heredity, an influence which, we do not hesitate to say, we think is overestimated."[93] Earp believed that the assumptions associated with vasectomy for the unfit "may rest upon no surer foundation than the dreams of faddists."[94]

Harry Sharp was well aware of the growing concern of intellectuals in the helping professions to do something about the problem of degenerate classes or the unfit. When he had read Ochsner's suggestion of using vasectomy as a less traumatic approach to cutting off the source of future generations of criminally degenerate classes, Sharp coupled both the therapeutic and the preventive medical goals of the procedure in suggesting to Clawson that his masturbation would respond to the vasectomy. Sharp was aware, of course, that a physician at a state institution was hired to treat the patients and not to initiate new social policy for the higher good of the state. He chose not to mention the vasectomy, nor did he list it in his table of illnesses treated, in the report of his cases covering 1 November 1898 through October 31, 1900 in the Second Biennial Report of the Board of Managers of the Indiana Reformatory at Jeffersonville. Even in the Third Biennial Report, for the period ending October 31, 1902, where he lists 34 vasectomies, not a word explains their sudden appearance in the report.

Sharp went public in another direction. In 1901 he spoke to the Mississippi Valley Medical Association meeting in Put-in-Bay, Michigan; and he published his first statement on the subject in the *New York Medical Journal* for 1902, using the title "The severing of the vasa deferentia and its relation to the neuropsychopathic constitution."[95] He had performed vasectomies on 42 patients ranging in age from 17 to 25. After his appeals to the governor in his 1904 and 1906 biennial reports, Sharp finally was given his opportunity. "When the Sterilization Bill was introduced in the legislature in 1907, I had all the men sit down in their cells and write a record of their experiences, in the form of testimonials, to impress upon the legislators that there was no kick from, and no subjection of the individual to 'cruel' and 'unusual punishment.'"[96] The bill was introduced by Dr. Horace D. Read, representative from Tipton County, and passed the House 59 to 22. The Senate concurred with a vote of 28 to 16. On March 9, 1907, the first compulsory sterilization law was signed and accepted by Governor Hanly three days after its passage.

RECEPTION OF INDIANA'S STERILIZATION LAW

The *Indianapolis Star*, using the headline "Surgeons to deal with criminals," interviewed Superintendent William H. Whittaker who spoke approvingly of the new law: "Men who have had experience of ten or twelve years in the handling of criminals can see the necessity of something being done along this line. This law will not only prevent the procreation, but in my judgment will be one of the best preventive measures to deter crime that has ever been placed upon the statute books."[97]

The Indiana law was broader than the *Star's* headline implied. The law read:

Be it enacted by the general assembly of the State of Indiana, That on and after the passage of this act it shall be compulsory for each and every institution in the state, entrusted with the care of confirmed criminals, idiots, rapists, and imbeciles, to appoint upon its staff, in addition to the regular institutional physician, two (2) skilled surgeons of recognized ability, whose duty it shall be in conjunction with the chief physician of the institution, to examine the mental and physical condition of such inmates as are recommended by the institutional physician and board of managers. If, in the judgment of the committee of experts and board of managers, procreation is inadvisable and there is no probability of improvement of the mental condition of the inmate, it shall be lawful for the surgeons to perform such operation for the prevention of procreation as shall be decided safest and most effective. But this operation shall not be performed except in cases that have been pronounced unimprovable. Provided, That in no case shall the consultation fee be more than three ($3.00) dollars to each expert, to be paid out of funds appropriated for the maintenance of such institution.[98]

THE FAILURE OF THE MICHIGAN AND PENNSYLVANIA CASTRATION LAWS

Although Sharp's campaign paid off and he convinced both the state and Governor Hanly of the worth of a compulsory sterilization law, Indiana was not the first state to attempt to put such a law into practice. Michigan in 1897 defeated such a bill, which was then limited to castration, a procedure many legislators looked upon as cruel or unusual and not in keeping with the constitutional prohibition of such acts. On March 30, 1905, Pennsylvania passed a law in both houses of its legislature, but Governor Samuel W. Pennypacker vetoed it. The bill was enacted in response to a campaign by Dr. Martin W. Barr, Director of the Pennsylvania Training School for Feeble-Minded Children at Elwyn, Pennsylvania. He had published his major book, *Mental Defectives; Their History, Treatment, and Training*, in 1904.[99] As early as 1901, Barr and his colleagues had written to the state legislature urging passage of a law to permit castration or vasectomy at the discretion of a surgeon. Barr pointed out this was not a punishment for crime but a method to prevent the spread of degeneracy and a treatment for the good of the patient. "Everyone who has paid thoughtful attention to the question knows how largely the element of heredity enters into the complex problems of degeneracy."[100] Barr claimed that the law was "returned by the Governor for the correction of some trifling technicality, was unfortunately lost, and thus failed to become a law."[101] Robert Reid Rentoul, the Birmingham physician had who labored in vain since 1903 to have a compulsory sterilization law by vasectomy for the degenerate classes in England, repeated, in a slightly modified form, this same interpretation: "the Governor wishing some slight alteration, did not have the Bill returned to him in time to sign it."[102]

GOVERNOR PENNYPACKER'S DENUNCIATION OF THE STERILIZATION LAW

Actually, Governor Pennypacker was scathing in his sarcasm as he returned the bill unsigned with a letter condemning it. The Pennsylvania law stated: "It shall be lawful for the surgeon to perform such operation for the prevention of procreation as shall be decided safest and most effective...." Governor Pennypacker noted that "it is plain that the safest and most effective method of preventing procreation would be to cut the heads off the inmates, and such authority is given by the bill to this staff of scientific experts."[103] He pointed out that the inmates would be subjected to the surgery without their consent and the law, in effect, allows the surgeons to experiment on them. "A great objection is that the bill would encourage experimentation upon living animals, and would be the beginning of experimentation upon living human beings, leading logically to results which can readily be forecasted."[104] He cited the disastrous results of the Norristown oophorectomies of 1893 and the implications of the bill that motivated its most zealous supporters. He quoted from a pamphlet on heredity written by one of the Elwyn physicians (probably Barr) who claimed that the "elimination of the weakling was the truest patriotism—springing from an abiding sense of the fulfillment of a duty to the state." Pennypacker took a dim view of giving scientists an unchecked authority to carry out their social philosophies and had no sympathy for the largely unproven thesis that the bill would do any good. His veto was sustained.[105]

SHARP ABANDONS MASTURBATION AS THE BASIS FOR VASECTOMY

Sharp dropped masturbation as the most important reason for performing vasectomies when the law allowed him to sterilize other degenerate classes. Whereas his first 176 cases were voluntary treatments for masturbation, his next 240 cases were primarily done for eugenic purposes. Sharp was vague about the medical or social problems of the reformatory inmates he now selected for surgery. "I shall not dwell on the various physical abnormalities that are found in the defective," he commented in 1909, "I do wish, however, to call attention to the two that are found most frequently, and the least dwelt on by writers on this subject, namely imperfect refraction and color blindness. It is very rare that we find a defective who does not have one or other of the above mentioned conditions, though possibly to so slight a degree that the defect may be entirely overlooked. There are persons of this class in whom the only indications are temperamental, the most common being selfishness, ingratitude, inconstancy, egotism, inability to resist an impulse or desire."[106] He described defectives as being "the most gifted as well as the most vicious, weakest, and ordinarily, the most unhappy of mankind."[107] He assumed that prenatal impressions and psychological depression during pregnancy harmed the embryo and made the newborn defective. His hopes were high that compulsory sterilization would be extended to other institutions and to other states. "There is a law providing for the sterilization of defectives in effect in Indiana," he told his fellow practitioners, "and it is being carried out at the Indiana reformatory. I regret very much that it is not being followed up in the other institutions of the state; but there is no doubt that it will come about in a very short time."[108]

Sharp was confident that his law "will never be rescinded for the simple reason that it is right, just to all, and humane."[109] He had dreams of further legislation and said he would "carry it a little further, and make provision in our marriage laws, that when one or both contracting parties suffer from a defect, or a chronic transmissible disease, the male should be sterilized. Then let them go and marry; and by this means there will probably be support given and a protectorate thrown about some feeble-minded woman, that in any event would become a public charge, or a prostitute, or more than likely the mother of illegitimate children."[110] He was also nervous. Governor Hanly was succeeded by Governor Thomas R. Marshall who was not happy with the law and ordered Sharp not to carry out any more vasectomies. The year's totals dropped from 119 in 1907–1908 to 39 in 1908–1909. In 1909–1910 there was only one vasectomy.

Hanly was a Republican and an ardent prohibitionist. His attempt to force through a law that would give each county an option to ban the sale of alcoholic beverages led to the defeat of the party in the 1908 election. Hanly went on to run for President of the United States in 1912 on the Prohibition Party ticket. Marshall, the Democrat who succeeded Hanly after the election of 1908, was sensitive to civil liberty issues. He became Vice President in the administration of President Woodrow Wilson.

INDIANA CEASES COMPULSORY STERILIZATIONS

Governor Marshall told Sharp that he was going to ask the legislature to rescind the law because he thought it was unconstitutional. Sharp per-

LEGISLATIVE STATUS OF EUGENICAL STERILIZATION IN THE SEVERAL STATES OF THE UNITED STATES, JANUARY 1935

STATES WITH STERILIZATION LAWS NOW IN EFFECT—1935	STATES WITH BILLS NOW PENDING —	BILLS VETOED SINCE 1907	STATES IN WHICH BILLS HAVE NEVER BEEN INTRODUCED	AMENDMENTS NOW PENDING	AMENDMENTS PASSED WITHIN ONE YEAR—	AMENDMENTS KILLED WITHIN ONE YEAR —
28	7	7	6	4	2	3
Alabama	Illinois	Georgia	Colorado	Iowa	Indiana	Kansas
Arizona	Maryland	Kentucky	Florida	Nebraska	South Dakota	Washington
California	Missouri	Massachusetts	Louisiana	Oklahoma		Wisconsin
Connecticut	New Jersey	Nevada	New Mexico	West Virginia		
Delaware	New York	Ohio	Rhode Island			
Idaho	Pennsylvania	Tennessee	Wyoming			
Indiana	Texas	Arkansas				
Iowa						
Kansas						
Maine						
Michigan						
Minnesota						
Mississippi						
Montana						
Nebraska						
New Hampshire						
North Carolina						
North Dakota						
Oklahoma						
Oregon						
*South Carolina						
South Dakota						
Utah						
Vermont						
Virginia						
Washington						
West Virginia						
Wisconsin						

The fate of sterilization bills. After 1907, some states enacted sterilization bills that were vetoed or amended, but most remained in force by 1935 if they abided by the Supreme Court restrictions cited in Buck v. Bell, 1927. (Reprinted, with permission, from Harry H. Laughlin Archives, Truman State University, courtesy of CSHL Eugenics Web site.).

suaded the Governor not to do so, and promised not to carry out vasectomies while he was in office. The Governor agreed, but did grant one exception, on a voluntary basis, when Sharp was prevailed upon to demonstrate the operation to a visiting Russian physician in 1910.[111] Although Indiana had ceased vasectomies, other states were adopting compulsory sterilization laws, and Sharp's initial success was becoming a national crusade.

FOOTNOTES

[1] There may be a magical expectation of science's successes by the general public. When science fails to deliver another marvel, a hopelessness can set in with the naive belief that if science can't solve a problem, nothing can. Thus, if prisons become "moral hospitals" and there are no cures, then treatment of criminals is not possible and only their isolation, exile, execution, or sterilization can prevent their presence and spread in society. This sense of science as magic I remember vividly about 1950 when an epidemic of polio was raging in Brooklyn and a local drugstore had a large hand-painted sign "We carry camphor," the 1950s magical equivalent to Booth Tarkington's Penrod, who wore a little sack of asafoetida to ward off disease. Five years later the same mothers who were desperate for camphor were lining up at their doctors' offices to have their children receive Salk vaccine. The mothers weren't more educated; they just followed the practical logic that if it's supposed to work, use it.

[2] A.E. Hamilton, "Pioneers in Eugenics," *Journal of Heredity* 5(1914): 370–372. Hamilton cites *Heredity: Its Relation to Human Development* by Loring Moody and Elizabeth Thompson (Institute of Heredity, Boston, 1882) as the source for Moody's views. The quotes that follow are Moody's.

[3] Ibid. p. 370.

[4] Ibid. pp. 371–372. Moody based his hereditary views on the work of Galton and some contemporary advocates of degeneracy theory, including "Sharpe, Anthon, Ribot, and Papilon."

[5] Ibid. p. 372.

[6] Philip R. Reilly, "Involuntary sterilization in the United States: a surgical solution," *Quarterly Review of Biology* 62(1987): 153–170. See also Reilly's *The Surgical Solution* (Johns Hopkins University Press, 1992). I thank John S. Haller, Jr. for sending me a copy of his manuscript "The role of physicians in America's sterilization movement, 1894–1925" which will appear in the *New York State Journal of Medicine*.

[7] Bernhard Heilbrun, Untitled and translated note from *Centralblatt fur Gynakologie* Sept. 22, 1883 in *Journal of the American Medical Association* 1(1883): 591.

[8] Wilhelm Tauffer, "The castration of women" translated from *Zeitschrift fur Geburtshulefe und Gynakologie* in *Journal of the American Medical Association* 2(1884): 632.

[9] Anonymous. "Castration in nervous and mental disease" editorial, *Journal of the American Medical Association* 7(1886): 547–549.

[10] Ibid. p. 549.

[11] E.W. Cushing, "Melancholia; masturbation; cured by removal of both ovaries," *Journal of the American Medical Association* 8(1887): 441–442.

[12] Ibid. p. 441.

[13] Ibid. p. 442.

[14] Anonymous. "Procreation of the criminal and degenerate classes," *Medical Record* 37(1890): 562.

[15] Ibid.

[16] Anonymous. "Castration for melancholia," *Medical Record* 42(1892): 736.

[17] Ibid.

[18] Ibid.

[19] Anonymous. "Castration recommended as a substitute in capital punishment," *Journal of the American Medical Association* 18(1892): 499–500.

[20] Ibid. p. 500.

[21] Robert Boal, "Emasculation and ovariectomy as a penalty for crime and the reformation of criminals," *Journal of the American Medical Association* 23(1894): 429–432.

[22] Ibid. p. 432.

[23] Anonymous. "Castration for neuroses and psychoses in the male," *Medical Record* 41(1892): 43.

[24] Ibid.

[25] F.E. Daniel, "Should insane criminals, or sexual perverts, be allowed to procreate?," *New York Medico-legal Journal* (1893): 275–292.

[26] Anonymous. "An experiment in castration," *Medical Record* 43(1893): 433–434.

[27] Ibid. p. 434.

[28] Daniel, *Insane Criminals*, p. 275.

[29] Ibid. p. 276.

[30] Ibid. p. 277.

[31] Ibid. p. 282.

[32] Ibid. p. 284.

[33] Ibid. pp. 284–285.

[34] Ibid. p. 286.

[35] Ibid.

[36] Ibid. p. 287.

[37] Ibid.

[38] Ibid. p. 288.

[39] Ibid. p. 289.

[40] Anonymous. "Castration of sexual perverts," *Medical Record* 45(1894): 479–480, p. 480.

[41] Harry H. Laughlin, *Eugenical sterilization in the United States* (Psychological Laboratory of the Municipal Court, Chicago, 1922), p. 351.

[42] Ibid. p. 351.

[43] In 1821, five years after the new state of Indiana passed its constitution, and several years after Governor William Henry Harrison urged the Territorial Assembly to finance the construction of a prison, an act was passed to locate it in Jeffersonville, on the Indiana side of the Ohio River, facing Louisville, Kentucky. The first group of prisoners was put to work to build the Ohio Falls Canal. As the state grew in population, buildings were added, and the crowding of prisoners spread infectious disease. The reform movements of the late 19th century extended to the prisons, and the state added dormitory bathrooms and sewage pipes, replacing the cell's pail and trenched latrines within the prison grounds. The more enlightened superintendents at Jeffersonville introduced vocational training, work projects, a school system, a library, and a hospital with its own surgical unit.

No biographical memoirs of Sharp exist. The *New York Times* of Friday, November 1, 1940, p. 25, column 4, provides a brief obituary. An interview by William M. Kantor provides his retrospective assessment of his career, "Beginnings of sterilization in America. An interview with Dr. Harry C. Sharp, who performed the first operation nearly forty years ago." *Journal of Heredity* 28(1937): 374–376. A photograph of Sharp may be seen on p. 24 of Paul Popenoe's "The progress of eugenic sterilization," *Journal of Heredity* 25(1934): 19–26.

[44] The biennial reports, later annual, were prepared and printed at the Jeffersonville Prison. The prison physician included an account of the health of the prisoners and recommendations for their welfare. Sharp was a concerned physician who fought for his prisoners' health and believed in strong public health and sanitary measures. Before Sharp became affiliated with the Jeffersonville Reformatory when these reports were first issued, there was little available information on the state of the prisoners' health prior to 1898. The prison's records were destroyed in a fire (Kantor, *Sterilization in America*, p. 376). The reports are individually bound and available at the Indiana University main library (Bloomington) under the call number HV 9105 I61. The titles vary somewhat from report to report. They begin with *First Biennial Report of the Board of Managers of the Indiana Reformatory, Jeffersonville: From November 1, 1896 to October 31, 1898, inclusive* (Reformatory Printing Trade School, 1898). A photograph of the prison is in the fifth report.

[45] Sharp had urged a more balanced diet and more calories for his prisoners. He also succeeded in replacing an open ditch, which ran through the prison carrying human wastes, with individual privies. Over the years he added an operating room, sterilization equipment, and other improvements to the prison dispensary. He was not always tactful in expressing his concerns to the governors. After he arrived in 1896, he claimed, "I recommended that two of the cell-houses be remodelled and

that the third one be torn down completely, and that a modern and larger one be built instead. This recommendation passed unheeded; the neglect was nothing short of criminal, and I wish to state here that the State of Indiana which has upon its statutes a crime known as involuntary manslaughter, is equally as guilty of that offence as any individual who has suffered the punishment prescribed by law." p. 27 in Sharp, *First Biennial Report, 1898.*

[46] Harry C. Sharp, "Neurasthenia and its treatment," *Proceedings of the Mississippi Valley Medical Association* (Nashville, Tennessee, October 11–16, 1898).

[47] Sharp was using a modified theme of Max Nordau's. In his *Degeneration*, Nordau claimed the rapid pace and complexity of civilization at the end of the 19th century was causing physical and mental degeneracy because of the overload of information and hectic life-styles urbanization and technology imposed on the individual. Sharp does not use degeneracy theory to support his views in this article.

[48] Albert John Ochsner, "Surgical treatment of habitual criminals," *Journal of the American Medical Association* 32(1899): 867–868.

[49] Daniel R. Brower, "Medical aspects of crime," *Journal of the American Medical Association* 32(1899):1282–1287.

[50] Ibid. p. 1282.

[51] Ibid.

[52] Ibid.

[53] Ibid. p. 1286.

[54] Ibid.

[55] *Second Biennial Report of the Board of Managers of the Indiana Reformatory, Jeffersonville: From November 1, 1898 to October 31, 1900, inclusive* (Reformatory Printing Trade School, 1900).

[56] Sharp identified him by last name only in the Kantor interview. But there is a record of death of a Harry Clawson about 1902 among the inmates who died in prison.

[57] Harry C. Sharp, "The severing of the vasa deferentia and its relation to the neuropsychopathic constitution," *New York Medical Journal* 75(1902): 411–414.

[58] Ibid. p. 413. Sharp used the various versions of Clawson's vasectomy to demonstrate his beliefs. In this first version, Clawson was the beginning of many allegedly successful therapeutic outcomes of the operation: "After having personally severed the vasa deferentia in forty two patients, whose ages range from seventeen to twenty five, I am prepared to speak favorably of the operation...." He goes on to claim they "improve mentally and physically."

[59] Sharp was embittered by the turn of events following Indiana's passage of the compulsory sterilization law. He didn't like the criticism of the state medical association and the resistance of Governor Marshall to his carrying out more sterilizations. He went into private practice in the midwest for several years and then did military service during World War I in France. He enjoyed that experience and eventually closed out his career in the veteran's hospital in Lyons, New Jersey, where he died.

[60] Harry C. Sharp, *Vasectomy: A Means of Preventing Defective Procreation* (Indiana Reformatory Print, Jeffersonville, 1908). Sharp is inconsistent with the numbers he cites. On page 10 he says "After observing nearly 500 males in whom I had severed the vas deferens...." and on page 18 he claims "I began this work in October 1899, and from 1899 to 1907 this operation was done on 176 men in the Indiana Reformatory on request." It is possible that he carried out nearly 300 involuntary vasectomies after the 1907 law was passed, but his prison reports do not support this. The recorded vasectomies are: 1898 = 0, 1900 = 0, 1902 = 34, 1904 = 42, 1906 = 30, 1908 = 119, 1910 = 1. Sharp did not report Clawson's case in the 1900 report. He also claims in 1909 (*JAMA* 53:1897–1902, p. 1899): "I have 456 cases that have afforded splendid opportunity for postoperative observation and I have never seen any unfavorable symptoms."

[61] Ibid. p. 18.

[62] I suspect tuberculosis because Sharp made it a point to talk about Clawson's weight gain and in some of his later articles he talks about the restored physical health and weight gain in some of the men he vasectomized. Loss of weight in tuberculosis was a common symptom. Tuberculosis was a major disease among his prisoners. In his report for 1900, Sharp claims 27% were from tubercular homes. Clawson's death was listed as tuberculosis.

[63] Harry C. Sharp, "Vasectomy as a means of preventing procreation in defectives," *Journal of the American Medical Association* 53(1909): 1897–1902, p. 1902.

[64] *Third Biennial Report of the Board of Managers of the Indiana Reformatory, Jeffersonville: From November 1, 1900 to October 31, 1902, inclusive* (Reformatory Printing Trade School, 1902).

[65] *Fourth Biennial Report of the Board of Managers of the Indiana Reformatory, Jeffersonville: From November 1, 1902 to October 31, 1904, inclusive* (Reformatory Printing Trade School, 1904).

[66] Sharp had begun his campaign to get the state to pass a compulsory sterilization law and his shift reflects this higher goal of preventive medicine rather than curative medicine. Sharp's stress on masturbation also began to fade as he convinced himself that vasectomies should be a public health issue.

[67] Sharp, *Fourth Biennial Report*, p. 52.

[68] *Fifth Biennial Report of the Board of Managers of the Indiana Reformatory, Jeffersonville: From November 1, 1904 to October 31, 1906, inclusive* (Reformatory Printing Trade School, 1906), p. 65.

[69] Sharp, *JAMA* 53, p.1900.

[70] David and Helen Wolfers, *Vasectomy and Vasectomania* (Mayflower Books, Frogmore, St. Albans, Herts, 1974). The Wolfers give a history of the use of vasectomies.

[71] Marc Goldstein and Michael Feldberg, *The Vasectomy Book* (J.P. Tarcher, Inc., Los Angeles, 1982). Reginald Harrison's ideas are in his book, *Selected Papers on Stone, Prostate, and Other Urinary Disorders* (Churchill, London, 1899). Harrison and Lennander may have been influenced by the work of Louis Auguste Mercier, who in

1841 attributed enlargement of the prostate to masturbation and venereal excess or to excessive sitting (one third of his patients were shoemakers). See J.T. Murphy's *History of Urology* (Charles Thomas, Springfield, Illinois, 1972), p. 383. Mercier used castration as a treatment for enlarged prostates.

[72] James M. Phalen, "Albert John Ochsner 1858–1925," *Dictionary of American Biography* 13: 616.

[73] Ochsner, *Surgical treatment,* p. 867.

[74] Ibid.

[75] Ibid.

[76] Ibid. p. 868.

[77] The National Union Catalog lists numerous medical books by Ochsner, including works on the proper design of hospitals. Ochsner's portrait is included in the obituary of the *Journal of the American Medical Association* 85(1925): 374.

[78] Robert Reid Rentoul, *Race Culture; or, Race Suicide (A Plea for the Unborn)* (Walter Scott Publishing Co., London, 1906).

[79] Sharp, *JAMA* 53, p. 1898.

[80] Harry C. Sharp, *The Sterilization of Degenerates* (National Christian League for Promotion of Purity, 1908), p. 2.

[81] Ibid.

[82] Quotation of Sharp in "Surgeons to deal with criminals," *The Indianapolis Star,* Thursday, March 7, 1907, p. 7.

[83] Sharp, *Sterilization,* p.1.

[84] Ibid.

[85] Ibid.

[86] Ibid. p. 7.

[87] Sharp is quoted for an address he gave in California, in Anonymous "Sterilization of human beings. The Indiana plan," *New York Medical Journal* 90(1909): 1241–1242, p. 1242. Sharp's views on the tonic effect of the seminal fluid at the untied end of the vasectomy are similar to the views later promoted by the Austrian physician E. Steinach, who used unilateral (occasionally bilateral) vasectomies in older men as a means of inducing rejuvenescence. The "Steinach procedure" was a popular medical fad between the two world wars, with prominent men like William Butler Yeats hoping for the restored vigor of years gone by.

[88] Sharp, *JAMA* 53, p. 1900.

[89] Of particular interest in that *JAMA* 53 article is the discussion section with his fellow physicians, most of them ecstatic in the hope that medicine would solve what charity, penology, and sociology could not do.

[90] Sharp, *JAMA* 53, p. 1900.

[91] Ibid. p. 1901.

[92] Samuel E. Earp, "Sterilization of the criminal," *Indianapolis Medical Journal* 14(1911): 136-137, quote on p. 136.

[93] An unsigned editorial. "The sterilization of criminals and other degenerates," *Indianapolis Medical Journal* 13(1910):163–165, quotes on p. 164.

[94] Ibid. p. 164.

[95] Harry C. Sharp, "The severing of the vasa deferentia and its relation to the neuropsychopathic constitution," *New York Medical Journal* 75(1902): 411–414.

[96] William M. Kantor, "Beginnings of sterilization in America," *Journal of Heredity* 28: 374–376, p. 375.

[97] "Surgeons to deal with criminals," *The Indianapolis Star,* Thursday, March 7, 1907, p. 10. The *Star* was supportive of the measure because they described the bill as having passed the senate "with few votes being recorded against it."

[98] *Laws of the State of Indiana Passed at the 65th Regular Session of the General Assembly 1907,* Chapter 215, pp. 377–378. The bill, H364, was approved on March 9, 1907.

[99] Martin W. Barr, *Mental Defectives: Their History, Treatment, and Training* (Blakiston, Philadelphia, 1904). Barr sent out questionnaires to 25 U.S. and 36 foreign institutions for the care of mentally defective individuals. There were 12 responses (9 from the U.S.), and only two said the state should not sterilize the defective and one said that it would be ineffective even if done. The rest endorsed the idea of compulsory sterilization. This suggests that in the period when Harry Sharp was obtaining his medical education and starting his career, there was already a growing movement among physicians for a legal backup to what many physicians perceived as an obvious remedy—the sterilization of the unfit.

[100] Ibid. p. 194. As Barr interpreted the title "the consideration of treatment naturally includes that of prevention" (p. 189). He based his proposal for a legalization of castration of defectives on a Connecticut marriage law (July 4, 1895) which read "No man and woman either of whom is epileptic, or imbecile, or feeble-minded, shall inter-marry, or live together as husband and wife when the woman is under forty five years of age. Any person violating or attempting to violate any of the provisions of this section, shall be imprisoned in the state prison not less than three years." (cited in Barr, p. 189).

[101] Ibid. p. 195. Barr was not prepared to abandon the effort for legalization, he states (p. 190) "For these, and against these—festering sores in the life of society—the only protection is that which the surgeon gives; and that which was spoken of behind closed doors, already begins to be the subject of open discussion and to appear in reputable journals." Note the metaphor of the body politic in Barr's plea.

[102] Robert Reid Rentoul, *Race Culture,* p. 167.

[103] Harry Hamilton Laughlin, *Eugenical Sterilization in the United States* (Psychopathic laboratory of the Municipal Court, Chicago, 1922). Laughlin discusses the Pennsylvania bill and its veto on pp. 35–36. Laughlin quoted the governor's remarks, I believe, in the hopes of showing the prejudice that existed against what he believed to be a sound medical practice.

[104] Ibid. p. 36.

[105] Ibid.

[106] One of the claims of those who saw masturbation as a disease was the deterioration of vision, especially nearsightedness, that accompanied the habit. It is not clear where the notion of color blindness as a degenerate trait originated. Since color blindness is found among males primarily, and since the more serious consequences of onanism were among males, perhaps this most prominent of sex-linked traits would have been associated with a degenerate heredity caused by masturbation. I have not seen such a claim in any of the works I've read on masturbation, but I doubt if Dr. Sharp originated this idea. The behavioral or temperamental traits Sharp used for selecting his prisoners for vasectomies is frightening and reminds me of the equally subjective criteria used by the SS in concentration camps for which Jews and other undesirable prisoners were to be gassed in the early stages of the Holocaust (see Chapter 20) when the gassings were justified among the staff as medical euthanasias.

[107] Harry C. Sharp, "Vasectomy as a means of preventing procreation in defectives," *Journal of the American Medical Association* 53(1909): 1897–1902, p. 1897. This shows that Sharp was more strongly influenced by the Continental school of degeneracy (from Quetelet to Nordau) than by Galton. Note, too, the Lamarckian origin of degeneracy which he attributes to maternal impressions.

[108] Ibid. p. 1899.

[109] Harry C. Sharp, *The Sterilization of Degenerates*, Pamphlet (National Christian League for Promotion of Purity, 1908), p. 9.

[110] Quoted in Hastings Hornell Hart *Sterilization as a Practical Measure* No. 11 (pamphlet), Department of Child Helping of the Russell Sage Foundation, Inc., New York, 1913, p.10.

[111] Kantor, *Sterilization in America*, p. 375.

13

The Emergence of Two Wings of the Eugenics Movement

THE USE OF STERILIZATION TO ENFORCE EUGENIC POLICY was an American invention. Although the term eugenics had not been coined when the Jukes kindred was converted from a potentially reversible population of misfits into an innately degenerate class, the intent of Sharp's campaign for sterilization was unmistakably eugenic. Ironically, the extreme libertarian views of Spencer provided the reasoning for protecting the public from the degenerate stock it nursed. Spencer's evolutionary ethics had, of course, also undergone transformation into social Darwinism. In this biologically justified form, the civil liberties of the unfit could be ignored and the health of the nation would demand protection.

Spencer provided the organic model of society, a view at least as old as Plato's authoritarian *Republic.* Spencer rejected the idea that society is designed or manufactured; he believed it grew or evolved, although its individual units, the people, did not form a permanently differentiated center of consciousness.[1] Although Spencer insisted in his *Social Statics* that the state cannot intervene to deny liberty to persons who did not violate the freedom of others, he became more ambivalent about the state's role in his later years. He accepted Malthusian checks as legitimate evolutionary processes to weed out the unfit. "The quality of a society is physically lowered by the artificial preservation of its feeblest member. ...The

231

quality of a society is lowered, morally and intellectually, by the artificial preservation of those who are least able to take care of themselves."[2]

SPENCER FAVORS STATE CONTROL OVER THE UNFIT

In his *Principles of Sociology*, Spencer saw a way out.[3] If the unfit harmed society, by somehow evading the checks of "misery and vice" even without public or private charity, then it would bestow on posterity an "increasing population of imbeciles and idlers and criminals."[4] Spencer had long argued that if the state takes away the wealth earned by honest effort through regulation or taxes and assigns benefits to those who are undeserving, its citizens have a right to say "Cease your interference. But when, in any way, direct or indirect, the unworthy deprive the worthy of their dues, or impede them in the quiet pursuit of their ends, then may properly come the demand, 'Interfere promptly; and be, in fact, the protectors you are in name.'"[5] No doubt the rejection of Spencer's unpopular views on war, colonization, the established church, public health, and education made it difficult, if not impossible, for him to influence Great Britain to enact legislation to restrict the degenerate classes. In the United States, however, Spencer was riding a crest of popularity, and those parts of his philosophy that appealed to the eugenically minded were isolated and endorsed while the rest of his beliefs were brushed aside as applicable only to his own country.

FRANCIS GALTON INTRODUCES EUGENICS

Eugenics as Great Britain experienced it was quite different in intent. It was invented and named by Francis Galton, whose work disproving pangenesis had embarassed Charles Darwin. Eugenics was introduced by Galton nearly 20 years before it was given a name and almost 40 years before he launched eugenics as a political and social movement.[6] Galton submitted an article to *MacMillan's Magazine* in 1865, bearing the title "Hereditary talent and character."[7] He was, perhaps, influenced by his relatedness to his cousin Charles Darwin and the fame that his relatives had achieved. He explored this by studying notables in a half-dozen source books and concluded that "talent is transmitted by inheritance in a very remarkable degree; that the mother has by no means the monopoly of its transmission; and that whole families of persons of talent are more common than those in which one member only is possessed of it."[8] Galton's rough estimate was that about one in six of the offspring of an eminent individual will inherit that talent to earn a place in history. This excited his imagination: "...how vastly would the offspring be improved, supposing distinguished women to be commonly married to distinguished men, generation after generation, their qualities being in harmony, and not in contrast, according to rules, of which we are now ignorant, but which a study of the subject would be sure to evolve."[9] Galton applauded the German tradition of professors marrying the daughters of other professors and he con-

Francis Galton presented by the artist in a scholarly pose. Galton was a generalist with contributions to meteorology, anthropology, statistics, geography, criminology, and psychology. He advocated and named the field of eugenics. (Reprinted, from F. Galton [1908] *Memories of My Life*, Methuen & Co., London [courtesy of CSHL Archives].)

demned religious celibacy, especially in the Middle Ages, "where almost every youth of genius was attracted into the Church."[10]

Galton expanded this first foray into selective breeding in humans and the heritability of talent by publishing several books, including *Hereditary Genius* (1869)[11] and *Natural Inheritance* (1889).[12] Galton was ecstatic when his cousin Charles Darwin wrote him after reading only 50 pages of *Hereditary Genius*[13]: "You have made a convert of an opponent in one sense, for I have always maintained that, excepting fools, men did not differ much in intellect, only in zeal and hard work; and I still think this is an *eminently important difference*."[14] Galton delayed starting a eugenics movement for several decades because of the hostility many people expressed to his article and his first book on the subject. Until the public was ready to accept the heritability of talents, a point much debated then, he was not prepared to advocate a social policy based on it. When he did advocate a eugenics movement, in 1901, he realized that in his earlier writings he had overrated the rate at which eugenic improvement would take place, but however modest the gain each generation, he felt the effort worthwhile.

Galton's crusade for "race improvement" or eugenics was launched upon his receiving a Huxley Medal from the Anthropological Institute. The title of his paper reflects the profound difference, as the 20th century began, between American values (e.g., Harry Sharp's) and British values (e.g., Galton's) in launching eugenics programs: "Possible improvement of the human breed under the existing conditions of law and sentiment."[15] Galton gave to the University of London a small bequest to establish a eugenics laboratory. He also broadened his definition of eugenics as "the study of agencies under social control that may improve or impair the racial qualities of future generations either physically or mentally."[16] Although the pessimistic side of eugenics is reflected in that updated definition, Galton's primary interest was in what later became known as "positive eugenics," the selective breeding of the healthiest, brightest, soundest, and wisest. He was also noncoercive and looked upon eugenics as a form of secular religion, with moral duty urging a couple to include eugenic attributes as part of a marriage decision. American eugenics suffered from the essentially nonelitist, democratic tradition that everything is fine as long as it's not corrupted from within or without. For this reason, American eugenics was dominated by a philosophy later called "negative eugenics," a holding operation with occasional purges of its allegedly weakest components.

HARRY LAUGHLIN AND NEGATIVE EUGENICS

The most notorious representative of negative eugenic thought in the United States was Harry Hamilton Laughlin (1880–1942), who was introduced in Chapter 11, for his work as superintendent of the Cold Spring Harbor Eugenics Record Office.[17] Laughlin was born in Oskaloosa, Iowa, the son of a preacher, professor of biblical languages, and a college president, first at Oskaloosa College and then in Ohio at Hiram College. Laughlin was one of the youngest of ten children, four of them brothers who all became osteopaths. His mother was a suffragette and was active in the Women's Christian Temperance Union. Young Laughlin grew up in a middle class and idealistic family. He studied history as an undergraduate at the Normal College for Teachers in Missouri and taught in rural schools in Iowa, where he learned quite a bit about agriculture and enrolled in Iowa State College to pursue this interest. In 1907 he went to the Brooklyn Institute to learn agricultural genetics and there met his future mentor and lifelong associate, Charles Benedict Davenport.[18]

It was Davenport who saw promise in this young man with a penchant for statistics and problem solving. He corresponded with Laughlin; sponsored him for membership in the American Breeder's Association; and after 1909, when Davenport's interests were strongly shifted to human genetics, he encouraged Laughlin to take up this new field. Davenport was fortunate that Mrs. E. H. Harriman, the mother of the future governor of New York state, provided funds to endow a Eugenics Record Office. Davenport asked Laughlin to join him and to manage the office. By 1916 the Cold Spring Harbor efforts on behalf of eugenics were nationally known, and Davenport arranged for Laughlin to gain academic credentials by going to Princeton University for a Ph.D. His project was a modest cytological study of mitosis in onion root tips, but his sponsors, E.G. Conklin and G.H. Shull, had impeccable records as first-rate biologists, and Laughlin returned in 1917 with the assurance that he would now be recognized as an expert in his new field of eugenics. Laughlin was shy, hesitant, opin-

EUGENICS

THE SCIENCE OF HUMAN IMPROVEMENT
BY BETTER BREEDING

BY

C. B. DAVENPORT

CARNEGIE INSTITUTION OF WASHINGTON
DIRECTOR, DEPARTMENT OF EXPERIMENTAL EVOLUTION,
COLD SPRING HARBOR, N. Y.
SECRETARY, COMMITTEE ON EUGENICS, AMERICAN BREEDERS' ASSOCIATION

NEW YORK
HENRY HOLT AND COMPANY
1910

Charles Benedict Davenport shifted from engineering to natural history before embracing genetics and eugenics as the central themes of his career. He planned the facilities and obtained the funding for the Cold Spring Harbor Laboratory complex that greatly benefited the field of genetics but proved embarrassing in its misguided eugenic zeal. Davenport's personality was complex. He was a good fundraiser and manager; some of his contributions to human genetics were solid (eye color, skin color), but most were marginal or reflective of his unexamined biases. He liked to be surrounded by yes-men, and he coveted power and influence. Also shown is the title page of his 1910 book on eugenics. (Both courtesy of CSHL Archives.)

ionated, and defensively intolerant of criticism. Like Francis Galton's, his marriage turned out to be sterile and he left no descendants.

Laughlin's scholarly research was on race horses. He hoped to use pedigrees to follow the thoroughbreds with the best records for speed and endurance. Unfortunately, the data in this field are notoriously unreliable because the matings are not always controlled and records are frequently forged or fraudulent to give horses a better ancestry on paper than they have in reality. Laughlin managed to publish seven papers on the genetics

Harry Laughlin (*right*), superintendent of the Eugenics Record Office, in 1918. Laughlin was a student of Davenport (*left*). His mathematical talents appealed to Davenport, who appointed him to the post of superintendent. Laughlin was shy, embarrassed by his late-onset epilepsy (then considered a sign of degeneracy), and a zealot for sterilization and restrictive immigration laws. His association with jurists made him an effective court witness and advisor to Congressional committees. Laughlin's contributions to basic genetics were minimal in his areas of training in cytology or in the complex genetics of thoroughbred horses. Like Davenport, he did not tolerate criticism. (Courtesy of CSHL Archives.)

of thoroughbred horses, most of them of marginal reputation among his peers. The bulk of his data on horse breeding was never published. Laughlin had far more personal, but not professional, success with his efforts in studying human pedigrees and using demography to identify the unfit.

The building used for the Eugenics Record Office was located just north of the Cold Spring Harbor Laboratory. (Courtesy of CSHL Archives.)

The Eugenics Record Office used eugenics field workers to interview families and prepare pedigrees and more detailed folders of family traits. A few of these involved physical traits both superficial (hair and eye color, skin pigmentation) and pathological (albinism, malformations). Most of the pedigrees focused on behavior and abilities. Davenport is fourth from the left in the front row. Davenport and some of his field workers went to Jamaica in 1913 to study human skin color in interracial families and established the inheritance of melanin as a quantitative trait. (Courtesy of CSHL Archives.)

EUGENICS RECORD OFFICE

ESTABLISHED OCTOBER 1, 1910

MRS. E. H. HARRIMAN, FOUNDER

---o---

BOARD OF SCIENTIFIC DIRECTORS •

ALEXANDER GRAHAM BELL, CHAIRMAN   WILLIAM H. WELCH, VICE CHAIRMAN

LEWELLYS F. BARKER,   IRVING FISHER,   T. H. MORGAN,   E. E. SOUTHARD

CHARLES B. DAVENPORT, SECRETARY AND RESIDENT DIRECTOR

H. H. LAUGHLIN, SUPERINTENDENT

---o---

THE FUNCTIONS OF THE OFFICE ARE:

1. To serve eugenical interests in the capacity of repository and clearing house.

2. To build up an analytical index of the inborn traits of American families.

3. To train field workers to gather data of eugenical import.

4. To maintain a field force actually engaged in gathering such data.

5. To co-operate with other institutions and with persons concerned with eugenical study.

6. To investigate the manner of the inheritance of specific human traits.

7. To investigate other eugenical factors, such as (a) mate selection, (b) differential fecundity, (c) differential survival, and (d) differential migration.

8. To advise concerning the eugenical fitness of proposed marriages.

9. To publish results of researches.

The Eugenics Record Office was one of Davenport's programs located at Cold Spring Harbor. Disagreement with eugenic policy led T.H. Morgan to resign as a member of the board of scientific directors. (Reprinted, with permission, from Harry H. Laughlin Archives, Truman State University, courtesy of CSHL Eugenics Web site.)

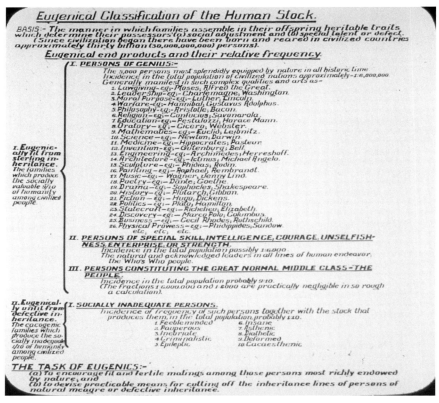

The value of eugenics represented in a classification of four levels of humanity. Note the exclusively hereditarian interpretation for all four classes. (Reprinted, with permission, from Harry H. Laughlin Archives, Truman State University, courtesy of CSHL Eugenics Web site.)

LAUGHLIN'S FIRST SPEECH ON EUGENICS

The year 1914 was very productive for Laughlin. He gave his first speech on eugenics to the National Conference on Race Betterment, held in Battle Creek, Michigan, and sponsored by the Kellogg family, enthusiasts for health food, exercise, and eugenics.[19] The Kellogg brothers had developed their health movement, from fitness centers to corn flakes, as a response to the dangers of masturbation and other practices leading to degeneration. W.K. Kellogg handled the business end of the enterprise and his brother, J.H. Kellogg, a physician, handled the research and scholarship. At the conference, thirteen scholars were invited to give speeches.

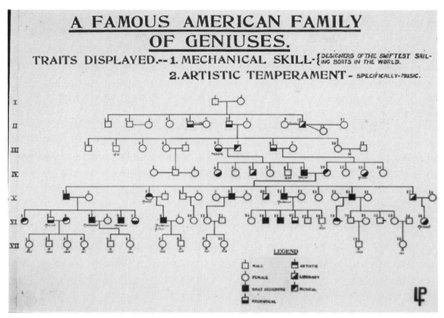

The pedigree studies carried out by the Eugenics Record Office were of poor quality. Few attempts were made to evaluate environmental influences on career choice. This pedigree ruled out X or Y linkage as well as autosomal recessive inheritance. If there were an autosomal dominant gene for mechanical talent, it is favored by males (7:1 ratio). More likely, sons mimicked their fathers' occupations and interests. (Reprinted, with permission, from Harry H. Laughlin Archives, Truman State University, courtesy of CSHL Eugenics Web site.)

Davenport and Laughlin represented the Cold Spring Harbor program. The conference included opponents and enthusiasts for segregation of the unfit, compulsory sterilization laws, and mass screening for defectives. Laughlin's paper was ambitious. He tried to calculate what it would take to make major inroads into the estimated 10% of the population who were unfit. He was convinced the only effective means was sterilization. "The recommended program would give ample opportunity for beginning on a very conservative scale. No mistakes need be made; for at first only the very lowest would be selected for sterilization, and their selection would be based on the study of their personal and family histories and the individual so selected must first be proved to be the carrier of hereditary traits of a low and menacing order."[20]

Laughlin tried to persuade his audience that "to purify the breeding stock of the race at all costs is the slogan of eugenics."[21] Whether Laughlin modeled his zeal on his father's exuberance as a preacher, or developed on his own a rigid belief in the righteousness of his views, he used this conviction effectively. He prepared tables of demographic information to show, decade by decade, the effects of a mass sterilization program as it gained momentum and acceptance over the years. His strategy was to begin with a nationwide education program, to lobby for legal restraints on marriage and habitation of the unfit, to agitate for the segregation of those identified as unfit, and finally to use sterilization, particularly on women, if the unfit are released to society. He singled out women because they carry the embryos; it would take too many male sterilizations to diminish the reproductive rate. He used the analogy of stray dogs; if the female is spayed she has no litters; sterilized male dogs are likely to be replaced by other strays not yet caught, who would readily copulate with females in heat.

LAUGHLIN'S CAMPAIGN FOR STERILIZATION LAWS

In 1914, Laughlin also prepared a report for the Eugenics Record Office on "Legal, legislative, and administrative aspects of sterilization."[22] Laughlin was enthusiastic about the prospects of using involuntary sterilization on a large scale. Davenport, equally concerned about the menace of the unfit, was less inclined to sterilization. He did not know how to respond to the charges that criminals and others of low repute, if sterilized, might lead a life of worry-free promiscuity. Laughlin's enthusiasm came from his association with a Committee on Sterilization set up by the Institute of Criminal Law and Criminology that year. Laughlin joined Harry Sharp, Bleeker van Wagenen (an admirer of Sharp's), and T.D. Carothers, all advocates of compulsory sterilization, in a committee whose remaining members were hostile to the idea.[23]

In Laughlin's report to the Eugenics Record Office, he carefully criticized the weakness of the existing state eugenic laws and the tests of their unconstitutionality. He found almost all of them flawed and virtually "all need amending or recasting."[24] They were, he claimed, a dead letter, and he suggested as a task for his office the preparation of a model sterilization law. In the meantime, he agreed that segregation of the unfit was the "most efficacious" procedure and that sterilizations should be few and selective until the defects in the laws were remedied. In his proposed model law (which was never adopted in full), he made sure that due process was followed, expert investigation and testimony were provided, and to avoid violation of equal protection of the law, Laughlin recommended that it should be used on those scheduled for release to society.[25]

The New York Bar Association was not happy with either the deliberations of the Committee on Sterilization or Laughlin's proposal for a model sterilization law. By 1917 the committee abandoned an effort to make a formal proposal to the Bar and disbanded. To make matters worse, the Carnegie Institution of Washington, the main funding source for the Cold Spring Harbor research programs in basic and applied genetics, was unhappy with the publicity it was getting.[26] It did not want its status as a foundation jeopardized by public identification of its staff as lobbyists for sterilization laws and other controversial legislation. Davenport was asked to curb Laughlin; he would not be permitted to lobby for compulsory sterilization laws. Fortunately for Laughlin, he found a way out of this difficulty. He succeeded in having himself identified as a consultant and as an expert witness so he could participate in court tests of the constitutionality of the sterilization laws and as a scholar for hire to those professional and legislative groups that needed his expert testimony. Laughlin continued his campaigns against the unfit by aligning himself with legal experts to draft more effective sterilization laws. He also found a new target, less controversial than America's homegrown tribe of defectives. He took on the hordes of immigrants from central and southern Europe whose alien names, religious beliefs, and troubled presence in American society led to a clamor for restrictive immigration laws.

FOOTNOTES

[1] For discussion of the biological model of society, see pp. 44–45 in Alan Swingewood's *A Short History of Sociological Thought* (St. Martin's Press, New York, 1984). Also see Walter M. Simon "Herbert Spencer and the 'Social Organism,'" *Journal of the History of Ideas* 21(1960): 294–299.

[2] Herbert Spencer, *The Study of Sociology* (Ann Arbor Paperback, Ann Arbor, Michigan, 1961), p. 313. The reprint is a compilation of themes from Spencer's *The Prin-*

ciples of Sociology, see footnote 20.

[3] Herbert Spencer, *The Principles of Sociology*, three volumes, 1876–1896.

[4] Spencer, *Study of Sociology*, p. 314.

[5] Ibid. p. 320.

[6] Francis Galton, *Memories of My Life* (Methuen, London, 1908). Galton started medical school but dropped out after inheriting an immense fortune from his father. He was one of the last amateur scientists who subsidized his own research. Galton, like Darwin, did not take any university appointments and enjoyed the freedom his wealth provided to make contributions to exploration, geography, anthropology, meteorology, fingerprint analysis, statistics, psychology, and heredity.

[7] Francis Galton, "Hereditary talent and character" *MacMillan's Magazine* 12(1865): 157–166, 318–327.

[8] Ibid. p. 157.

[9] Ibid. p. 163. Note the recognition that Galton gives to women; he was not, like some of his contemporaries, a sexist who only saw intellect transmitted by males.

[10] Ibid. pp. 164–165. Galton's thesis in this first part of his article is clearly one of positive eugenics. The brightest and healthiest should marry each other and establish an ever-increasing population of talented people. Galton was rhapsodic in his enthusiasm: "If a twentieth part of the cost and pains were spent in measures for the improvement of the human race that is spent on the improvement of the breed of horses and cattle, what a galaxy of genius might we not create" (p. 165). Galton also recognized the possibility of a degeneracy requiring some corrective, but he saw that remedy through natural selection rather than through a planned negative eugenics movement. Galton used the ideas of Prosper Lucas to justify the hereditary nature of many diseases and behavioral traits such as "a craving for drink, or for gambling, strong sexual passion, a proclivity to pauperism, to crimes of violence, and to crimes of fraud" (p. 320) in much the same way that Emile Zola accepted Lucas's theories.

[11] Francis Galton, *Hereditary Genius* (MacMillan, London, 1869; second edition 1892).

[12] Francis Galton, *Natural Inheritance* (MacMillan, London, 1889).

[13] Galton, *Memories*, p. 290. I am not sure Darwin would have been as persuaded if Galton were not his cousin. Darwin was not a distinguished student and yet had achieved eminence. He worked hard to gather information and dedicated his life to his scholarly studies. Darwin's view, that except for fools (biologically retarded individuals) it is motivation and commitment that count most in success, is still held by critics of biological determinism. Victor and Muriel Goertzel's *Cradles of Eminence* (Little, Brown & Co., Boston, 1962) supports that theory.

[14] Some of Galton's social views, in fact, were reflective of the racism of his day (see Galton, *MacMillan's* 12, p. 321 for his descriptions of American Indians and blacks). His views of the United States were also not very flattering. He believed North America was settled by restless idealists and rebellious criminals and schemers. "If

we estimate the moral nature of Americans from their present social state, we shall find it just what we might have expected from such a parentage. They are enterprising, defiant, and touchy; impatient of authority; furious politicians; very tolerant of fraud and violence; possessing much high and generous spirit, and some true religious feeling, but strongly addicted to cant" (p. 325). Galton may have felt that the culling of defectives was best left to nature and that only a positive eugenics scheme was worthy of promotion because of its simplicity of encouraging the exceptionally talented to have larger family sizes.

[15] Francis Galton, "The possible improvement of the human breed under the existing conditions of law and sentiment," *Nature* 64(1901): 659–665. See also Galton's "Eugenics: its definition, scope and aims" *Sociological Papers* (MacMillan & Co., London, 1905).

[16] Galton, *Memories*, p. 321.

[17] An excellent account of Laughlin's life may be obtained in Frances Janet Hassencahl's doctoral dissertation *Harry H. Laughlin, 'Expert Eugenics Agent' for the House Committee on Immigration and Naturalization, 1921–1931* (University Microfilms, International, Ann Arbor, Michigan, 1971).

[18] A detailed and critical evaluation of Davenport's life and work is found in E. Carleton MacDowell's "Charles Benedict Davenport 1866–1944: A study of conflicting influences," *Bios* 17(1946): 1–50.

[19] John H. Kellogg was a physician; he founded the Race Betterment Foundation at Battle Creek, Michigan. His brother, W.K. Kellogg, managed the family's cereal business. Vernon L. Kellogg was a biologist at Stanford University who served on David Starr Jordan's Committee on Eugenics for the American Breeder's Association in 1906.

[20] Harry Laughlin, "Calculations on the working out of a proposed program of sterilization," *Proceedings of the National Conference on Race Betterment January 8–12, 1914* (Race Betterment Foundation, Battle Creek, Michigan), pp. 478–494, quotation from pp. 490–491.

[21] Ibid. p. 478.

[22] Harry Laughlin, "Legal, legislative, and administrative aspects of sterilization," *Eugenics Record Office Bulletin No. 103*, February 1914, Cold Spring Harbor, New York.

[23] Hassencahl, *Harry H. Laughlin*, p. 153.

[24] Laughlin, cited in Hassencahl, p. 98.

[25] Hassencahl, *Harry H. Laughlin*, p. 98.

[26] The Carnegie Institution of Washington was a philanthropy devoted to promoting knowledge, especially in the sciences. Many of its staff members were nervous about the use of the Institution's name in press reports on the activities of the Eugenics Record Office. Although there was separate funding for the various branches of the Cold Spring Harbor enterprises, the public did not always make such fine distinctions.

14

Europe's Undesirables Replace the Domestic Unfit

THE INDIANA LAW, PASSED ON THE 9TH OF MARCH 1907, aroused public interest, mostly sympathetic, from lawyers and physicians concerned about the habitual criminal and other degenerate classes of humanity. In 1909 three more states adopted sterilization laws: Washington on March 22, California on April 26, and Connecticut on August 12. Iowa, Nevada, and New Jersey joined the list in 1911, and New York passed its law in 1912. By the mid-1930s, the majority of Americans were subject to such laws, 30 states having passed them.[1]

The Indiana Plan, as it was sometimes called, was appealing because it seemed rational and required no more than an office visit to carry out the operation. The *Indianapolis Medical Journal* acknowledged that some physicians and medical journals approved vasectomy as a humane way to prevent degenerates from breeding, and cited Dr. W.T. Belfield who justi-fied the law because the "startling increase of crime, especially in our large centers of population, by men of inborn criminal tendencies, demands some means of protection."[2] The editorial asked for a wait-and-see atti-tude before going national with this surgical solution to social ills and described the new law, at best, as an experiment and not a proven treat-ment; and, at worst, the editor saw the new law as a danger for which he hoped "to live to see the day it will be stricken from the Indiana statutes."[3]

The state sterilization laws varied in intent. Most were eugenic and addressed the problem of sterilizing males and females who were epileptic, insane, or mentally retarded. Some states targeted all three categories of mental defect; others singled out only one. Still other states combined mental defect with criminal behavior and permitted sterilization as a punishment for certain crimes such as rape, indecent exposure, or repeated arrest and conviction. The inconsistencies of these laws led to challenges in the courts. New Jersey's law was struck down in 1913, two years after its passage. It provided sterilization of "certain defectives in State institutions."[4] An epileptic female was ordered sterilized. The New Jersey Supreme Court declared the law unconstitutional because it violated the woman's protection under the equality clause of the Fourteenth Amendment of the U.S. Constitution. The epileptics in institutions were being treated differently from those who were living at home and who were thus not subject to sterilization. Justice Garrison, in rendering a decision on the case, stated "the force of the statute falls wholly upon such epileptics as are 'inmates confined in the several charitable institutions in the counties and state'. It must be apparent that the class thus selected is singularly narrow when the broad purposes of the statute and the avowed object sought to be accomplished by it are considered."[5] The state had arbitrarily created two different classes and applied the sterilization law only to one of them, thereby establishing a procedure Justice Garrison had to reject.

The following year, a poorly worded sterilization law in Iowa that mixed eugenic and punitive goals was struck down because it did not provide adequate opportunity for appeal or appointment of a guardian to serve the inmate scheduled for sterilization. The law violated "due process," and its use of sterilization for specific criminal acts was regarded as a "cruel and unusual punishment." Failure to abide by the due process guarantees of the Bill of Rights and the inflicting of a cruel or unusual punishment made the law unconstitutional.[6]

VIRGINIA'S LAW BECOMES A TEST CASE

Five more state laws were struck down by 1925. Nevada, New York, and Michigan had their laws ruled unconstitutional in 1918, Oregon's was struck down in 1921, and Indiana's law of 1907 finally went to its defeat in

1925, fulfilling a prediction that Governor Marshall had made in 1908 when he ordered Dr. Sharp to discontinue sterilizations at the Jeffersonville Reformatory. Those who favored compulsory sterilization laws recognized the defects in these earlier laws and looked upon them as something of a legal experiment. Harry Laughlin, Director of the Eugenics Record Office at Cold Spring Harbor, was the nation's most ardent advocate of these laws and after the armistice ending the First World War, he worked as a consultant with states enacting new compulsory sterilization laws. He was particularly happy with the law passed by the state legislature of Virginia. The Senate approved it 30–0 on February 22, 1924, and the House of Delegates, by a vote of 75–2, approved it on March 8. The Governor signed it on March 20. The law applied only to hereditary forms of "insanity, idiocy, imbecility, epilepsy, and crime"[7] and provided the protections that Laughlin felt would make it constitutional.

The superintendent of a state institution would appoint a board of directors who would recommend to him those inmates whose sterilization would permit them to leave the institution and return to their communities. Laughlin argued that sterilization eliminated the need for institutionalization of those inmates, especially feebleminded women, who would either end up marrying or finding employment in unskilled work such as serving as a domestic. The purpose of the sterilization was solely to prevent a recurrence of the condition among potential children the inmate might intentionally or unintentionally procreate. The Virginia law also required that a proper guardian be given to the inmate to appeal the process once the superintendent signed an order for sterilization on recommendation of the board of directors. This would cover the due process clause of the constitution. Only those inmates who were likely to benefit from being released and who were not capable of safeguarding themselves against pregnancy were subject to sterilization. This made the law apply to the individual and not to a class, and Laughlin believed it met the objections of the "equal protection of the laws" clause of the Fourteenth Amendment. The law did not punish any person for the condition that deprived the inmate of personal freedom. It only applied to the hereditary nature of the defect and was intended for the benefit of society. This last aspect, Laughlin believed, made it immune to the "cruel and unusual punishment" clause of the Bill of Rights.

CARRIE BUCK SELECTED TO TEST VIRGINIA'S NEW LAW

Once Laughlin felt satisfied that the Sterilization Act of Virginia met the criticisms of earlier laws, he asked Dr. A.S. Priddy, Superintendent of the State Colony for the Insane or Feeble-minded in Amherst County, to select a patient to test the case and pursue it, if need be, to the U.S. Supreme Court.[8] Priddy was pleased to oblige, and selected Miss Carrie Buck, claiming "I arrived at the conclusion that she was a highly proper case for the benefit of the Sterilization Act, by a study of her family history; personal examination of Carrie Buck; and subsequent observation since admission to the hospital."[9] He also asked Laughlin to prepare her family history and serve as the state's consultant to establish that Carrie Buck suffered from a hereditary form of feeblemindedness.

Carrie Buck was born July 2, 1906 in Charlottesville. Her mother, born Emma Harlow (who later used the name Addie Emmitt), had a troubled marital history. She was married to Frank Buck, but they had no children, and Emma had several affairs that led to the collapse of the marriage. Carrie was the product of one such affair.[10] Two additional males fathered half-siblings of Carrie, a brother named Roy Smith and a sister named Doris Buck. All three of the children (there may have been an additional child, a boy) were put up for adoption. Carrie was adopted by Mrs. J.T. Dobbs when Carrie was four years old. She attended school through the sixth grade and showed "no physical defect or mental trouble" in her early years. She was also described by Mrs. Dobbs as being "fairly helpful in domestic work" especially when supervised. As she reached puberty she began to become more difficult to raise and no longer obeyed the Dobbs family.[11] She stayed away from home and became pregnant, giving birth to a daughter, Vivian Buck, in 1924. Her pregnancy made the Dobbs's despair about her future and they complained that they could not "be responsible for her self control." They requested that she be committed to a state institution. A hearing at the Juvenile and Domestic Court of Charlottesville ruled that Carrie was "epileptic and feeble-minded" and she was committed January 23, 1924.[12]

Dr. Priddy signed the order for sterilization after asking the board of directors to document her case. Mr. R.G. Shelton was appointed her guardian, and he agreed to appeal the case if the Circuit Court of Amherst County sustained the law. Priddy was prepared to appeal the case if the law

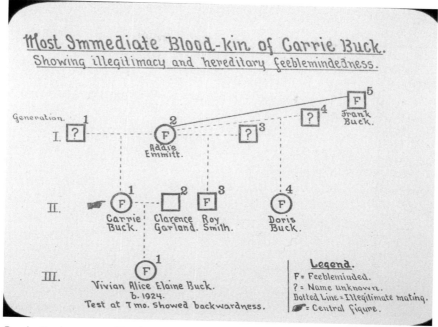

Most Immediate Blood-kin of Carrie Buck.
Showing illegitimacy and hereditary feeblemindedness.

Generation. I.

II.

III.

Carrie Buck.
Clarence Roy Garland. Smith.
Doris Buck.
Frank Buck.
Addie Emmitt.

Vivian Alice Elaine Buck.
b. 1924.
Test at 7 mo. showed backwardness.

Legend.
F = Feebleminded.
? = Name unknown.
Dotted Line = Illegitimate mating.
⬛ = Central figure.

Carrie Buck was sterilized in 1927. Her biological father is unknown (I-1); her mother (I-2) is listed as feebleminded, as are her illegitimate children (II-3 and II-4). Frank Buck (I-5) had no genetic contribution to Carrie or her two half-siblings. Carrie's daughter (III-1) is classified spuriously at 7 months as feebleminded. The multiple illegitimacies (*dotted lines*) may have been more influential than the alleged feeblemindedness in the Supreme Court's ruling upholding the sterilization law. (Reprinted, with permission, from Harry H. Laughlin Archives, Truman State University, courtesy of CSHL Eugenics Web site.)

were declared unconstitutional. Judge Bennett T. Gordon presided and Harry Laughlin submitted a deposition for the "hereditary analysis of Carrie Buck." On April 13, 1925, Justice Gordon rendered his verdict; he found no fault with the procedures and sustained the writ for Dr. Priddy to go ahead with the sterilization. Mr. Shelton and his court-appointed attorney then appealed the case. It went to the Virginia Supreme Court of Appeals, Judge Jesse F. West presiding, but Dr. Priddy died before the decision was rendered and he was replaced as superintendent by Dr. J.H. Bell, who agreed to continue the case.[13] On November 12, 1925, Justice West sustained the law permitting its appeal by Carrie Buck's guardian to the U.S. Supreme Court.

Hospital Form No. 131

VIRGINIA:

BEFORE THE STATE HOSPITAL BOARD

AT

_____
(Institution)

In re

_____, Register No._____

)
) Order for
) Sexual Sterilization
)
Inmate)

Upon the petition of_____

Superintendent of _____

and upon consideration of the evidence introduced at the hearing of this matter, the Board finds that the said inmate is

{ insane
  idiotic
  imbecile   }   and by the laws of heredity is the probable potential parent of socially
  feeble-minded
  epileptic   }

indequate offsprings likewise afflicted; that the said inmate may be sexually sterilized without deteriment to {his her} general health, and that the welfare of the inmate and of society will be promoted by such sterilization.

Therefore, it appearing that all proper parties have been duly served with proper notice of these proceedings, and have been heard or given an opportunity to be heard, it is ordered that_____

_____   {perform
     (Superintendent)   {have performed

by Dr._____, on the said inmate the operation of {vasectomy salpingectomy}

after not less than thirty (30) days from the date hereof.

_____
(Designated Member of Board)

Dated _____

Note: Make two copies; one for guardian or committee and one for Record.

Form used for the sterilization of the unfit in the state of Virginia. Carrie Buck was the first to be ordered sterilized, an order upheld by the U.S. Supreme Court in 1927. (From Paul Lombardo, Ph.D., J.D., courtesy of CSHL Eugenics Web site.)

Since Carrie Buck was to be the first person to undergo compulsory sterilization under the Virginia law, the State Colony wanted to have a strong case in which the legal procedures were faithfully carried out. Dr. Bell and his consultants from the Eugenics Record Office were happy that the law had twice been sustained, and they looked forward to the challenge in the U.S. Supreme Court. Paul Popenoe reviewed "The progress of eugenic sterilization" in a 1934 article in the *Journal of Heredity*. He claimed that "it was desirable to have the Virginia law passed on by the courts before it was put into effect and the litigation in which Carrie Buck was the plaintiff, and which is now historical under the title 'Buck vs. Bell', was arranged as a test case by the officials of the State Colony."[14]

LAUGHLIN AND ESTABROOK TESTIFY AGAINST CARRIE BUCK

The case against Carrie Buck was developed at each trial by Harry Laughlin and his associate at the Eugenics Record Office, Arthur H. Estabrook. Laughlin, after consulting with Dr. Priddy in 1924, described Carrie as feebleminded with a mental age of 9 (at the time of her commitment, when she was 18 years old). He described her as having a "record during life of immorality, prostitution, and untruthfulness"; that she "has never been self sustaining; has had one illegitimate child" who was "supposed to be mentally defective."[15] This assessment, made originally by Laughlin when Carrie's daughter Vivian was only 7 months old, is surprising, but he tried to support it in Carrie's later court trials by relying on the judgment of a nurse who believed Vivian to be abnormal. Laughlin also relied on the recollections of Miss Caroline E. Wilhelm, a Red Cross nurse who related to him that Carrie's mother also gave birth to two boys and a girl, all from different lovers. Laughlin was not able to trace the records of these siblings because the adoption laws of Virginia provided anonymity to the adoptive families. He attributed the lack of useful family records to the Harlow and Buck stock, both of which were known to produce pauper elements in Albemarle County where Charlottesville is located. "These people belong to the shiftless, ignorant, and worthless class of anti-social whites of the South."[16] Because they were a "moving class of people" documentation on their vital statistics was difficult to obtain. Similarly, Carrie's mother Emma Harlow, born in 1872 in Charlottesville, had a mental age of 7 years, 11 months, when she was tested at the age of 52. She too had a "record during life of immorality, prostitution and untruthfulness; has never been self-sustaining, was maritally unworthy; having been divorced from her husband on account of infidelity; has had a record of prostitution and syphilis; has had one illegitimate child and probably two others inclusive of Carrie Buck."[17]

Estabrook, who testified at the Virginia Supreme Court of Appeals, attempted to prove that the type of feeblemindedness seen in Carrie Buck and her family is a simple Mendelian recessive trait. "Where feeble-mindedness is found in two strains, the two strains meeting, feeble-mindedness will show up in one-fourth of the children."[18] In cases like Carrie's where her mother was feebleminded, he assumed the risk was 50% if Carrie's father was essentially normal. He explained that two feebleminded people only produce feebleminded children. "The rule, so far as we can find, has no exception."[19] Estabrook had to go over these points several times to clear up the genetics for the judge and attorney for Ms. Buck. Estabrook tried to show how a feebleminded individual could arise from two essentially normal parents, a bit of a puzzle in those days to most educated people. Using his argument about two strains coming together, Estabrook declared: "That gives the explanation of where the feeble-minded child comes from in families that are apparently normal. The blood is bad. They carry the defective germ plasm, and where two defective germ plasms meet, the effect again appears."[20]

Dr. Priddy was asked: "So far as patients are concerned, do they object to the operation or not?" He replied confidently, "They clamor for it,"[21] because it permits them to leave the institution and they value their freedom. He also stated that the Dobbses would take Carrie back if she were sterilized.

THE SUPREME COURT'S 8–1 DECISION FOR STATE STERILIZATION LAWS

The case that would then become known as "Buck versus Bell" was accepted for consideration for the October 1926 term of the Supreme Court of the United States. A decision to sustain the sterilization of Carrie Buck was made on May 2, 1927. The vote was 8–1 in favor of the Virginia law, Justice Pierce Butler (1866–1939) dissenting. Oliver Wendell Holmes, Jr., son of the famed essayist, delivered the majority opinion. Holmes reviewed the legal process and noted that Carrie Buck had benefitted from due process through her right to appeal through her court-appointed guardian, who had exercised that right. "There can be no doubt," Holmes asserted, "that so far as procedure is concerned the rights of the patient are most carefully considered, and as every step in this case was taken in scrupulous compliance with the statute and after months of observation, there is no doubt that in this respect the plaintiff in error has had due process of law."[22]

Holmes rejected the arguments, put forward after the New Jersey law was declared unconstitutional, that sterilizing Carrie Buck denied her equal protection of the law. He noted that the law applies to "any inmate of a state institution for the insane, feeble-minded or epileptic, who is afflicted with hereditary recurrent insanity, idiocy, imbecility, feeble-mindedness or epilepsy, and who, if sterilized could be paroled or discharged and could become self supporting."[23] That restricted usage, he argued, was within the state's powers for looking after the welfare of its citizens and did not constitute an unequal application of the laws. He also argued that the request made of Carrie Buck was not a punishment and not a major personal burden. "We have seen more than once that the public welfare may call upon the best citizens for their lives. It would be strange if it could not call upon those who already sap the strength of the State for these lesser sacrifices, often not felt to be such by those concerned, in order to prevent our being swamped with incompetence. It is better for the world, if instead of waiting to execute degenerate offspring for crime, or to let them starve for their imbecility, society can prevent those who are manifestly unfit from continuing their kind. The principle that sustains compulsory vaccination is broad enough to cover the fallopian tubes. Three generations of imbeciles are enough."[24]

Although the Buck versus Bell decision is often singled out as a blunder by the Supreme Court, the story is even more depressing in what it reveals about the way justice was done in this case. Paul Lombardo, in a devastating analysis in the New York University Law Review (1985), pointed out some significant facts. All three of the principals in the legal case, Albert Priddy (the superintendent of the asylum), Irving Whitehead (the appointed defendant's advocate in court), and Aubrey Strode (Priddy's legal arm for the asylum), were friends who knew each other well and who worked together as a team to see the Virginia law through to the Supreme

Court. Strode wrote the new Virginia law based on Laughlin's model eugenic law that he had worked out with Judge Olson in Chicago. Strode and Whitehead were boyhood friends who had grown up on adjacent farms. Priddy had prevailed on Strode to write a more effective law after an awkward lawsuit which he was fortunate to win after sterilizing a woman and her daughter on flimsy charges that at best could be described as motivated by Priddy's sexual prudery. Whitehead had served as a board member for the asylum and had served on committees with Priddy to promote sterilization at other institutions in Virginia. Strode served in the Virginia Assembly that passed his new sterilization law. Strode's wife studied eugenics at Cold Spring Harbor with Arthur Estabrook, Laughlin's colleague at the Eugenics Record Office. It was Estabrook who came down to present the case for Carrie Buck's alleged hereditary mental deficiency. There is no evidence that Carrie Buck, her mother, or Carrie's daughter were either feebleminded or seriously deficient in intellect. Both Carrie and her daughter did well in school and neither was ever left back because of poor grades. Complicating the case was the information (withheld from the courts) that Carrie's unplanned pregnancy arose (according to Carrie) from a rape by her adoptive parent's nephew, a fact the Dobbses tried to hide when they asked that she be committed. Surely the Supreme Court was unaware of these compromising circumstances that should have given, but did not give, Carrie a much needed legal defense. Also complicating the decision was Justice Holmes's filial duty to his father's memory; the elder Holmes was a staunch hereditarian long before there was a eugenics movement. One could also be charitable in interpreting Holmes's pejorative use of "imbeciles" instead of "feebleminded" as a lack of sophistication about the technical vocabulary of educational psychology in those days. However one does interpret the case, it is clearly a gross miscarriage of justice, poisoned by cronyism and conflict of interest, supplemented with shared prejudices about Carrie Buck and her family.[25]

CALIFORNIA'S EXTENSIVE USE OF ITS STERILIZATION LAW

Of all the sterilization laws, California's was most vigorously applied. Dr. F.W. Hatch was Secretary of the State Lunacy Commission and later General Superintendent of State Hospitals. He was urged to enforce the law through the efforts of E.S. Gosney, an industrialist and ardent supporter of the eugenics movement, and by Paul Popenoe, Jordan's student from Stanford University, who wrote extensively on eugenic sterilization as a humane measure. Almost half of the 38,087 sterilizations carried out by these laws through 1942 were performed in the state of California.

LAUGHLIN'S INVOLVEMENT IN IMMIGRATION LAWS

Before 1920, Laughlin did not distinguish national origin for the various categories of unfit who were subject to state sterilization laws. They were homegrown culls or degenerates who had to be isolated or sterilized to keep the American population healthy. Laughlin shifted his attention from the sterilization movement to the inferior quality of immigrants pouring into the United States. This transition came from Davenport's initial interest in the problem and Laughlin's frustration that he could not make lobbying a central focus of his eugenic research.

Davenport's classmate at Harvard was Prescott Hall, a lawyer with a love for German philosophy and Wagnerian opera (which he found particularly helpful for his chronic insomnia). In 1894 Hall and Robert deCourcy Ward, a meteorologist who became a Harvard professor, founded a club-like organization, the Immigration Restriction League. Hall and Ward feared the dilution and eventual loss of the Anglo-Saxon heritage in the United States.[26] The Immigration Restriction League never had more than 30 members and they were usually recruited through personal contact. In 1895 it urged Senator Henry Cabot Lodge to introduce a literacy test for foreign immigrants; this was vetoed by President Cleveland in 1897. In 1911 the league requested materials from Davenport on eugenics, and Davenport reciprocated by appointing Hall to the Immigration Subcommittee of the American Breeder's Association Eugenic Committee.

The Immigration Subcommittee also included Franz Boas, an internationally respected anthropologist at Columbia University.[27] Boas rejected the claims of racial and ethnic inferiority associated with the immigrants. He opposed the proposal of the subcommittee to urge Congress to increase the number of Nordics and diminish the number of Southern and Eastern Europeans as part of a national immigration policy. After writing a minority report to the American Breeder's Association, Boas resigned. He was replaced by a person more sympathetic to the goals of the Immigration Restriction League.[28]

RACISM AS A MOTIVE IN RESTRICTING IMMIGRATION

Davenport had made it clear to Hall, as early as 1912, that the Eugenics Record Office could not lobby Congress for changes in immigration law. He suggested using the American Breeder's Association and its *Journal of Heredity* as a suitable arena to inform the public of its concerns. Hall and Ward found additional support from a wealthy New York lawyer, active in the conservation movement, who was also a major contributor to the founding of the Bronx Zoo. Madison Grant (1865–1937) cannot be faulted for his contributions to the culture of New York City, but he became notorious for his racism. In 1916 he published *The Passing of the Great Race*.[29] Grant lamented the deterioration of the Nordic character of American life and saw it being swallowed up by inferior stocks brought in through a self-destructive immigration policy.

The public sympathy for restrictive immigration increased yearly through the first two decades of the 20th century. Americans who could trace their ancestry to the 18th century looked with distrust at the new Americans speaking strange languages, practicing alien religions, and allegedly clinging to their ancestral customs. They were accused of not entering the melting pot; of living in self-imposed ghettoes; of importing the foreign ideologies of Marxism and Anarchism; of agitating laborers to join labor unions; and of ruthlessly replacing American workers by taking low wages no native-born American could possibly accept.[30]

The anti-immigration movement found support in Congress from Representative Albert Johnson of the state of Washington. He came to Congress in 1912 on a campaign against the evils of foreigners. He started out as a printer and journalist, rising to the editorship of the *Seattle Times*. He feared Japanese immigration in the Pacific coast states, and he identified foreigners with subversion after the Industrial Workers of the World tried to organize the lumber mills in the Northwest.[31] In 1919 Johnson succeeded in getting immigration slowed down to 355,000 per year. The 1910 census was used and 3% of each national origin in that census was permitted entry.

Johnson chaired the Congressional Committee on Immigration and scheduled hearings for a more restrictive law that would effectively reduce to a trickle those immigrants who came from the eastern and southern nations of Europe. Laughlin was delighted in 1921 to be asked to serve as an "expert witness" for the hearings that Johnson proposed. He hoped to amass an overwhelming amount of evidence to confirm the inferiority of

Immigrants landing at Ellis Island in New York City, early in the 20th century. (Courtesy of the Library of Congress.)

the Russians, Poles, Italians, and smaller nationalities found among these two regions of Europe that sent so many of their impoverished and unhappy citizens to the New World. Laughlin believed the immigrants and their children would outnumber the native-born in commitments to mental institutions, illnesses recorded at public hospitals, arrests and convictions for crimes, and failures in the public school systems. Where the data did not fit his preconceived notions, he tried to explain it away, and occasionally he omitted it or used a system of classification favorable to his views. Thus, when immigrant Jews did better than the native-born in the public schools he no longer listed them as a separate category and instead tucked them in among the larger majorities of non-Jews for a nationality-by-nationality listing.[32]

Laughlin hoped that only eugenically sound immigrants would be admitted, with inspection teams examining the mental and physical abilities of potential immigrants in their native lands. This was prohibitively

Harry Laughlin pictured in 1935, at the end of his career. (Courtesy of CSHL Archives.)

expensive and politically impractical. Instead, Congress chose to use the 1890 census when few Southern or Eastern Europeans were among those settled in America. Congress considered its bill non-eugenic, and it passed without difficulty, 62–6 in the Senate and 323–71 in the House.[33]

As in his role in propagandizing for sterilization laws, Laughlin got into trouble with the Carnegie Institution for his racist-sounding interpretations of other nationalities. After the public debate on the immigration law revisions was over, Laughlin was discreetly told to resume his horse pedigree analysis and to see whether his studies of human pedigrees and family histories could lead to anything useful for the medical and social sciences.[34] To compound his distress, the collapse of the economy after the 1929 Wall Street panic made Congress turn to other problems. The election of majorities of Democrats to Congress in 1930 weakened conservative causes, and by 1933, Johnson was voted out of office by his constituents. Laughlin was no longer wanted as a government witness and he no longer had like-minded or sympathetic friends in Congress. Without that support from his congressional friends, without support from the geneticists he knew, and with no private funding in sight, Laughlin was forced to retire and returned to obscurity in Missouri.

FOOTNOTES

[1] Rev. Edgar Schmiedeler, *Sterilization in the United States* [pamphlet, 38 pp.] (Family Life Bureau, National Catholic Welfare Conference, Washington, D.C., 1943). Schmiedeler claims that by 1942 there were 15,780 males and 27,307 females who were sterilized under state laws, mostly feebleminded, insane, epileptic, habitual criminals, or moral degenerates (sexual perverts).

[2] Anonymous editorial, "The sterilization of criminals and other degenerates," *Indianapolis Medical Journal* 13(1910): 163–165, p. 163.

[3] Ibid. p. 164.

[4] Harry Laughlin, "Legal status of eugenical sterilization," Supplement to the *Annual Report of the Municipal Court of Chicago*, 1929. Laughlin cites *Smith vs. Board of Examiners*, 85 Law 46 on p. 40 of his supplement. New Jersey's law was passed on April 21, 1911.

[5] Ibid. p. 46. Garrison argued that even if the law were rigorously applied, it would have little eugenic effect because most epileptics were cared for at home and the law did not apply to them.

[6] Few of the court rejections criticized the eugenic argument itself, perhaps because the justices deferred to the expert testimony that feeblemindedness (and other medical degeneracies) followed Mendelian laws and the pedigrees used may have seemed persuasive.

[7] Laughlin, Supplement, p. 10. Virginia's law was Chapter 394, SB 281, signed March 20, 1924. The issue is complicated because there are some instances of hereditary insanity (e.g., certain porphyrias), idiocy (e.g., phenylketonuria [PKU]), imbecility (e.g., mannosidosis), and epilepsy (e.g., congenital absence of the corpus callosum). There are no known biological inherited syndromes associated with criminality. The overwhelming majority of insane people and probably of those called feebleminded do not owe their defect to single-gene disorders. It is interesting that Virginia's law lists "idiocy" and "imbecility" but not "feeblemindedness" in its preamble, yet Carrie Buck, the person whose mental competence was to be challenged under this law, could hardly be called an idiot or imbecile by the standards of that day.

[8] The Buck versus Bell case is a good example of how myths form. Many people who have a vague idea of the case believe that Carrie Buck, the victim, was represented by civil-liberty-minded liberals who took a dim view of the sterilization laws. In

fact, Laughlin and the Institution's superintendent initiated the whole thing to force the Supreme Court to rule on what they thought was a law that defended all the civil rights of Carrie Buck. It is not easy to recognize that many liberals and civil libertarians supported the eugenics movement as a progressive and humane change for the better (for example, birth control advocate Margaret Sanger).

[9] Laughlin, Supplement, p. 25.

[10] Ibid. p. 16. The father is not known and thus Frank Buck's family history is of no validity to her alleged hereditary deficiencies. She carries the name of her mother's husband at the time, as is the tradition of marital law.

[11] Ibid. It is strange that none of the justices picked up this early normal childhood to question the validity of Laughlin's and Estabrook's claim that she was genetically feebleminded. The rebelliousness of her teens is not surprising when middle-class homes with professional parents often have rebellious sons or daughters.

[12] Ibid. The complaints against Carrie are "immorality, prostitution, and untruthfulness" and that is what made her adoptive parents commit her. She was not committed because she lacked the ability to do domestic work, go to school, or function at a normal or subnormal level of intelligence. Stephen Jay Gould has investigated some of those achievements and rightfully questions the diagnosis of "feeble-minded and epileptic." (See Gould, *The Mismeasure of Man* [W.W. Norton, New York, 1981], pp. 335–336.)

[13] Priddy, of course, could not legally be the defendant in the appeal to the Supreme Court and Bell gets the historical credit (or infamy) for this landmark case. One irony, from the perspective of hindsight, is the anomalous blemish on Justice Holmes's record as a civil libertarian.

[14] Paul Popenoe, "The progress of eugenic sterilization," *Journal of Heredity* 25(1934): 19–26, p. 22.

[15] Laughlin, Supplement, p. 16.

[16] Ibid. p. 24.

[17] Ibid. p. 16.

[18] Ibid. p. 24.

[19] Ibid. Carrie's father would have been heterozygous (a carrier of the recessive gene) in the model Estabrook presented in court. It is not clear what evidence Estabrook used to justify this claim. The one instance where a retarded individual always produces retarded children is maternal PKU (phenylketonuric women, if pregnant, will cause a toxic damage to the nervous system of the developing embryo or fetus, regardless of the genotype or genetic constitution of the fetus).

[20] Ibid.

[21] Ibid. p. 27.

[22] Popenoe, *Progress of sterilization*, p. 23. Justice Butler did not write a dissenting opinion. His conservative political outlook and his observant Catholic beliefs may have been at odds, and he may have chosen to vote his conscience without comment. Holmes's views on eugenics may have been influenced by his father's writings

and comments (see Chapter 11). Pierce Butler was a Democrat from Minnesota, but his law career defending railroads from government regulation made him popular with conservatives. He was nominated by President Harding and surprised his backers by his staunch defense of the rights of criminals and those accused of crimes. He was one of the "four thorns" that led President Roosevelt to attempt to pack the Supreme Court. Butler voted against every New Deal legislative effort that came to the Supreme Court. See pp. 110–111, *The Supreme Court of the United States*, edited by Kermit L. Hall, James W. Ely, Jr., Joel B. Grossman, and William Wiecek (Oxford Univeresity Press, New York, 1992).

[23] Laughlin, Supplement, p. 8.

[24] Popenoe, *Progress of Sterilization*, p. 23.

[25] Paul A. Lombardo, Three generations, no imbeciles: new light on Buck v Bell [1985] (New York University Law Review 60(1985)(1): 31–62.

[26] Frances Janet Hassencahl, *Harry H. Laughlin 'Expert Eugenics Agent' for the House Committee on Immigration and Naturalization, 1921 to 1931*. Ph.D. dissertation (University Microfilms International, Ann Arbor, Michigan, 1971), p. 163.

[27] Boas (1858–1942) was a German-born anthropologist who settled in the United States, becoming a professor at Columbia University and curator of anthropology at the American Museum of Natural History. He was a life-long critic of racist theories.

[28] Irving Fisher, an economist sympathetic to the restrictive immigration objectives, took his place.

[29] Madison Grant, *The Passing of the Great Race* (Scribner's, New York, 1916). Grant (1865–1937) was a bachelor and gave his money and time to conservationist movements, such as "Save the Redwoods." Grant's views were not considered outrageous in the 1920s when many Americans feared the changes from what had been essentially a familiar Anglo-Saxon heritage and ethnicity to an ethnic pluralism with profoundly different cultural values and traditions.

[30] The labor unions were supporters of restrictive immigration acts because union workers feared that immigrants would take nonunion jobs at any wage out of desperation and this would crush the union movement.

[31] The IWW or Industrial Workers of the World were socialist in political philosophy, unlike most of the labor unions that formed in the 20th century. The IWW was founded in 1905, and because of its radical philosophy, the union was often falsely accused of instigating violence that frequently was brought about by company owners who requested police and state militia to break up strikes. In 1919 the backbone of the union's organizers were arrested and 249 aliens were deported to the USSR.

[32] Hassencahl, *Harry H. Laughlin*, pp. 283–301.

[33] The lop-sided votes in its favor reflect the wide support for restrictive immigration laws. Many of those who supported the legislation did not consider themselves racists. They believed that the United States had an Anglo-Saxon heritage and that tradition should be preserved. Others believed that the United States had matured, completed its "manifest destiny," and did not need any immigrants other than a trickle of highly competent people.

[34] The Johnson Act of 1924 remained in force until modified in 1952 by the Walter-McCarren Act, which reaffirmed the racial composition restrictions of the Johnson Act. A less racial policy has been used since the 1970s. Immigration is allowed for refugees from political oppression, a standard for admission that has allowed immigrants from the USSR, China, Central America, and South East Asia, especially if the political oppression is identified as Communist.

15

Eugenics Becomes an International Movement

E
UGENICS WAS NOT LIMITED TO THE UNITED STATES and England. The idea
swept through Europe and found sympathy in other continents. By
1912 the first International Eugenics Congress was meeting in London.
There were competing terms such as human betterment, race hygiene,
puericulture, and the neo-Malthusian movement to address different ways
each country looked on eugenics, or sometimes there were several competitive movements within a larger country like Great Britain or France.
What they had in common was a belief that human or societal control
over reproduction was essential to preserve the heredity of the nation.

The two major wings of eugenics as the 20th century began were
called negative eugenics and positive eugenics. Galton's sympathies were
mostly for positive eugenics. He believed the brightest, healthiest, and
most talented males and females in Great Britain had a moral duty to have
more children. His views were based on the heroic model of history—a
civilization owes its existence to the works and ideas of its most talented
few—the philosophers, writers, musicians, artists, jurists, scientists, engi-
neers, militarists, statesmen, and other gifted leaders who put a stamp on
history and culture. He hoped that eugenics would become a secular religion, practiced by the best and brightest. That fit Great Britain well. It was
a largely closed society of layered classes with the nobility on top, the landed aristocracy immediately below, a large wealthy and middle class (large-

ly nouveau riche) following in third place, and a substantial white- and blue-collar working class in fourth place. At the bottom of the pyramid of hierarchy were the failures—the unemployable, the psychotic, the retarded, the paupers, the vagrants, the beggars, and the most detested, the criminal. People knew their places and accepted their status as an act of fate or God's will. Mobility was difficult, and even the most talented from the white-collar classes were limited in how far they could advance. It was said, less with envy than admiration, that the future leaders of Great Britain were tested and shaped "on the playing fields of Eton."

Negative eugenics was more suited to the American tradition. It was a land of immigrants and largely a classless society with no hereditary titles; the holder of the highest office was called Mr. President; and there were opportunities for fortunes, land, and mobility spurred by the Constitutionally mandated "pursuit of happiness." American culture was filled with phrases reflecting this attitude. The United States was "the land of opportunity," despite its racist outlook (blacks being excluded from "the American Dream"). The advice given to most American males before the civil rights and feminist movements was to strive for success because "the world is your oyster; you're free, white, and 21." Financial success was made possible by a growing population, lots of cheap land in the west, a government willing to shove native Americans out of the way, and leaders who took pride in their humble origins. There were three productive components of American society: the nouveau riche, always changing as new technologies brought new business tycoons into millionaire status; the middle class consisting of farmers who owned their own land, shop owners, clergy, teachers, and other professionals; and the white-collar and blue-collar working class. At the bottom were the paupers, criminals, psychotics, and other failures of society. In contrast to Great Britain, there was plenty of movement up or down the acceptable classes. If a young family was disappointed with opportunities on the east coast, they could become pioneers and establish a new life in the midwest (to the 1840s) and far west (mostly after the Civil War). The railroads made mobility convenient and less hazardous for those pulling up their roots and seeking a new life. The basic attitude in America was that you were as good as anyone else and, with a lot of ambition and effort, you could obtain almost anything you desired.

This sense of freedom and opportunity led the overwhelming majority of Americans to look with suspicion on the failures of society, especially in the last 20 years of the 19th century. Some failure of physical ability, of motivation, of values, or of mental capacity was assumed to be present in a degenerate layer of society. Preserving the heredity of the majority was then a reasonable response to an alleged contaminating minority. This was especially driven home after Weismann's theory of the germ plasm fixed degeneracy in the reproductive tissue. There was little need for positive eugenics when so many talented people in the middle and working classes were constantly finding their way upward to success through their energy and talents.

THE FIRST INTERNATIONAL CONGRESS OF EUGENICS

Galton's death in 1911 made his most sympathetic followers seek an international gathering of eugenics movements around the world. Major Leonard Darwin took the lead and organized a consortium of leaders to invite their counterparts in other countries to send delegates to an international congress to be held in London.[1] It is an astonishing roster of consultants: Alexander Graham Bell, Winston Churchill, Charles B. Davenport, Charles Eliot (emeritus president of Harvard University), David Starr Jordan, Sir William Osler, Gifford Pinchot (founder of the national forest conservation movement), August Weismann, Alfred Ploetz, J. Langdon Down, Havelock Ellis, R.C. Punnett, and William Bateson. The gathering of delegates and presentation of papers took place July 24–30 at the University of London. Leonard Darwin served as president.[2]

In his address, Darwin stated his views. The meetings were held "...to endeavor both to study the laws of heredity and practically to apply the knowledge thus acquired to the regulation of our lives..." It was "...a paramount duty which we owe to posterity."[3] He acknowledged as a high priority an on-going need for a "...filling up of the blanks in our knowledge... ."[4] Despite the ignorance of how heredity worked in humans, he felt enough was known to permit some applications to society. "As an agency making for progress, conscious selection must replace the blind forces of natural selection." That sentiment was shared by many others in the audience, and he encouraged their participation by offering a challenge and a goal: "...May we not hope that the twentieth century will ... be known in future as the century when the Eugenic ideal was accepted as part of the creed of civilization?"[5] The published papers included presentations from Spain (1), Italy

Major Leonard Darwin, Charles Darwin's son, embraced eugenics, but unlike his American counterpart, Charles Davenport, he failed to make basic changes in society through eugenics. In Great Britain, eugenics remained an unrealized aspiration for the privileged. (Reprinted from *A Decade of Progress in Eugenics: Third International Congress of Eugenics*, 1934, courtesy of CSHL Archives.)

Alexander Graham Bell, who was honorary president of the Second International Congress of Eugenics, is best known for his invention of the telephone. His interest in deaf people led him to study hereditary forms of that condition as well as human heredity in general. (Reprinted from *Eugenics, Genetics and the Family: Second International Congress of Eugenics*, 1923, courtesy of CSHL Archives.)

(8), Denmark (1), the United States (9), Great Britain (5), France (6), Belgium (1), Norway (1), and Germany (1). They covered many aspects of human abilities and disabilities. Some papers gave data on infant mortality, some on criminality, some on racial differences in body morphology, some on alleged weaknesses of miscegenated children, some on marriage laws, some on the hereditary damage allegedly caused by alcoholism, some on venereal diseases, and some on the growing movement to use compulsory sterilization for the unfit. A favorite theme was the sudden decrease in family size in the industrial nations and the fears that this would lead to "race suicide."

THE SECOND INTERNATIONAL CONGRESS OF EUGENICS

The organizers of the First International Congress agreed to hold annual meetings (using as its name the International Federation of Eugenic Organizations) to plan the next International Congress, to accept new member organizations, and to keep contacts among nations open. The original plans for a second congress in 1915 in America were set aside by the outbreak of World War I, and that meeting was rescheduled after the war for 1921 in New York, at the American Museum of Natural History. Henry Fairfield Osborn, who was president of the American Museum of Natural History, was the president of that congress which met September 22–28, 1921.[6] The organizers felt they needed an education in the new field of genetics that was rapidly developing around the world. Half the invited papers were in basic genetics (many of the presenters having little interest and some even having open contempt for the social policies of the negative eugenics wing). The rest were eugenics papers, many of them presenting pedigrees as evidence for families of degenerates or successes.

American geneticists, now the dominant leaders in that field due to the successes of Morgan and his school at Columbia University, as well as other distinguished geneticists at the many universities that gave high priority to plant and animal genetics, were happy to have a vehicle to publicize some of their new findings. Fifteen of the major presentations were on first-rate contributions to genetics. These included work by L. Cuenot (on Mendelism in mice), H.S. Jennings (on paramecia genetics), C.E. McClung (on Oenothera cytology), C.B. Bridges (on fruit fly chromosome aberrations), A.F. Blakeslee (on Datura chromosome aneuploidy), J. Belling (on meiosis), G.H. Shull

The exhibit hall of the Second International Congress of Eugenics (1921). Poster exhibits were (and still are) common at scientific meetings. Some of these represent basic genetic studies and others purely eugenic exhibits. At the Third International Congress of Eugenics, all of the exhibits were devoted to eugenics. (Courtesy of CSHL Archives.)

Henry Fairfield Osborn was a staunch evolutionist and president of the American Museum of Natural History. Osborn favored the conservative, upper-class values of his peers who feared contamination of the American stock by social failures, either home-grown or among its immigrants. (Reprinted from *A Decade of Progress in Eugenics: The Third International Congress of Eugenics*, 1934, courtesy of CSHL Archives.)

(on Oenothera balanced lethals), R.R. Gates (on cytoplasmic factors in Oenothera inheritance), H.J. Muller (on the nature of mutation), C. Zeleny (on reverse mutation in fruit flies), R.A. Fisher (on a mathematical analysis of mutation and population genetics), P.W. Whiting (on haploidy in wasps), A.F. Shull (on rotifer genetics), C.H. Danforth (on the mathematical fate of genes in a population), S. Wright (on inbreeding in guinea pigs), and a few others less well known but also focused on basic science.

The eugenic papers were less memorable in their contributions either to medical genetics or human genetics, with a few exceptions such as papers by R.A. Fisher on twins, C.C. Little on the genetics of cancer, and L. Howe on the inheritance of eye defects. The range of the eugenics papers was varied and focused on topics such as inbreeding, race mixture, pedigree histories of degenerate kindreds (especially pauper stocks and an update on the Tribe of Ishmael), inheritance of music, and adolescent runaways.

THE THIRD INTERNATIONAL EUGENICS CONGRESS

For the Third International Eugenics Congress, the organizers decided to exclude basic genetics papers because the geneticists were invited to make their presentations at the Seventh International Congress of Genetics, which was also scheduled for 1932, but in Ithaca, New York. At this Third (and last) International Eugenics Congress, the president was Charles B. Davenport, director of Cold Spring Harbor Laboratories. It was well attended and hosted again by the American Museum of Natural History.[7] It had several sections, including the welcoming and thematic addresses, anthropometrics, race mixture, society and eugenics, positive and negative eugenics, disease and infertility, differential fecundity, and human genetics. The majority of the papers were by American authors, but other contributions came from England, France, Germany, Holland, USSR, Italy, Hungary, Poland, Cuba, Norway, Denmark, India, and Canada.

One paper (see also Chapter 19) captured a lot of attention from the press. H.J. Muller presented "The dominance of economics over eugenics," despite the efforts by Davenport to block its presentation. When that failed, Davenport's committee tried to curtail its length.[8] Muller prevailed to let his views be known. He accused the American (and most of the world's) eugenic efforts as a failure. He did so because he believed that societies were largely unequal, with limited opportunities for women,

An exhibit at the Third International Congress of Eugenics in 1932. (Reprinted, with permission, from Harry H. Laughlin Archives, Truman State University, courtesy of CSHL Eugenics Web site.)

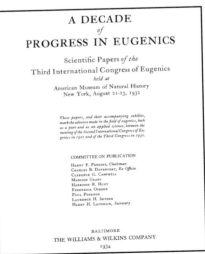

Charles Benedict Davenport as the president of the Third International Congress of Eugenics in New York, 1932. Also shown is the title page of the proceedings of this meeting. (Courtesy of CSHL Archives.)

lower classes, and minorities. As long as such overt bias existed, no eugenics for social traits was practical. Muller argued "...There is no scientific basis for the conclusion that socially lower classes, or technically less advanced races, really have a genetically inferior intellectual equipment, since the differences found between their averages are to be accounted for fully by the known effects of environment."[9] Muller insisted that instruments such as IQ tests have no merit when environments are grossly unequal. His strong stand against racism, sexism, and spurious elitism made the front pages of the newspapers covering the congress, and the international wire services sent the story abroad. It was a message that resonated with the American public, which was engulfed in a great economic depression. The unemployed did not see themselves as inferior in ability but just unlucky and victims of an economic system that failed them. Muller was unabashedly Marxist in his analysis, a sentiment that cost him dearly in later years, but one that made him at home with the new eugenics shaping up in Great Britain.

EUGENICS IN EUROPE

The geneticists of Muller's generation rejected the assumptions that prevailed in the first and second eugenic congresses. They did not accept a simple monogenic basis for social traits, especially those assigned to the unfit—the paupers, beggars, criminals, feebleminded, and psychotic. They argued that sterilization was ineffective and either did not change gene frequencies or did so at such a slow rate that any one generation would have little reduction in the incidence of these failure classes. They felt that the gene–character relationship was complex, with many interacting genes and a strong environmental component for almost all traits of social significance. They argued that a socialist system was more likely to provide equality of opportunity by eliminating class privileges (but at the time they were naive and did not realize that in the USSR it would be party loyalty and not merit that served as a basis for opportunity). These views were shared by Julian Huxley (who had recruited Muller to Rice Institute in 1916 for his first teaching job), J.B.S. Haldane (one of the leading mathematical biologists in the world and a theoretician of first rank in the study of human genetics), Joseph Needham (an eminent embryologist, historian of science, and Sinologist), and Lancelot Hogben (another brilliant mathematical geneticist). Their outlook was one of Galtonian positive eugenics for health, intelligence, talents, and personality rather than the negative eugenic obsessions with real and alleged single-gene traits. They assumed the future of eugenics resided in the study of polygenic traits.[10]

In France a very different tradition in eugenics developed. Eugenics was an umbrella term embracing many smaller and conflicting schools of thought. It included classical negative eugenics (championed by Alexis Carrel) with its sterilization programs, racist and anti-Semitic organizations, birth control advocates, anti-birth control natalists who favored a larger family for everyone to avoid race suicide, puericulturists who believed in a Lamarckian favorable environment prior to and during pregnancy to bring about superior babies, and Lamarckian socialists who saw good environments as the answer to bad heredity. Their net effect was to endorse a marriage law that required a certificate of good health (not severely retarded or syphilitic or alcoholic) as a basis for marriage. During the Second World War, Vichy France, the southern remnant of the country that was allowed to function after Northern France was kept as occupied German territory, embraced the racist and the negative eugenics wings of the eugenics movements. With the war's end, eugenics collapsed, as it did among almost all countries engaged in the war.[11]

Scandinavia had both conservative (like the American negative eugenics movement) and progressive (like the British new eugenics with a socialist slant) eugenics movements. Otto Mohr in Norway was progressive; Jon Mjoen was conservative and more inclined to German race hygiene with its baggage of racism. A similar division existed in Sweden. Sterilization was carried out at modest levels for problem classes, mostly for those with conditions having serious enough defect to merit institutionalization.[12]

Germany had the most robust of the eugenics movements, both before and after Hitler took power. Race hygiene had its origins in Germany when Alfred Ploetz introduced it in 1908. Anti-Semitism had flourished for centuries in Germany. The hygiene movement owed its origins to the efforts of Louis Pasteur in France; Joseph Lister in Great Britain; and, in Austria and Germany, Ignatz Semmelweis, Robert Koch, and Rudolf Virchow, whose work led to reforms in childbirth, child care, purification of water supplies, removal and sanitizing of wastes and garbage, antiseptic surgery, pasteurization of milk, school nurses, and mandatory examinations of school children for diseases. Race hygiene reflected Ploetz's fusion of two movements in Germany: a growing belief in a Germanic people and the admiration of the hygiene movement for diminishing infectious diseases. The self-image of a growing number of German people can be seen in this newspaper report of Berlin, December 11, 1908:

> The Emperor is reported to be interested in a plan proposed by professor Otto Hauser for the propagation of a fixed German type of humanity—a type which will be as fixed as the Jewish in its characteristics, if the suggestions of the professor can ever be carried out. The fixed type is to be produced as follows: —Only 'typical' couples are to be allowed to mate. The man is to be not more than thirty years old, the woman not over twenty-eight, and each have a perfect health certificate. The man should be at least five feet seven inches tall; the woman not under five feet six inches. Neither the man nor the woman should have dark hair. Its tint may range from blonde to auburn. The eyes of the pair should be pure blue without any tint of brown. The complexion should be fair to ruddy without any sign of heaviness or 'beefiness.' The nose ought to be strong and narrow, the chin square and powerful, and the skull well developed at the back. The man and the woman must be of German descent and must bear a German name and speak the language of Germany. These 'mated couples' are to get a wedding gift of $125 and an additional grant for each child born. The couples may settle in the United States if they prefer."[13]

Most curious was the short-lived eugenics movement in Bolshevik USSR. In the 1920s, A.S. Serebrovsky proposed positive eugenics programs for the USSR, sharing with Muller across the Atlantic a vision of a Communist society where all are given equal opportunities for work, education, and the resources of society to keep them clothed, fed, and healthy. In this perfected socialist environment, idealistic Communists believed, the major differences among individuals would be genetic. The talented would show their abilities in school and head to the universities. The ablest athletes, musicians, artists, scientists, and writers would enter the ranks of those professions. It was to these future leaders, the cherished assets of the USSR, that Soviet eugenics sought a Galtonian growth in population through use of their sperm by women other than their wives and by having larger families of their own. When artificial insemination entered the agricultural world and later entered the world of infertility among sterile couples, this new technique was favored as a portal for bringing eugenic worth into being.[14] Soviet society, even more than French society, rejected hereditarian models based on Weismann and Mendel and favored an environmentalist approach. Throughout the later 1920s and early 1930s, two sides began to emerge in the USSR: those with values like Muller, Haldane, and Huxley among the Soviet geneticists and those with Lamarckist sympathies who saw the work of Paul Kammerer and I.V. Michurin as the basis for an environmental training or good nurturing of less able individuals and populations into healthy and robust populations. That fight, using agriculture as its chief battleground, would intensify during 1935–1948 and culminate in the political dictate of the Central Committee to crush genetics in the USSR. Long before that fight between agricultural (led by Trofim D. Lysenko after 1935) and basic genetics erupted, the environmentalists had routed the eugenicists on matters of human heredity.

What characterized eugenic thinking in early 20th-century intellectual thought in almost every country was the belief that the state had higher interests and rights than the individual. The state had the right to raise an army by a draft; it had the right to send its young men into battle and risk death to protect national interests; and it had the right to protect the healthy public from the infected few and keep those with infectious disease in quarantine against their will. It just didn't bother the consciences of liberals or conservatives, negative or positive eugenicists, socialists or capitalists, that individuals had rights to live and reproduce without being sterilized if they had genetic defects. In general, positive eugenicists hoped that natural selection would do the job to cull the weak and the unfit from the population. They hoped that education and moral suasion would lead to a reform of social conscience and the procreation of more talented, bright, and healthy children. In general, negative eugenicists preached the reverse, and they hoped to hasten the process of evolution by eliminating the unfit and to allow evolution to take its course in leading to a superior strain of humanity.

FOOTNOTES

[1] For a very thorough account of the three congresses of eugenics, see Harry H. Laughlin, "Historical background of the Third International Congress of Eugenics," pp. 1–14, in *A Decade of Progress in Eugenics: The Third International Congress of Eugenics* (William and Wilkins, Baltimore, 1934).

[2] The various delegates and committees are listed in *Problems in Eugenics: First International Eugenics Congress, Held at the University of London, July 24–30, 1912* (Charles Knight and Company, London, 1912).

[3] Ibid. p. 3. Leonard Darwin Presidential Address, pp. 3–6, *Problems in Genetics*.

[4] Ibid. p. 5.

[5] Ibid. p. 6.

[6] The papers are in *Eugenics, Genetics, and the Family: The Second International Congress of Eugenics, volume I, Held at American Museum of Natural History, New York, September 22–28, 1921* (William and Wilkins, Baltimore, 1923).

[7] The publication of the papers is in *A Decade of Progress in Eugenics*.

[8] H.J. Muller. Ibid. The dominance of economics over eugenics, pp. 138–144.

[9] Ibid. p. 142.

[10] See Garland E. Allen, "Julian Huxley and the eugenical view of human evolution," pp. 193–222, *Julian Huxley: Biologist and Statesman of Science. Proceedings of a Conference Held at Rice University 25–27 September 1987.* Edited by C. Kenneth Waters and Albert Van Helden (Rice University Press, Houston, Texas, 1992). Also see Daniel Kevles, *In the Name of Eugenics: Genetics and the Uses of Human Heredity* (Knopf, New York, 1985), for an account of the reform movement in eugenics in the 1920s.

[11] An excellent and thorough account of French eugenics may be obtained from William Schneider, *Quantity and Quality: The Quest for Biological Regeneration in Twentieth-Century France* (Cambridge University Press, Cambridge, England 1990).

[12] For an overview of international eugenics, see Mark B. Adams, editor, *The Well Born Science* (Oxford University Press, New York, 1990).

[13] Cited in Horatio Hackett Newman, *Evolution, Genetics and Eugenics* (The University of Chicago Press, Chicago, Illinois 1925), pp. 599–600.

[14] For an account of A.S. Serebrovsky's eugenic proposals, see David Joravsky, *Soviet Marxism and Natural Science, 1917–1932* (Routledge and Kegan Paul, London, 1961). See pp. 304–307. Serebrovsky was the founder of the Morganist (western) school of genetics and a staunch foe of the older Lamarckist school of heredity in Soviet biology. Serebrovsky published his eugenic views, promoting positive eugenics through artificial insemination for married couples (as a socialist duty). Serebrovsky believed sex and reproduction would come to be seen as separate activities, the former for intimacy, the latter for the good of society. The negative response in the popular press was overwhelming.

Part III: Racism, the Holocaust, and Beyond

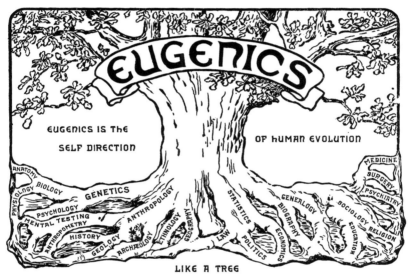

EUGENICS

EUGENICS IS THE SELF DIRECTION OF HUMAN EVOLUTION

MEDICINE SURGERY PSYCHIATRY ANATOMY BIOLOGY PHYSIOLOGY GENETICS PSYCHOLOGY MENTAL TESTING ANTHROPOMETRY ANTHROPOLOGY HISTORY GEOLOGY ARCHÆOLOGY ETHNOLOGY GEOGRAPHY STATISTICS LAW POLITICS ECONOMICS BIOGRAPHY GENEALOGY SOCIOLOGY EDUCATION RELIGION

LIKE A TREE
EUGENICS DRAWS ITS MATERIALS FROM MANY SOURCES AND ORGANIZES THEM INTO AN HARMONIOUS ENTITY.

16

Racism and Human Inequality

T HE FAILURE TO TREAT ALL HUMAN BEINGS as social equals includes certain practices that are almost universally accepted as reasonable or understandable and others that fall under the name of bigotry or racism. Members of a family treat themselves with a different set of rules and expectations than do unrelated persons. Sometimes laws are necessary to prevent abuses based on family ties, such as nepotism as a basis for hiring in government jobs. Here the discrimination against an unrelated person for a position to be filled by a possibly less qualified family member is recognized as a wrong, but it is not perceived as bigotry or racism against the unrelated person.[1] A similar distinction must be drawn between the rights of citizens of a nation and those who are aliens. Almost all nations have laws that limit what aliens can do. They usually cannot earn a living in the host country unless they request and receive permission to do so. They cannot reside permanently in the host country unless granted permanent immigrant status. They usually cannot vote in the host country or hold elective office. In some countries, they may be prevented from buying property or they may find their travel limited to restricted parts of the country. It is not bigotry that generates these rules governing alien life, but national economic and social interests that favor the country's own citizens over those whose livelihood and rights are assumed to be the responsibility of the alien's home country.

Jews as a Race, Seed, or People

Somewhat more difficult to interpret are restrictions imposed on those who belong to a different religious faith.[2] In Ezra, very clear prohibitions on intermarriage are stated: "Now therefore let us make a covenant with our God to put away all the wives, and such as born of them, according to the counsel of the Lord, and of those that tremble at the commandment of our God; and let it be done according to the law" (Ezra 10:3). The Jews had to divorce their non-Jewish wives and send them and their children back to their former homes if they were to enter Jerusalem and reestablish a Jewish community and temple.[3]

One of the undesired consequences of intermarriage is described by the phrase "… the holy race has mixed itself with the peoples of the lands" (Ezra 9:2). The term "holy race" in the Revised Standard Version is translated as "holy seed" in both the King James Bible and in Judah Slotski's *Daniel, Ezra, and Nehemiah*. The term "mixed itself" is translated as "mingled" in Slotski's version and it is rendered as "become contaminated" in the Anchor Bible. Post-Holocaust translators would probably use the phrase "holy people" rather than identify the Jews as a race or seed (with its implications of being a genetic stock).[4] Whatever the original meaning of the passage, it implies a different reason from the practical concerns of Ezra and Nehemiah that assimilation will lead to a gradual loss of faith. No one can fault the desire of a religion to survive, but if a religion defines its people as holy or set apart or as being a special seed, it thereby excludes other people, and the potential for resentment, disapproval, and bigotry may arise. More likely, the bigotry arising from unrelated causes, especially economic competition, may seek its justification through this interpretation of God's covenant with the Jewish people.

Biological Concept of Race

The concept of races as biologically, as well as culturally, distinct populations is based on the perceived real differences in skin color, body shape, and other physical features assocated with geographically remote peoples.[5] Added to this is the false assumption that the behavioral differences of these people must also be of the same biological nature as their physical traits.

The classification of plants, animals, and humans is both practical and difficult. It is necessary for people to talk about the kinds of animals and plants they encounter to distinguish at least those that are beneficial and those that are harmful. In a small community it may be possible to identify several hundred plants and animals by name, and for most of humanity and most of history this was sufficient. As the voyages of navigators increased and new animals and plants were introduced, the task of naming, describing, and characterizing these new forms of life became more difficult, with no international standards available for either their names or the criteria for classifying them. In biblical times, organisms ("creatures") were vaguely described as belonging to a "kind."[6]

The successful unification of classification was largely due to the effort of Carl Linnaeus (1707–1778), who took an interest in flowering plants

Carl Linnaeus, the Swedish naturalist, introduced a system of classifying organisms that founded the field of taxonomy or systematics. (Reprinted from Blunt W. [©1971] *The Compleat Naturalist: A Life of Linnaeus*, The Viking Press, New York, with permission of Curtis Brown Group Ltd., London on behalf of the Estate of Wilfrid Blunt).

while still a youth in Latin school.[7] He began his medical studies in Lund and Uppsala and completed them in Holland. While pursuing his degree, he began classifying plants on the basis of their sexual organs, with the number and location of pistils and stamens serving as a precise means of classifying them. In 1735, he published his classification of animals, plants, and minerals in his *Systema Naturae*.[8] By 1749 he had clarified the naming of organisms by using a distinction between genus (an inferred concept of relatedness) and species (the actual organism that lives and can be described). The use of a two-word sequence of Latin names for the genus and species, or binomial nomenclature as biologists call it, has been the international standard since. Linnaeus believed, at first, that the species was a fixed type, as readily identifiable in his day as on the assumed day of creation in Genesis. Later he studied reproduction or crosses of different species and accepted the modification of species through hybridization.[9]

LINNAEUS'S VIEWS ON HUMAN RACES

Linnaeus recognized the similarity of humans to the monkeys and apes and put humans among the primates rather than in some unique taxonomic unit separate from all other animals.[10] He acknowledged that there were racial differences among humans, but none of the characteristic differences was sufficient to merit species status, and all humans were put in a single species, *Homo sapiens*. It is not that Linnaeus was without the biases of his generation. He described aboriginal Indians as persevering, content, and free; Europeans were light, active, and ingenious; Asians were severe, haughty, and miserly; and Africans were crafty, lazy, negligent, and governed by whim. Race or variety was not a fixed characteristic in Linnaeus's system. Species characteristics were fixed and found among all populations of humans.[11]

Although Linnaeus did not offer a theory on the origin of races or their future fate, many other scholars of that period proposed models for racial differences. Buffon and Rousseau both believed climate and diet caused racial variation. Debate focused on the perfectibility of races. If, as most European thinkers believed, whites were the most successful of the races, then proper climate, diet, and circumstances could lead to a single human race. The assumed inferiority or superiority of any one race, even if hereditary, would be subject to environmental modification, a view strengthened by an almost universal acceptance in the early 19th century of Lamarck's theory of evolution through the inheritance of acquired characteristics.[12] Voltaire and Jefferson were not as optimistic about the unlimited modifications made possible by environmental means and accepted some hereditary differences in behavior, personality, and talents among races.

Geneticists have recognized, since the 1950s, that there is more variation within any one alleged race (white, black, yellow, brown, and red in the broadest classification by color) than there is between any two of them. This was first shown by use of blood groups (ABO, Rh, etc.) and increasingly by studies of the sequences of human proteins and genes. The Human Genome Project confirms that out of tens of thousands of genes, only a few dozen are involved in the physical differences (skin color, hair texture, facial features) that characterize most racial classifications.

GOBINEAU'S RACIAL DOCTRINES

Modern racism has its origins, in part, from the views of Joseph Arthur, Comte de Gobineau (1816–1882).[13] Gobineau's father, Louis, was a captain in the Royal Guards and remained a monarchist after the French Revolution; his mother, Magdeleine de Gercy, considered herself the daughter of Louis XV's illegitimate son. Gobineau was a commoner who adopted the title from his uncle, who also had a vague and disputed claim to past royalty.[14] Gobineau was tutored by a German and then attended college in Redon, where he took up Oriental studies. He dropped out of college to support himself as a postal clerk and later as a journalist. His views were initially liberal, but after the 1848 revolution he became a conservative in his political belief, favoring the monarchy and abhorring democracy. Gobineau's interest in politics was stimulated by his association as the private secretary for Alexis de Tocqueville (1805–1859), who was Minister of Foreign Affairs.[15]

Gobineau was an unhappy person who did not get along well with others, thought little of his fellow conservatives, and retained an aloof, pessimistic attitude toward life. He served as a diplomat in Germany, Persia, Greece, Brazil, and Sweden. He felt contempt for the French nobility because he believed they were impotent to restore France's grandeur and suffered from a lost vitality brought on by miscegenation with inferior stocks of Europeans.

The racial doctrines of Gobineau are set forth in *The Inequality of Human Races*, published in 1853.[16] Gobineau proposed an elitism based on the alleged virtues of a specific population of whites he described as the Aryans.[17] The term "Aryan" was originally applied to one of the Indo-European language groups that stretches from India to the British Isles. In the 19th century, European scholars variously associated the origin of the Aryans with those who developed Sanskrit, with those who lived in central Asia north of the Himalayas, or with those who lived on the Scandinavian peninsula itself, or at least were of European origin west of Russia. All of these diffusion models (from a center of origin) were difficult to assess because they were based on linguistic analysis. Late 20th-century scholarship adds genetic diffusion to these language models and favors a central Asian model for the Indo-European languages. Nazi ideology favored a Teutonic origin for the Aryan languages and shared Gobineau's belief in Aryan cultural superiority.

His idealized version of their role throughout history is sentimental and unsupported: "Recognizing that both strong and weak races exist, I preferred to examine the former, to analyze their qualities, and especially, to follow them back to their origins. By this method I convinced myself at last that everything great, noble, and fruitful in the works of man on this earth, in science, art, and civilization, derives from a single starting point, is the development of a single germ and the result of a single thought; it belongs to one family alone, the different branches of which have reigned in all the civilized countries of the universe."[18]

The decline of civilizations, he claimed, "lies hidden in the very vitals of a social organism."[19] Gobineau identified the cause: "...Nations die when they are composed of elements that have degenerated." The dying civilization or nation perishes "because it has no longer the same vigour as it had of old in battling with the dangers of life; in a word, because it is degenerate."[20] Gobineau rejected the idea that nations fall because of sinfulness, fanaticism, corrupt morals, or abandoned faith. Using the biological model of F.X. Bichat's discovery of tissues as the components of body organs and a basis for pathological anatomy, Gobineau endorsed the idea that the corruption can be seen within the society, although the source of the corruption was external and biological. "The word *degenerate*, when applied to a people, means (as it ought to mean) that the people has no longer the same intrinisc value as it had before, because it has no longer the same blood in its veins, continual adulterations having gradually affected the quality of that blood. In other words, though the nation bears the name given by its founders, the name no longer connotes the same race. In fact, the man of a decadent time, the *degenerate* man, properly so-called, is a different being, from the racial point of view, from the heroes of the great ages."[21]

Gobineau uses the metaphor of the human body for a race; he describes some of the Polynesian and African races as "embryo societies" permanently arrested at that stage. He enlarges the metaphor, allowing for conquest, fusion, enlargement, and assimilation of nations. Although empires are built in this way, he makes the pessimistic assumption that this leads to their own destruction because "...The human race in all its branches has a secret repulsion from the crossing of blood...."[22] Foremost in his belief is the beneficial contribution of the Teutons, "...Where the Germanic element has never permeated our special kind of civilization does not exist."[23] The lowest race in Gobineau's scale belongs to Oceania, which "has the privilege of providing the most ugly, degraded, and repulsive specimens of the [black] race, which seems to have been created with the express purpose of forming a link between man and brute pure and simple."[24]

GOBINEAU'S MUTATION THEORY OF RACIAL ORIGIN

Gobineau believed some races were stable, like the "...Jewish type" who has "...remained much the same; the modifications it has undergone are of no importance and have never been enough in any country or latitude, to change the general character of the race."[25] Gobineau believed the progenitor of a race was a sudden mutation, like the case he cites of a father (born in 1727) and his four sons who had an unusual scaly or horny skin (possibly ichthyosis).[26] He speculated that conditions may have been more rigorous in the years after the Creation, and this led to the origin of new racial types. Gobineau accepted only 3 races—white, black, and yellow. Among the whites he included Caucasians, Semites, and Japhetics; the blacks were the Hamites; and the yellow race included Altaic, Mongol, Finn, and Tatar variants. He rejected Blumenbach's classification of 28 races and Prichard's assertion of 7 races because "both of these schemes include notorious hybrids."[27]

Among Gobineau's other racist beliefs was his claim that white people have beauty and other races at best approach this ideal; that whites are more muscular than other races; that Italians are more beautiful than Germans, Swiss, French, or Spaniards; and that the English are more beautiful than Slavs. He attributes to Benjamin Franklin the racist remark that the negro is "an animal who eats as much, and works as little, as possible."[28] At the same time he recognized that individual blacks may surpass average Europeans and denied that "every negro is a fool." He justified this more flexible statement on blacks because if he did not do so "I should have to recognize, for the sake of balance, that every European is intelligent; and heaven keep me from such a paradox."[29] Gobineau's biases are typical of the sterotypic bigot. He describes the negro as "careless of life," the yellow race as "mediocre in everything," and the white race as energetic, courageous, idealistic, and having an "instinct for order."[30] He justified intermarriage for the development of artistic talent which was otherwise missing in the three races. "Artistic genius," he claimed, "arose only after the intermarriage of white and black."[31]

Gobineau's prejudices are a mixture of cultural bigotry and racism. Although he did not originate many of his racial stereotypes, he has the distinction of having formalized a racist system and popularized it with the self-deceptive pretense that his findings were based on scholarly study.

INTOLERANCE IN NORTH AMERICA

In the New World, bigotry, more perhaps than tolerance, characterized the colonial history of North America. Although many Protestant denominations fled persecution in Europe and sought religious freedom in the New World, that freedom was seldom extended to those of other faiths who settled in the same colony. As Gustavus Myers perceived the state of religion in America in the 17th century, the various denominations were "engaged in contention among themselves, the various Protestant sects in American colonies sought in common the suppression of Catholics; and Protestant and Catholic alike desired the elimination of freethinkers."[32]

The European past was filled with the horrors of religious persecution. From the time of the launching of the Crusades by Pope Gregory and his instruction to convert the infidels, Jews had been expelled, massacred, deprived of property, and banned from professional occupations in different parts of Europe. There were massacres of Catholics by Protestants, Protestants by Catholics, and heretic Protestant sects by those Protestants claiming exclusive control over religious belief. Henry VIII's break with the Catholic church led to persecutions of Catholics, confiscation of monasteries and convents, and the establishment of an Anglican church. Catholic retaliation against Protestants, especially in France, was severe. On St. Bartholemew's day, 23 August 1572, about 30,000 Huguenots were massacred by French Catholic armies under Charles IX. Hysteria against Catholics in England reached a fever pitch on 5 November 1605 with the alleged gunpowder plot of Guy Fawkes to restore the papacy to Great Britain.

Prior to the American Revolution, those persecuted for their religious beliefs fled or were expelled and set up new colonies. When Roger Williams, a Puritan minister, preached in 1635 for the separation of church and state and the right of individuals to exercise their own religious consciences, at a time when Massachusetts was a theocracy, he was expelled. He founded Rhode Island, but even that colony failed to follow his religious and political beliefs. At the time of the ratification of the U.S. Constitution, only 3 of the 13 states allowed Catholics to vote (Pennsylvania, Delaware, and Maryland).[33] Anti-Catholic prejudice remained strong throughout the 19th century. Samuel F.B. Morse, celebrated as an artist and the inventor of the telegraph, was the author in 1835 of *Foreign Conspiracy against the Liberties of the United States.*[34] Morse believed there was a Jesuit plot to take over the United States through increased immigration and subsequent voting as a Catholic bloc to elect Catholics to as many political offices as possible.

THE NATIVE AMERICAN PARTY AND THE KNOW-NOTHINGS

Bias was not limited to religion. The Native American party, founded in 1835, sought exclusion of foreigners from elected office; they wanted a 21-year residency for citizenship (instead of 5). They looked upon the bulk of immigrants, most of them English, as a lower class of competitors who

Samuel F.B. Morse's theories of a Jesuit plot to overtake the United States supported anti-Catholic prejudice in the 1830s. (Courtesy of the Library of Congress.)

were willing to take lower wages and the jobs the Native American party supporters wanted.[35] By 1841 the Native American party had become anti-Catholic, and in 1843 the Irish were singled out as America's major threat. The bigotry gave way to violence in 1844, with the burning of Catholic churches, schools, and homes in Philadelphia by rioting supporters of the Native American party.

A more sinister group fueled by bigotry was the Know-Nothings, a secret society whose members had to be Protestant of Protestant birth. They sought election of Protestants and were angered by President Pierce's appointment of a Catholic to the cabinet position of Postmaster General. In 1854 the Know-Nothings had their strongest following and elected nine governors, eight Senators, and 104 members of the House of Representatives. Samuel Morse embraced the Know-Nothings, and later, when some of its members supported the abolition of slavery, he broke the momentum of the society by forming, and serving as president of, the American Society for Promoting National Unity. Morse believed that slavery was ordained by God.[36]

THOMAS WATSON'S RACISM AND THE GROWTH OF THE KU KLUX KLAN

While anti-Catholicism was endemic throughout the latter half of the 19th century, it was fused to racism by political bigots. One of the notorious bigots in American politics was Thomas E. Watson (b. 1856), a lawyer, publisher, and later a congressman and senator from Georgia.[37] At his peak popularity he was the Populist party nominee for vice president in 1896 and presidential candidate in 1904. His editorials grafted anti-Catholic bigotry with racism against blacks. From 1913 on he added Jews to his list of despised people. A Jewish businessman, falsely accused of the rape of a young female employee, was kidnapped from jail and lynched by a mob. Watson's editorial commentary in his magazine reveals the intensity of his bigotry: "Leo Frank was a typical young Jewish man of business who lives for pleasure and runs after Gentile girls. Every student of sociology knows that the blackman's lust after the white woman is not much fiercer than the lust of the licentious Jew for the Gentile."[38]

Watson's views were not very different from those of the Ku Klux Klan. The Klan was founded in 1866 in Tennessee, an offshoot of pre-war pro-slavery, anti-North, and pro-secession societies. The original Klan was disbanded by federal troops in the 1870s after a campaign of terror to intimidate blacks from voting or participating in civic government. The Klan's revival was initiated by Joseph Simmons, in Georgia, in 1915. It embraced the views of Watson, and throughout the South it was candid in its bigotry against Catholics, blacks, and Jews. In the 1920s the Klan extended its recruitment program to northern states, but there its bigotry was more muted and it portrayed itself as a patriotic organization promoting American virtues, white supremacy, restrictive immigration, and anti-Communism. By 1924 the Klan had about 5 to 8 million members and its power was felt with the election of Klan-supported candidates in Indiana, Maine, and Colorado. The Klan campaigned against "evolution, atheism, modernism, and Communism."[39]

The Grand Dragon of Indiana, D.C. Josephson, disgraced himself and the Klan through a sex scandal resulting in the death of a mistress. Josephson's arrest, and the subsequent investigations of his activities by the press, tarnished the image of the Klan in the northern states. The Klan's difficulties were compounded by revelations of misuse of dues from members, and by 1927, membership had fallen to 321,000. Despite repeated attempts to make a comeback, the Klan remained a small, but potentially violent group, largely restricted to the southern states.

Racism directed against blacks did not require the Klan for its spread or perpetuation. Those views had been generated as part of the rationalization for the slave trade and the long history of slavery in the New World.

FOOTNOTES

[1] Nepotism is discriminatory in both ways. It is blatantly so when one relative hires another at the expense of a better-qualified person, especially in government positions. It is less obviously so in the university where a talented spouse might be forbidden a job at the same institution because it might appear to be nepotism to hire that person. Until the 1970s, many universities rigorously enforced that rule, fearing the loss of two faculty if one should receive an attractive offer from elsewhere and fearing the recruitment of a mediocre candidate as the price of getting a much desired one. After the 1970s, so many husbands and wives were both professionals that it became difficult to hire a candidate without finding a job for the professional spouse. Nepotism would only apply in conflict of interest situations (it would not be good practice for a spouse, chairing a department, to hire his or her mate in that same department).

[2] In the Old Testament, marriages between Jews and non-Jews were quite common in the first two millennia after the Jews were identified as a separate people who had made a covenant with God. Joseph married an Egyptian (Genesis 12:45), Judah a Canaanite (Genesis 38), Moses a Cushite (Numbers 12:1), Mahlkon and Chilion took Moabite women (Ruth 1), and any woman captured in war was acceptable for marriage (Deuteronomy 21:10). This tolerance for intermarriage came to an end after the return to Jerusalem from a long period of exile after Babylonian conquests.

Most of the land was owned by non-Jews, and intermarriage would have led to assimilation and the disappearance of the Jews, the potential compromising of their religious beliefs, and the feared adoption of idolatry.

[3] Note that this demand by Ezra is made on behalf of a minority religion of its own people in a land where they would comprise a small proportion of the inhabitants. In traditional bigotry, it is a majority religion that makes demands on a minority, usually inhibiting its practice of religion. The request to avoid intermarriage is repeated in Nehemiah 10:32: "...We would not give our daughters unto the peoples of the land, nor take their daughters for our sons." Nehemiah's justification is based on the consequences of assimilation that he witnessed: "In those days also I saw the Jews who had married women of Ashdod, of Ammon, and of Moab; and half their children spoke the language of Ashdod, and they could not speak the language of Judah, but the language of each people. And I contended with them and cursed them, and beat some of them, and pulled out their hair; and made them take oath in the name of God: 'You shall not give your daughters unto their sons, nor take their daughters unto your sons, or for yourselves" (Nehemiah 13: 25).

Note the reference to Ashdod, the major center of the Amalekites. By showing that intermarriage is tolerant even to the point of ignoring God's sworn enemy, the prophets can denounce the habit as a prohibited act (the Amalekites were referred, at least in anger, as mamzerim). See Chapter 1 for a discussion of the unfit peoples in the Bible.

[4] The translations reflect the way present-day cultures influence the words selected for equivalent meaning. The terms race, seed, and people have very different connotations. Even in context, the precise meaning of the Hebrew term "Jews" as a population in antiquity may not convey the same meaning as these modern words.

[5] Biologists differ on the interpretation of the term race when applied to humans. Some prefer the term variety because the term race is used for "incipient species" or "isolates" or geographically separated populations of the same species that have compiled enough differences that they can be distinguished, but not quite enough that they have formed barriers to breeding when brought together. All humans can breed with one another and thus do not fall into the category of "incipient species." Breeds, as in domesticated animals, would have been appropriate, but because of its animalistic image, the phrase "human breeds" is rarely used. Breed is self-explanatory—the differences are genetic, but they are not of species rank; they are temporary and subject to change. Racism makes its impact on people by elevating the status of human breeds to races, implying that there are fixed traits that should remain with those groups called races. "Breeds" implies that there is one species (e.g., *Felis domestica*, the common house cat), but the breeds are arbitrary collections of traits established by human selection or accidental accumulation.

[6] It is one of the difficulties fundamentalists face when they are forced to take vaguer terms of translation (like biblical "kinds") and equate them to modern equivalents. Thus "kind" can be a species, a genus, or a family (as in the cat family, which includes lions, leopards, tigers, ocelots, and domesticated house cats). Some fundamentalists use the narrowest (species) and others the broadest (family) meaning when interpreting what the original creation included.

[7] Heinz Goerke, *Linnaeus: A Modern Portrait of the Great Swedish Scientist* (Scribner's, New York, 1973) gives an account of Linnaeus's life and career.

[8] Carl Linnaeus, *Systemae Naturae* (Leyden, 1735).

[9] Goerke, *Linnaeus*, p. 95. Linnaeus interpreted hybrids as evidence of the continuing process of creation by the Creator.

[10] Many critics of the Linnaean system felt that evolutionary theory would not have been as persuasive as it was in the 19th century if Linnaeus had put humans in a separate category from the primates.

[11] Marvin Harris, "Race," *International Encyclopedia of the Social Sciences* 13(1968): 263–268, p. 265.

[12] Ibid. p. 265.

[13] D.R. Stevenson, "Joseph Arthur Gobineau," *Historical Dictionary of the French Second Empire 1852–1870*, edited by W.E. Echard (Greenwood Press, Westport, Connecticut, 1985). Also R. Thenen, "Joseph Arthur, Comte de Gobineau" in *International Encyclopedia of the Social Sciences* 6(1968): 193–194.

[14] Stevenson, *Gobineau*.

[15] Tocqueville was different in philosophy and political outlook, but he enjoyed Gobineau's work habits and companionship.

[16] Arthur de Gobineau, *The Inequality of Human Races*, translated from the 1853 French by Adrian Collins (Howard Fertig Co., New York, 1967).

[17] The term Aryan was applied to a language relation (Indo-European), a culture, and an anthropological population (variety, race, tribe, or breed), often with the three usages being confused.

[18] Gobineau, *Inequality of Human Races*, p. xiv.

[19] Ibid. p. 23.

[20] Ibid. p. 24.

[21] Ibid. p. 25.

[22] Ibid. p. 29.

[23] Ibid. p. 93.

[24] Ibid. p. 107.

[25] Ibid. p. 122.

[26] Curt Stern explored the Lambert pedigree, used in the 18th century as an example of human heredity. Some of the Lamberts were exhibited in freak shows. The family history was "improved" to make it look as if fathers passed this to their sons. It was used as a possible example of Y-linked inheritance until Stern, in 1957, showed this was not true. The Lamberts had an extreme form of ichthyosis hystrix. See Stern, "The problem of complete Y linkage in man," *American Journal of Human Genetics* 9(1967): 147–166.

[27] Gobineau, *Inequality of Human Races*, p. 146.

[28] Ibid. p. 180.

[29] Ibid.

[30] Ibid. p. 207.

[31] Ibid. p. 208.

[32] Gustavus Myers, *History of Bigotry in the United States* (Random House, New York, 1943).

[33] Ibid. p. 111.

[34] Ibid. p. 163. He published it in 1834 in the *New York Observer* under the pseudonym Brutus, but used his own name in 1835 when he expanded it to book form.

[35] Ibid. p. 166.

[36] This was a common theme of racists. Making human inferiority a religious mandate helped remove the guilt of illegal dominion over a fellow human being.

[37] The success of a demagogue is often based on the frustration of voters with problems they cannot solve. The appeal to racism, religious bigotry, and political suspicion is often a highly successful technique when the demagogue's opponent cannot provide answers that seem like a quick, cheap, and effective solution. Watson, like Hitler, found this appeal to prejudice to be very effective.

[38] Myers, *History of Bigotry*, p. 259.

[39] Ibid. p. 303.

17

Jews as People, Race, Culture, Religion, and Victims

B IOLOGISTS USUALLY SEE LIVING THINGS, including humanity, through an evolutionary perspective. Historical time is a brief moment in the much longer geological time that contains the past history of the remains of all preceding life. Biologists, in hope of interpreting the past, have amassed immense quantities of data from the fossil record, from comparative anatomy, from genetics, and from the sequences of proteins and nucleic acids of plants, animals, and microbes. The pace of evolution is slow, even in those moments of rapid evolution (involving hundreds or a few thousands of generations) that occasionally burst forth when the earth's environment undergoes dramatic change. A species, to an evolutionist, represents a frozen moment in a continuum that fades into prior ancestors and future descendants who merit new species names. The break between one species and another may not be sharp and, in any one time period chosen, the look back and forward reveals the same illusion that not too far in either direction there are closely related ancestors and descendants.[1]

The term race, to the biologist who appreciates the dynamic state of a species, is at best a temporary collection of variations that reflect different adaptations of a single species. The races may fuse, blend, develop new races, or, given the right circumstances and sufficient time, lead to new species. All humans belong to one species, *Homo sapiens*, and a flow of genes trickles back and forth among Inuit and Hottentot, Ainu and Papuan, Swede and Thai. The genetic bridges that link continents may have an excruciatingly slow movement of genetic traits across them, but for humanity, that is relatively fast compared to the far greater isolation between continents of most plants and animals. Humans migrate; they sail; they explore; they trade; they exchange goods; they conquer. In historical times the flow of genes, compared to prehistoric times, is a torrent.[2]

RACE AS AN ANTHROPOLOGICAL CONCEPT

The concept of human races was given its anthropological legitimacy by Johann Friedrich Blumenbach (1752–1840), a physician, who identified five major races of humanity—white, black, yellow, brown, and red. These roughly corresponded to the populations of Europe, Africa, Asia, Malaysia, and the Americas before the 15th century. Blumenbach did not assign inferiority or social status to his races and rejected the bias that "black men were on a lower level of humanity than white men."[3]

The term race in popular culture has ranged from narrow ethnic identification to broad collections of diverse people sharing a common trait, such as skin color. The term has been assigned by a larger group to a smaller one to set them apart, or it has been self-assigned to create a feeling of community and identity. The Jews represent such a group, at times identifying themselves as a race and at other times, despite their attempts to function as citizens of another country, to be looked upon as an alien race no matter how many generations they had resided there.

JEWISH CULTURES AND POPULATIONS

Jews trace their identification as a people to Abraham, who resided in Ur, a city of Mesopotamian culture that was located between the Tigris and Euphrates Rivers in present-day Iraq. Abraham's two wives and two concubines produced 12 sons, 11 of them forming scattered tribes and the 12th, Joseph, founding the state of Israel. Most scholars identify the Jews as a people with a common culture, language, religion, and history. Although

Hebrew may have been lost in the dispersion of Jews across Europe and the Middle East, the Hebrew alphabet was retained in Yiddish (Germanic), Ladino (Spanish), Judeo-Franco, Judeo-Italian, Judeo-Greek, Judeo-Arabic, Judeo-Persian, and many other Jewish isolates.

The Jews are today classified in three major groups. The largest, 82% of all Jews, are the Ashkenazic or European Jews who migrated from the Middle East to Rome and across the Alps, settling in France and Germany in the early Middle Ages.[4] After the persecutions of the Crusaders in the 11th century, they moved east to areas today designated as the Baltic states, Poland, the Ukraine, Russia, Rumania, and Hungary. Most of the Jews who associated with the classical Greek and Roman Empires moved west across Africa and into Spain to become the Sephardic Jews. Those Jews who moved East into Iran, Iraq, India, Afghanistan, Kurdistan, and other countries nearby constitute the Oriental Jews. Each of these three broad categories of Jews has experienced dispersions that occurred later as the religious and political affiliations of the nations they lived in underwent change. Sephardic Jews were sent into exile from Spain in 1492, and most headed back to North Africa, Egypt, or Turkey.

There were about 16.5 million Jews at the onset of World War II. The ravages of that war, especially the mass murders in the Holocaust, reduced the population to 10 million Jews. Although the population of Jews has fluctuated over the past three millennia, it has never fallen below 1 million since the days of King David and King Solomon. The circumstances that led some Jewish groups to form isolates, however, has created both cultural and biological differences among the Jews. The smallest recognized Jewish community is the Samaritans, only about 500 of whom still survive. Some of the Oriental Jews, like the Karaites (Iraq and Persian Jews), do not have a talmudic tradition. The Falashas (Ethiopian Jews) are also non-talmudic and have developed their own unique religious practices that differ from both the Palestinian and Babylonian traditions.[5]

THE ORIGIN OF JEWISH ETHNIC DISEASES

The dispersion of Jews by pogroms in the Middle Ages and the settlement of the survivors in central and eastern Europe led to expanding populations from initially small isolated groups. Geneticists recognize that such conditions lead to "founder effects," in which rare gene mutations may be present in one isolated group but not in neighboring groups. The mutation then increases as the population increases, leading to an artificially high percentage of the population carrying it compared to the original population prior to its dispersal. A number of ethnic genetic disorders have been identified among Ashkenazic Jews. The best known of these is Tay-Sachs disease, a central nervous system disorder leading to blindness, paralysis, loss of intellectual function, and death usually before the age of four. It is a classical recessive trait, the parents being normal and carriers (heterozygous) for the trait and each of their children having a 25% risk of receiving the two mutant genes from their parents. The gene is present in about 1 of every 25 Ashkenazic Jews.[6] It is much rarer in non-Jewish populations or among Oriental and Sephardic Jews. Other Ashkenazic genetic disorders that are rare among other Jews or non-Jews are Gaucher syndrome, Niemann-Pick syndrome, spongy degeneration of the central nervous system, Bloom syndrome, and dysautonomia.[7] The Ashkenazic Jews who emigrated back to western Europe and to the Americas and South Africa also brought these genes along. In the 1990s, the identification of a gene, *BRCA1*, associated with breast cancer was quickly shown to be disproportionately more frequent in Ashkenazic Jews and is probably a genetic founder effect.

Oriental and Sephardic Jews have their own collection of rare ethnic disorders in higher proportion than among Ashkenazic Jews or among non-Jews. These include the Louis-Bar syndrome (mostly among Moroccan Jews), congenital adrenal hyperplasia (North African Jews), cystinuria (Libyan Jews), Dubin-Johnson syndrome (Iranian Jews), glucose-6-phosphate dehydrogenase syndrome (Kurdish, Iraqi, Iranian, Afghan, Yemenite, and Indian Jews), ichthyosis vulgaris (Indian Jews), and several others.[8]

The existence of ethnic disorders is not unique to Jews. Any isolated population can show these, and those settled in recent times, like the Amish in North America, have several documented recessive traits that appear in disproportionate number because of the small number of Mennonite founding families that came over after the Reformation. Even more recent isolates, such as the inhabitants of Pitcairn's Island (survivors of the mutiny on the HMS *Bounty*) in the South Pacific Ocean or the shipwrecked survivors on Tristan da Cunha (an island in the Atlantic Ocean midway between southern Africa and Brazil) show unique collections of ethnic disorders.[9]

The occurrence of different ethnic diseases among the various Jewish isolates, however, does show how biological variation can arise within a group and establish different combinations and frequencies of genes. Complicating the genetic history are infusions of genes from neighboring, non-Jewish, peoples through rape, intermarriage, conversions, or out-of-wedlock pregnancies over the hundreds of years that a minority population resides in a geographic area.

EARLY HISTORY OF THE PERSECUTION OF THE JEWS

The relation of Jews to their conquerors and host countries has rarely been stable. They were given permission to practice their own religion by Caesar because they supported his Egyptian campaign. They prospered as a separate colony under King Herod. When Augustus became the emperor, he continued that support. Tiberius forbade conversions of non-Jews to Judaism in most of Rome and Jews remained relatively undisturbed through the reigns of Caligula and Claudius. A confrontation with Caligula, who wanted his statue placed in the temple in Jerusalem, was avoided after Caligula's assassination. Conditions deteriorated after that, with Jerusalem put under Syrian rule, and by 70 A.D., the Romans under Titus destroyed the temple in Jerusalem and the Jews were scattered. Whereas some of the conquerors only sought political loyalty, others, like Caligula in his madness, sought universal worship, a demand that the Jews could not meet. Attacks on the Jews after the spread of Christianity were mostly on religious grounds.

The migration of Jews to cities, like Alexandria in Egypt, was often followed by a resentment and distrust that native inhabitants feel to outside minorities coming in. When that distrust is fed by accusations of disloyalty, anti-Semitism develops. This is described aptly in the Bible when Haman, the prime minister of the Persian King Ahasuerus, sought vengeance against Mordecai and his fellow Jews for refusing to bow before him.[10] Mordecai's adopted daughter was Esther, the second wife of King Ahasuerus. Haman used a classical anti-Semitic approach to shake Ahasuerus's confidence in Mordecai: "There is a certain people scattered abroad and dispersed among the peoples in all provinces of your kingdom; their laws are different from those of every other people, and they do not keep the King's laws, so that is not for the king's profit to tolerate them. If it please the king, issue a decree that they be destroyed" (Esther 3: 8–9). Fortunately for the Jews, Haman's plot was eventually exposed and Esther saved her father and the Jews.

It was the exception for Jews, or any minority religion, to be free to practice their faith in most countries of the world until the 18th century. Most countries equated citizenship with a state religion. Jews, Christians, and Moslems alike shared a belief that theirs was the one true religion sanctioned by God. Their coexistence in any one state was always frail and usually short-lived. The Jewish claim to a unique relationship with God was described in Deuteronomy 7: 6–8. "For you are a holy people, dedicated to the Lord your God. He has chosen you from all the people on the face of the whole earth to be his own chosen ones. He didn't choose you and pour out his love upon you because you were a larger nation than any other, for you were the smallest of all! It was just because he loves you and he keeps his promises to your ancestors."[11]

BIBLICAL INTOLERANCE OF HEATHENS

Intolerance against religious foes is as old as the Bible, which describes many wholesale massacres, imposed conversions, expulsions of non-Jewish spouses and children, and the executions of idolatrous priests. These acts are not only carried out by Jews, but Jews were sometimes ordered to do so by God in his rage against Jewish lapses into heathen rituals.[12]

The long history of intolerance of Jews as an alien religion was compounded by the centuries of persecution of Jews as deicides. The anger of the Jewish high priests against Jesus is described in Matthew 26: 65–66 and Mark 14: 63–64. Luke 23: 13–24 adds Herod to those who condemn Jesus, and John 18:12-16 specifically uses Jews in the global sense rather than their priests or council as the group seeking the crucifixion of Jesus. The attack on Jews as killers of Christ was frequently used to fan the persecution of Jews and the confiscation of their property, as well as to justify the exclusion of Jews from the professions and trades practiced by the non-Jewish inhabitants of the cities and countries the Jews lived in. It was not until 1965 that the Vatican rejected collective responsibility of the Jews for Christ's death.

MYTHS USED TO SLANDER JEWS

Among the myths associated with Jews as enemies of Christianity was the belief that the "host" or wafer used in the Catholic mass was tortured by Jews. The red spots seen on these wafers were bacterial colonies, but in the eyes of a superstitious people these were drops of blood shed by the re-tortured body of Christ.[13] Jews accused of this crime were massacred near Berlin in 1243. The Popes usually favored conversion of the Jews rather than their massacre, and their human status and souls were affirmed by Pope Martin V. Despite these official statements of tolerance, Jews were locally accused of fattening and killing babies (by the Greeks and later by Christians).[14] They were also accused of spreading the Black Plague in 1348 and massacred for that alleged crime. But none of these was as malicious in its intent to inflame prejudice against Jews as the belief that ritual murders of Christian children was carried out to make matzos.[15]

Jews served as moneylenders during the Middle Ages because that was an occupation forbidden to Christians, and it served a need for the many loans that medieval rulers sought to finance their state affairs. They were forced to live in ghettos in Venice, and they were forced to wear a yellow or red badge in the 13th century.[16]

CHAMBERLAIN'S TEUTONIC THEORY OF CIVILIZATION

Modern anti-Semitism stems from Gobineau's hierarchy of racial types with a superiority of the Aryan race and an implied or outright assigned inferiority to all other races. Jews are identified as a race, but no special hostility is directed at them by Gobineau. Similarly, Houston Stewart Chamberlain, an English-born but naturalized German and son-in-law of Richard Wagner, endorsed the Teutonic theory of racial superiority. In his *Foundations of the Nineteenth Century*, Chamberlain attributes to Jews "indifference and unbelief" and "historical narrowness,"[17] but "what gives the Jew a special claim to our honest admiration is his unceasing struggle against superstition and magic...."[18] Despite these mixed impressions, Chamberlain was firm that "the less Teutonic a land is, the more uncivilized it is."[19] The Teuton's special gift is imagination which, Chamberlain claimed, arises from individualization. "Every individual person reveals progress and degeneration, every individual thing likewise—whatever its nature—the individual race, the individual nation, the individual culture; that is the price that must be paid for the possession of individuality."[20] Chamberlain's Teuton embraces discovery, science, industry, economy, politics, the church, and culture in the broad sense. The Teutonic temperament, he asserted, is restless, curious, and filled with love of nature as the source of knowledge.

Modern anti-Semitism is often associated with the publication, in 1879, of the first anti-Semitic bestseller, *The Victory of Jewry over Germany* by Wilhelm Marr. Marr was also the first person to coin the term "anti-Semitism" as a synonym for a dislike or hatred of Jews. Marr, curiously, was sympathetic to Jews in his early adult life, and his second wife was Jewish. His third wife was half-Jewish. At the end of his life, Marr regretted his anti-Semitism and renounced it, but although he was penitent, the name of Scrooge from *A Christmas Carol* will always conjure up the miserly, hardened, and dyspeptic life he led rather than his near-deathbed conversion to a generous benefactor of the poor. So too, Marr's name will inescapably be associated with the virulent anti-Semitism his book stimulated and reinforced.[21]

Anti-Semitism became widespread in Germany after 1870, but it was in the early 20th century that a worldwide anti-Semitic movement began. Jews were becoming more successful in commerce, politics, the arts, and the sciences as their opportunities to become full-fledged citizens in their host countries increased. They were frequently seen by envious non-Jews as being unfair competitors in England, France, Germany, Italy, Russia, or the United States. In this form of anti-Semitism, Jews were seen as clannish, driven by a monomania for wealth and power, manipulative, and conspiratorial.

JEWISH RESPONSES TO ANTI-SEMITISM

Anti-Semitism was used to accuse the Jews of betraying France during the Franco-Prussian War.[22] It was used to portray the Jews as an alien nation, loyal only to fellow Jews and subversive of the nations they lived in. The hostility that flared up in pogroms in Russia and Poland and the discriminatory practices in Europe's most cultured countries made it difficult for Jews to respond effectively. Some chose the path of assimilation, hoping that their outward European clothes and habits would make them be seen as loyal citizens differing only in religion. Some chose the possibility of emigration from lands of terrible persecution to lands where persecution was less intense, such as the United States, Canada, or Argentina. Still others began thinking of a homeland of their own, preferably in Palestine in the land once ruled by Jewish kings. Max Nordau, the acid-tongued critic of modern European culture, chose Zionism as the best response to anti-Semitism in his adopted country, France. He co-authored with Gustav Gottheil *Zionism and Anti-Semitism*[23] and gave his reasons for wanting a country where Jews could establish their own nation. "The new Zionism has grown in part only out of the internal impulsions of Judaism itself, out of the enthusiasm of modern educated Jews for their history and martyrology, out of the awakened consciousness of their racial qualities, out of their ambition to save the ancient blood, in view of the farthest possible future, and to add to the achievements of their forefathers the achievements of their posterity."[24]

The combination of nationalism throughout Europe and the growing anti-Semitism that accompanied it were the reasons Nordau gave for his own interest in the Zionist movement. Nordau's vision was idiosyncratic. He wanted a Zionism without mysticism; he repudiated the Reform movement, in which Jews assimilated in their adopted countries; but most of all he saw a Jewish state as one founded on a racial theory of the Jewish people. He pointed out how assimilated Jews in Munich protested and succeeded in preventing the first International Zionist Congress from meeting there (they met in Basel in 1897 instead). Whatever the religious and spiritual intensity of the new nation would be, Nordau favored it because "only the return to their own country can save the everywhere hated, persecuted, and oppressed Jewish nation from physical and intellectual destruction."[25]

A Fraudulent Pamphlet Defames Jews

Gottheil identified the anti-Semitism movement as a reaction to the Jews' refusal to be converted to Christianity; their dispersion as a minority in every land; and their success in business, science, law, politics, arts, press, and finance; their sobriety, character, and energy.[26] That reaction against the Jews was soon expressed in print. An anonymous pamphlet, *The Protocols of the Learned Elders of Zion*,[27] appeared in Russia about 1905; it was reprinted in many languages and used as evidence of an international plot by Jews to conquer the world.

The *Protocols* were prepared by the Czar's secret police and used to offset the world criticism of the pogroms against the Jews in Russia. The document was prepared from a satire (non-anti-Semitic) that had appeared a generation earlier lampooning the French emperor, Louis-Phillipe. The major theme of the protocols was the conspiratorial nature of Jews who had an agenda for world conquest through a takeover of the world's newspapers and magazines, its financial centers, the real estate of its major cities, and the positions of power in Europe and the Americas. Each protocol stressed one of the anti-Semitic beliefs then current in late 19th-century Europe. Thus:

The seventh protocol: "To each act of opposition we must be in a position to respond by bringing on war through our neighbors of any country that dares to oppose us, and if these neighbors should plan to stand collectively against us, we must let loose a world war."[28]

The ninth protocol: "We have misled, stupefied, and demoralized the youth of the Gentiles by means of education in principles and theories patently false to us, but which we have inspired."[29]

The tenth protocol: "To wear everyone out by dissensions, animosities, feuds, famine, inoculation of diseases, want, until the Gentiles see no other way of escape except an appeal to our money and power."[30]

The fourteenth protocol: "When we become rulers we shall regard as undesirable the existence of any religion except our own, proclaiming One God with Whom our fate is tied as The Chosen People, and by Whom our fate has been made one with the fate of the world. For this reason we must destroy all other religions. If thereby should emerge contemporary atheists, then, as a transition step, this will not interfere with our aims."[31]

THE *DEARBORN INDEPENDENT* PROMOTES ANTI-SEMITISM

The *Protocols* were still being used in the mid-1950s by Gerald L.K. Smith, the national director of the Christian Nationalist Crusade.[32] In

1957, he republished excerpts of the *Protocols* with the repudiated anti-Semitic articles of the *Dearborn Independent* published by Henry Ford, Sr. and his staff. This newspaper, published in 1921, was based on Ford's alleged fear of competition from Jewish-owned businesses or Jewish-inspired labor unions. Ford eventually signed a letter repudiating the anti-Semitic issues of the *Dearborn Independent* after an outcry of protest and to avoid the possible damage that might arise from a boycott of his automotive products.

Smith republished the long-banned and scarce copies of Ford's newspaper in *The International Jew: The World's Foremost Problem.*[33] In his introduction, Smith claims he interviewed Ford in 1940 and Ford denied having apologized for his *Dearborn Independent.* Ford, according to Smith, blamed the caving in to the public outcry on his associate, Harry Bennett, who signed Ford's name without his permission because Bennett feared the company would be damaged. Smith was presented a Morocco leather-bound set of the *Dearborn Independent* by one of Ford's "inner circle staff members."[34]

In these articles, the themes of the *Protocols* are extended and the major industries (entertainment, cotton, oil, steel, alcohol, finance) "are in control of the Jews of the United States either alone or in association with Jews overseas."[35] The success of the Jews is attributed to their "adhesiveness," and their control over the money markets creates ill will. Jews are accused of being materialistic, wanting to "get" money rather than "make" money, the former an act of avarice and the latter a legitimate product of a "sense of service." The "getting" of money breeds an artificial division between capital and labor and leads to loss of worker loyalty.[36] The authors of the articles deny being anti-Semitic. They claim that the Jews in the International Conspiracy are not Old Testament Jews but a "tribe of Judah" who have perverted the authentic religious tenets of biblical Judaism.[37] Jews are accused of inventing liberalism to subvert governments with a long-range goal of becoming a "super nation." They are also blamed for causing wars. In one of the most infamous claims of the *Protocols*, the authors state that Jews have never been persecuted, such accusations being a coverup for their "virulent attacks upon any and all forms of Christianity."[38] The authors assert there is no such thing as anti-Semitism; "There is however, much anti-Goyism." To offset incredulity even among unsophisticated readers, they follow this charge with the claim that "there is only a very little and a very mild anti-Jewish" feeling among non-Jews critical of the International Jew.[39]

Heaped on the Jews are the corruption of public morals, the control of gambling, the liquor business, prostitution, and popular music and movies that destroy family values. Jews are accused of living off people and not off the land.[40] They are described as parasites on society. Bolshevism, the collapse of Germany in the First World War, and the rise of troublemaking labor unions are blamed on the Jews. The pessimistic conclusion of the authors of the *Dearborn Independent* is that "Jews conquered Germany, England, France, and Russia; they are now taking over the United States."[41]

The United States had and continues to have its anti-Semitic sentiments which emerge occasionally in public. Fortunately, the courts and the Constitutional protections for religious belief have held firm despite the viciousness of anti-Jewish ideas promoted by movements like those headed by Gerald L.K. Smith and publicized by articles like those appearing in the *Dearborn Independent*. Equivalent propaganda against the Jews had far more success, with calamitous consequences, in Germany with the rise of the Nazi party.

FOOTNOTES

[1] There is general agreement among biologists that evolution is real (in the same sense that history is real) and that life on earth, no matter how many years it extends into the past, has left some of its record in the earth's sediments. There is also general agreement that any theory of evolution involves some form of natural selection (in genetic terms, the survival of certain genotypes over others), but there is disagreement about both the pace of evolution and the biological mechanisms that lead to differences above the species level (in taxonomic terms, new genera, families, orders, classes, phyla, or kingdoms). Some biologists (neo-Darwinians of the 1920s to 1960s such as Julian Huxley, R.A. Fisher, Sewall Wright, J.B.S. Haldane, Ernst Mayr, H.J. Muller) believe there is an imperceptible gradation from one species into another. Others (proponents of "punctuated equilibrium" such as Stephen Jay Gould and Nils Eldridge) believe there is little speciation over long periods of time and then certain favored environments or circumstances lead to rapid change. Even the rapid change, however, is measured by hundreds or thousands of generations rather than the hundreds of thousands of generations often required for the gradual process of the neo-Darwinians. Few biologists believe in a de novo origin of species.

[2] In historical times there has been some shifting of human populations to form what some anthropologists look upon as new races. The highly miscegenated Latin

American is such a population. See *Race Mixture in the History of Latin America* by Magnus Morner (Little, Brown & Co., Boston, 1967). Simon Bolivar solved the vexing problem of racial classification (white versus black in North American classification of miscegenated people) by using the definition Latin American for anyone with a Hispanic ancestor, regardless of skin color and appearance.

[3] Walter Baron, "Johann Friedrich Blumenbach (1752–1840)," *Dictionary of Scientific Biography* 2: 203–205. See p. 204.

[4] This is the theory most commonly accepted among Jewish historians and scholars. There is a curious interpretation resurrected and popularized by novelist Arthur Koestler. In his book, *The Thirteenth Tribe*, he argues that the Ashkenazic Jews arose from an Aryan group in the Caucasus, the Khazars, who were converted to Judaism to avoid conflict with Christian neighbors on one border and Arabic neighbors on another. The Jews at that time (about the 8th century) were tolerated by both these major religions. There is little direct evidence to support this theory.

[5] The talmudic tradition includes commentaries by rabbis over the past two millennia. Rabbinic courts frequently ruled on controversies brought to the religious elders for settlement. The views of different rabbis who commented on Jewish law and traditions are often consulted when a decision is made in such difficult cases. Because of the diaspora of the Jews over the past three millennia, quite different traditions arose.

[6] In genetic terms, the normal allele ts^+ is dominant and the mutant allele, ts, is recessive. The parents of a child with Tay-Sachs disease are both heterozygotes (the parental cross being represented by P $ts^+/ts \times ts^+/ts$). Their affected child is ts/ts and normal siblings will be ts^+/ts or ts^+/ts^+. If the gene is present in 1 among 25 Jews, then the odds of 2 such Jews encountering each other are $1/25 \times 1/25$, and the chances they will have a child with Tay-Sachs disease is $1/25 \times 1/25 \times 1/4 = 1/2500$. These are relatively frequent events compared to most recessive mutations in the human population. Tay-Sachs is one of many dozen disorders in which an enzyme is missing from the cellular organ of digestion, the lysosome, and these disorders are called lysosomal storage disorders because the product that should be digested remains inside the lysosome and forms engorged crystalline masses that eventually lead to cell death. The defective enzyme, in this case, is hexoseaminidase A, and tests for the presence or absence of this enzyme can be used to detect heterozygotes as well as homozygotes lacking the enzyme altogether in the newborn or in embryos.

[7] Gaucher's is a lysosomal storage disorder that affects the bone and vascular systems. It is not usually fatal, but it can cause debilitating fractures and massive growth of the spleen. Niemann-Pick is a lysosomal storage disease of the nervous system with symptoms similar to Tay-Sachs disease, causing an early death among affected infants. This is also true for spongy degeneration of the nervous system. Bloom syndrome affects an enzyme that repairs broken chromosomes, and its absence leads to stunted growth, a high incidence of childhood cancers, and skin that is extremely sensitive to sunlight. Dysautonomia is a nervous disorder in which slurring of speech with inappropriate autonomic nervous system functions, including excessive sweating, reduced production of tears, reduced sense of pain, reduced

sense of taste, and choking when swallowing are commonly encountered. It is not known whether it is a coincidence or causally significant that four of the Ashkenazic disorders are lysosomal storage diseases. For a detailed account of the clinical genetics of these disorders, see Richard M. Goodman's *Genetic Disorders Among the Jewish People* (Johns Hopkins University Press, Baltimore, 1979).

[8] The Louis-Bar syndrome is also called ataxia telangiectasia; it includes a paralysis and high incidence of skin cancer and sensitivity to the sun. Like Bloom syndrome, it is a DNA repair enzyme defect. The adrenal hyperplasia causes XX embryos (who are normally female) to be born with ambiguous or male genitalia, but they have ovaries and not testes. Cystinuria is an amino acid disorder that causes severe kidney stone formation and renal damage by the second or third decade of life. The Dubin-Johnson syndrome is a chronic jaundice with an onset about the second or third decade of life. Glucose-6-phosphate dehydrogenase deficiency is an X-linked defect that is ordinarily benign and may confer a milder course of certain forms of malarial disease. It causes severe destruction of red blood cells when cyclic compounds, including aspirin, are introduced into the blood. Certain foods, like fava beans, can also cause severe anemia among males with this defect. Ichthyosis vulgaris is a scaly condition in which the skin looks like dried fissured mud, the skin crusts being about a half inch (5 mm) in length or width. There are many other such disorders discussed in Goodman's monograph. What is of significance is the striking difference in the collection of disorders of each major branch of the Jews. This is consistent with a founding effect from a dispersed number of isolated small groups who served as progenitors of present-day populations (see footnote 9).

[9] The probable origin of most ethnic diseases is the founder effect, in which small isolates grow to a very large size, and the chance inclusion of a rare recessive allele in the founding group becomes disproportionately amplified compared to other populations of similar large size that had no such origin from a small isolated group. Selection may play a role, as in the allele for sickle cell anemia, which can confer better survival to those who are infected with malaria. Attempts to relate Jewish ethnic disorders with extinct or once prevalent diseases (e.g., Tay-Sachs heterozygotes are better at surviving leprosy or tuberculosis) have not been successful. For a nontechnical account of human genetics, see my non-science-major's text, *Human Genetics* (D.C. Heath & Co., Lexington, Massachusetts, 1984).

[10] Haman is the son of Hammedatha, an Agagite and descendant of the Amalekites, the condemned Philistine enemies of God (see Chapter 2). Ahasuerus, the Persian king, is probably Artaxerxes I, the son of Xerxes. The book of Esther describes how this adopted daughter of Mordecai, the King's advisor, gained knowledge of Haman's evil plot to kill the Jews and eventually convinced the king of Haman's malice and duplicity. This brought about the quick execution of Haman and his ten children instead of the massacre of Jews that Haman had hoped for.

[11] The special love that God has for the Jews was the basis for their self-perception as a "chosen people" (variously associated with a holy seed, a holy nation, or a holy people). The translation of the Hebrew into English thus can convey a biological (holy seed), political (holy nation), or religious (holy people) status as the special relationship with God.

[12] In Numbers 25: 16, the command is direct: "Then the Lord said to Moses, 'Destroy the Midianites, for they are destroying you with their wiles. They are causing you to worship Baal, and they are leading you astray....'" Later, in Deuteronomy 7:1–4, Moses instructs his people, "When the Lord brings you into the Promised Land, as he soon will, he will destroy the following seven nations, all greater and mightier than you are: the Hittites, the Girgashites, the Amorites, the Canaanites, the Perizzites, the Hivites, the Jebusites. When the Lord your God delivers them over to you to be destroyed, do a complete job of it—don't make any treaties or show them mercy; utterly wipe them out. Do not intermarry with them, nor let your sons and daughters marry their sons and daughters. That would surely result in your young people's beginning to worship their gods. Then the anger of the Lord would be hot against you and he would surely destroy you."

The brutality of God's command is difficult to interpret in today's values. To "utterly wipe them out" has the same genocidal ring to it as massive retaliation in a nuclear war or the obsessed desire of the Nazi leadership for a "Jew-free" Europe. Those who reject a vengeful God who would include innocent children among the victims of these Jewish wars soften the implications by restricting the "wiping out" to the heathen religion itself and not to the younger children who have not yet practiced it.

[13] Several species of bacteria form red colonies. *Serratia marscesens* is probably the most widely known because of its use as a research organism. These bacteria can also simulate blood drops on wooden sculptings of Christ on a crucifix. In the days before microbiology, such outbreaks were assumed to be miracles or signs of divine warnings or forecasts of historic events.

[14] Vamberto Morais, *A Short History of Anti-Semitism* (W.W. Norton & Co., New York, 1976). See p. 94.

[15] Ibid. p. 104.

[16] Ibid. p. 111. It is remarkable how the prejudices and symbols used in the persecution of Jews have endured for millennia (ritual slaughter of children) and centuries (wearing badges on clothing). For a theological history of Christian anti-Semitism, see William Nichols, *Christian Antisemitism: A History of Hate* (Jason Aronson, Inc., North Vale, New Jersey, 1993).

[17] Houston Stewart Chamberlain, *Foundations of the Nineteenth Century* (John Lane, The Bodley Head, Ltd., London, 1910, reprinted by A. Fertig Co., New York, 1968). See p. 45.

[18] Ibid. p. 123.

[19] Ibid. p. 188.

[20] Ibid. p. 215.

[21] Wilhelm Marr, *The Victory of Jewry over Germany*, 1879. An account of Marr's influence in promoting 19th century anti-Semitism is discussed critically by Albert S. Lindemann "Anti-Semitism: banality or the darker side of genius?" *Religion* 18(1988): 183–195. Lindemann contrasts the theory of anti-Semitism as banality (Hannah Arendt's view in her books on Eichmann and the Holocaust,

discussed in Chapter 19) with the theory of anti-Semitism as a pathology of influential and creative people (Jacob Katz's *The Darker Side of Genius: Richard Wagner's Anti-Semitism* [Brandeis/University of New England Press, Hanover, New Hampshire, 1986]).

[22] This led to the Dreyfus affair. Alfred Dreyfus (1859–1935) was a career military officer associated with the Ministry of War. He was falsely accused, in 1894, of supplying information to the German military command. The anti-Semitic bias of his accusers allowed spurious testimony to convict Dreyfus of espionage. He was sent to prison for life on Devil's Island off French Guiana. The actual letter alleged to have been written to the Germans by Dreyfus was written by a Colonel Esterhazy, but this evidence was initially suppressed until Emile Zola brought the case to national and international attention in a full-page article with the banner headline "J'Accuse!" [I accuse] in the liberal paper, *l'Aurore* (January 13, 1898). The Dreyfus case led to a new trial. The French court martial reduced the sentence to 10 years; eventually Dreyfus was pardoned and later the sentence was repealed. As a result of the Dreyfus case, many intellectuals rallied to denounce anti-Semitism. It also helped Theodor Herzl and Max Nordau in their founding of the Zionist movement.

[23] Max Nordau and Gustav Gottheil, *Zionism and Anti-Semitism* (Scott-Thaw Co., New York, 1904). Nordau's first speech to the founding session of the Zionist congress made him one of the most popular and effective speakers for the Zionist movement. The political crises of two world wars deflected the successful resolution of creating a homeland for Jews in Palestine (then part of Turkey). Some Zionists were willing to compromise with a staged progression (starting with land in Uganda), but Herzl and Nordau were firm in their commitment to a Palestine Jewish state that had the guarantees of the British and other European powers for its security. Nordau's Jewish state would not be founded on either the assimilationist cultural model of Reform Jews nor on the mystical traditions of some of the Orthodox Jews (such as the Hasidic Jews). Nordau believed the Jews to be a race and not just a religious community.

[24] Ibid. p. 17.

[25] Ibid. p. 39.

[26] Ibid. p. 57.

[27] Although the *Protocols* were translated into many languages and millions of copies were sold or distributed, copies are relatively scarce in libraries because those offended by "hate literature" often destroy such documents. Similar charges of an international conspiracy to subjugate the world have been made by American Protestants about Roman Catholics and by European Roman Catholics about Freemasons. The willingness of free individuals in a democracy to surrender that freedom for the security of authoritarian personalities is discussed in Erich Fromm's *Escape from Freedom*.

[28] Anonymous, *Protocols of the Learned Elders of Zion*, in Gerald L.K. Smith, editor, *The International Jew: The World's Foremost Problem* (reprint of the 1921 edition with additions, Christian Nationalist Crusade, privately printed, 1957). I found this copy in a used book shop in Colorado Springs. The spine and the front cover were

sealed with black electrician's tape. The translated title is sometimes given as *Protocols of the Wise Men of Zion*, or *Protocols of the Elders of Zion*.

[29] Ibid. p. 186.

[30] Ibid. p. 144.

[31] Ibid. p. 68.

[32] Anti-Semitic groups accused the Jews of running the country through their alleged control of the press, banks, Wall Street, heavy industry, the universities, and other symbols of power in society. In their extreme form they have even denied that Jesus was a Jew. During World War II these "hate groups" laid low out of fear of being arrested for sedition. After the war, they resumed their anti-Semitic campaign. Many of the arguments used against Jews were weakened when many assimilated Jews (most American Jews) showed support for the many of the same values that the anti-Semitic groups championed: a strong military defense against Communism, opposition to quotas and other compensatory programs that were not based on merit, and a preference for capitalist rather than socialist economic policies.

[33] It is not clear how much Henry Ford himself supported the anti-Semitism of the *Dearborn Independent*. Some of his critics claimed that he feared competition from Jewish industrialists entering the automotive business and used anti-Semitism as a means of discrediting his competitors rather than out of any strong personal bias against Jews. Others claim he, like many Protestants in the 1920s, accepted anti-Semitism as a shared belief that Jews were what the propaganda against them claimed them to be. His most dedicated supporters claim he was duped and did not realize what was being published in his newspaper.

[34] *Protocols*, p. 5.

[35] Ibid. p. 16.

[36] Ibid. p. 24.

[37] Ibid. p. 25.

[38] Ibid. p. 41.

[39] Ibid. p. 111.

[40] Ibid. p. 201.

[41] Ibid. p. 231.

18

The Smoke of Auschwitz

T HE PATH TO THE HOLOCAUST has many tortuous and ancient routes. None of these individual components can serve as a predictor of the tragedy that was to befall Jews, gypsies, homosexuals, the retarded, the infirm, Communists, Slavs, and the outspoken voices of conscience in Germany and its conquered lands against the single-minded racist vision of the Nazi party leadership. As already detailed, the Nazis did not invent euthanasia, eugenics, racism, anti-Semitism, totalitarianism, ethnocentrism, the Aryan myth, sexism, military conquest, or genocide. Because of its authoritarian traditions, its anti-Semitism, and its painful defeat and humiliation after World War I, Germany was the right place for all these tributaries to flow together and create a nightmare state that plunged the world into its bloodiest war and threatened the survival not just of the political character of conquered nations, but of their entire people.

Adolf Hitler (1889–1945) shaped Nazism from many sources as he emerged embittered from the defeat of Germany in World War I. Hitler's father, Alois Schicklgruber, was born out of wedlock in 1837 to Maria Schicklgruber, who later married Johann Heidler. Alois did not receive the Heidler name until 1876, but he preferred using the spelling Hitler. Alois married his third wife, Klara Pozl (his second cousin), and Adolf was their third child but the first to survive.[1]

The young Hitler grew up in an era of sanctioned anti-Semitism. The candidate of the anti-Semitic Social Christian Union had won the election for mayor of Vienna during Hitler's childhood and held that post until

315

1913. Hitler thought he might become an architect or artist but his career was frustrated because he did not pass the entry examinations for the academy he hoped to attend in Vienna. During the war, Hitler was a non-commissioned officer who fought with enthusiasm in the front-line trenches and was struck down by poison gas. Like many young German soldiers, he was dismayed, as he recuperated in a hospital, when Germany was forced to admit defeat. Hitler was sympathetic to a widespread belief that Germany was betrayed by special interests. He included as enemies of Germany the Jews, priests, the Hapsburg monarchy, and Social Democrats. He developed a contempt for weaker organizations and political philosophies that favored pacificism, civil rights, and the protection of the vulnerable classes of society. He distrusted the labor unions then in force because of their ties to the liberal Social Democrats.[2]

Hitler was a free-lance artist and casual laborer after the war, and he spent many hours, while unemployed, in the library reading political and racist literature. He fused elements of pan-German nationalism, authoritarianism, anti-Communism, anti-Semitism, and anti-Hapsburg movements into a political party that would avoid their alleged errors. He recognized the importance of appealing to the masses; he despised parliamentary disorder with coalitions of weak parties; he recognized the importance of using smaller minorities as scapegoats; and he knew that he could not win middle-class support by attacking Catholic or Protestant religious beliefs. The Jews were a prime scapegoat because he saw them not as a social class or a different religion but as a malignant or parasitic race.

HOW HITLER FOUNDED THE NAZI PARTY

While he was in the army during the war, Hitler made the acquaintance of two soldiers who would later be among his closest allies in the new National Socialist German Worker's party (NSDAP) he founded. Rudolf Hess later became his secretary and Max Amann became his publisher. The NSDAP grew out of a German Worker's party that Hitler had infiltrated for army intelligence. He was impressed by many of the goals he himself shared and he left the army in 1920 to devote full time to the party's development. He adopted the swastika as the party's symbol, aligned himself with Ernst Rohm, an early member of the German Worker's party, who had developed a street army from veterans of the war. Hitler was an effective orator and learned he could hold an audience and win them over. He applied his theories of power, hierarchy, obedience, and mass appeal through symbolism and found this combination rapidly swelled the membership of the NSDAP. In 1921 he demanded to run the party or resign. He got his way and appointed Amann to run the party's business affairs and organize a newspaper, the *Volkischer Beobachter* (The People's Observer) as its official voice.[3]

Hitler led 15,000 of his supporters in an attempt to overthrow the Bavarian government in Munich in 1924. The "putsch" failed and Hitler served nine months in jail for his insurrection. During this time, he wrote *Mein Kampf*, a savage attack on democracy, the Jews, the Versailles Treaty, Slavs, Communism, and other enemies he hoped to destroy. He sketched his dream for a Germanized Europe and the triumph of his party's favored Aryan culture over all others.

Hitler's political fortunes were greatly assisted by the chaotic economy created by a runaway inflation in the early 1920s and the collapse of the world economy in 1929. The Weimar Republic, as Hitler hoped, proved ineffective with its many splinter parties representing polarized goals. Hitler laid Germany's woes to those enemies he had held in contempt, and as the problems of unemployment and turmoil increased, Hitler's appeal as the person who could take charge of the nation and restore it to its former glory found an ever larger electorate favorable to his views.

ORIGINS OF MODERN GERMAN ANTI-SEMITISM

The rise of the Nazi party attracted anti-Semitic, pan-Germanic, and eugenics movements that had long predated Hitler's political career.[4] German anti-Semitism had been endemic for centuries, and Jews suffered periodic persecution or expulsion, reaching a peak in 1096 during the Crusades, when zealous Christians identified all non-Christians as infidels. Conversion was usually sufficient to stay the loss of livelihood or life itself. Modern, politicized, anti-Semitism arose during the 1870s. Jews were perceived not as a religion to be converted to Christianity but as a people and culture to be shunned or expelled. They were identified as a racial strain whose mixture with German "blood" would lead to degeneration, as portrayed in Gobineau's racial views. This racial theory of Jews received popular acclaim after the publication in 1873 of Wilhelm Marr's pamphlet *The Victory of Jewry over Germany*, discussed in Chapter 17. Marr accused Jews of strangling German finance and monopolizing the media. Marr made much of the rise of Benjamin Disraeli (1804–1888), who later became Earl of Beaconsfield. Disraeli was a Jew by birth who was converted to the Church of England in 1817 after his father had a falling out with the leadership of London's Portuguese (Sephardic) synagogue. Disraeli became a leader of the Tory party (conservatives) and served as prime minister of Great Britain. Marr also envied the Jewish successes in France and Germany as they entered the professions and assumed middle-class respectability and power. Marr's anti-Semitic book was reprinted in twelve editions, and Marr enjoyed both the success and the income his book achieved. In 1878, he founded the Anti-Semitic League, the first organized group publicly committed to this prejudice. The term anti-Semitism he introduced has survived ever since. Marr's movement attracted both Roman Catholic and Protestant conservatives, who identified Jews with liberal movements. As is often the case in irrational prejudice, the Jews were simultaneously attacked as capitalists exploiting workers and as socialists destroying the fabric of society. Anti-Semitism was embraced by teachers and scholars, including the eminent historian Heinrich von Trietschke, who popularized the slogan "Die Juden sind unser Ungluck" (The Jews are our misfortune).[5]

The political opportunism of anti-Semitism was exploited by Prince Otto von Bismarck, who attracted the conservative and pro-military segments of German society and tolerated their anti-Semitism. Bismarck was not an ideologue, and, after having won the power he sought, he did not pursue the efforts by anti-Semitic organizations to ban the immigration of Jews nor to deprive German Jews of their rights as citizens.

ORIGINS OF GERMAN EUGENICS

German eugenics had its formal origins in 1904 when Alfred Ploetz (1860–1939) founded the German Society for Racial Hygiene. Ploetz's term was based on the earlier movement for public health or hygiene that had been pioneered by Rudolf Virchow (1821–1902). Virchow was a

Alfred Ploetz founded race hygiene shortly after the start of the 20th century. Ploetz's views were complex and resembled a mixture of the views of Herbert Spencer, Marxism, and the American eugenics movement. His unfortunate choice of race, rather than the individual or humanity as a whole, made it an easy target for incorporation by racist ideology. (Reprinted, with permission, from R. Proctor [1988] *Racial Hygiene: Medicine under the Nazis*, Harvard University Press, Cambridge, Massachusetts.)

prominent physician who supported the germ theory of infectious diseases and who sought public programs to improve the quality of life among the lower classes, who were most seriously ravaged by disease. He is best known to biologists for his contributions to pathology. He correctly identified cancer as a cellular disease, and he formulated the cell doctrine, which holds that all cells arise from preexisting cells. Virchow

Rudolf Virchow had an enormous influence on the field of medicine. He promoted the cell doctrine, pioneered the field of pathology, and became politically active to get government involved in public health issues. His students (many were from the United States) brought back the hygiene movement to their own communities and colleagues. (Courtesy of the National Library of Medicine.)

favored higher wages, better housing, and better working conditions for the Silesian miners whose epidemic of typhus he studied in 1848. He was forced to leave Berlin because of his socialist sympathies, and he resided in Wurzberg until 1856.[6]

In the last decades of the 19th century, Virchow pioneered effective sewage disposal, water purification, and other hygienic measures. Virchow was a popular figure in German science and served in the Reichstag. He was also an enthusiastic anthropologist and archaeologist and went on expeditions with Heinrich Schliemann in the search for ancient Troy. In 1871 Virchow studied the distribution of hair color among school children throughout Germany and concluded that only the northernmost provinces, constituting a minority of the German population, would be considered Teutonic or Aryan in the manner that German anthropologists described Teutons. The Teutonic form was described by anthropologist Hans Gunther, a colleague of Ploetz. They were depicted as "... tall, long headed [with] narrow face, well-defined chin, narrow nose with very high root, soft and fair golden-blond hair, receding light (blue or grey) eyes, and pink white skin color."[7]

RACE HYGIENE AS THE GERMAN EUGENIC MODEL

Ploetz's idea of racial hygiene (rassenhygiene) appealed to biologists and physicians in Germany. His views are difficult to classify by the traditional conservative or liberal political views of the late 20th century. Ploetz believed that war, revolution, and welfare contributed to racial degeneration. He favored welfare for those past the reproductive age. He hoped to discourage early or late marriages. He recognized alcoholism and sexually transmitted (venereal) diseases as threats to marriage and children. He favored preventive medicine as a social philosophy. He was progressive in his support for social security, accident insurance, shorter working hours, profit sharing, and user- or worker-owned enterprises (cooperatives) but felt these reforms could only occur when accompanied by differential breeding to offset the dysgenic effects that social welfare would cause. A vigorous racial hygiene program would necessarily lead to the social reforms sought by liberals and socialists.[8] Ploetz was also influenced by

Gobineau's racial typology, with the German Volk as the "last bastion of the Nordic race."[9]

Also active at the turn of the century in shaping German eugenic philosophy was Wilhelm Schallmayer, a social Darwinist, who warned of degeneracy from the medical care of the weak and unfit. Schallmayer echoed the views of American hereditarians who feared the unfit were out-reproducing the fit.[10] Schallmayer favored state intervention to prevent degeneracy in the German stock.

Ploetz's Society for Racial Hygiene had 32 members in 1905, but the movement grew rapidly, opening branches in Berlin (geneticist Erwin Baur joining), Freiburg (human geneticists Fritz Lenz and Eugen Fischer joining), and Stuttgart (geneticist Wilhelm Weinberg joining).[11] Liberal support for the society came from Alfred Grotjahn, father of the socialized medicine movement, and also from the Communist party, which in 1931 favored the compulsory sterilization of psychotic patients.[12] Ploetz was not an anti-Semite. He denounced anti-Semitism in 1895 and considered European Jews a cultured race, more Aryan than Semitic. Ploetz ranked the Aryans highest in their biological potentials and, with Fritz Lenz, developed the "Nordic Ideal" with the "German Volk [as the] last bastion of the Nordic race."[13]

FORMATION OF NAZI EUGENICS

The race hygiene movement shifted from a liberal to a conservative membership after World War I. The architects of Nazi race hygiene were scholars with excellent academic credentials. Fischer, Lenz, and Schallmayer were students of August Weismann, one of the great theoretical biologists of the 19th century. The first textbook in human genetics appeared in 1921, co-authored by Baur, Fischer, and Lenz.[14] Ploetz, who was never happy with the growing anti-Semitic direction of race hygiene, did not join the Nazi party until 1938. Baur was probably the most academic and least political of these early founders of the race hygiene movement. Baur edited an internationally respected series of monographs on the genetics of laboratory organisms, *Bibliotheca Genetica*. Thomas Hunt Morgan contributed a volume on the *Genetics of Drosophila* to the series.

Baur's work on the genetics of snapdragons (*Antirrhinum*) and cytoplasmic inheritance made him internationally respected. He founded the leading German journal of genetics, *Zeitschrift fur induktive Abstammungsund Vererbungslehre* (ZIAV, The Journal for the Analytical Study of Ancestry and Descent), and he was considered the most influential classical geneticist on the continent. Baur did not consider himself a Nazi, and after the war he was cleared of those charges and resumed his career as a botanical geneticist. His membership in Ploetz's society, his acceptance of the race hygiene movement, and his silence on the racism it embraced after 1930 were costly to his reputation after the war ended.

Wilhelm Schallmayer's work appealed to the fledgling Nazi party because he favored racial models of eugenics. He was Bavarian and thus influential in the early Nazi party's ideological development. His book *Inheritance and Selection in the Life of Nations* (1910) won the Krupp Prize for its social significance. He asserted that the role of the woman was to be a wife and a mother; her reputation would be enhanced in proportion to the number of children she raised. He wanted early marriage, family values, and a eugenic basis for marriage, including compulsory sterilization of those unfit for marriage.[15]

Fritz Lenz (1887–1976) was far more influential in developing a race hygiene philosophy that was assimilated by the Nazi party in its formative years. His first publication on race hygiene was in 1917. He believed that the state does not serve the individual but the race.[16] His enthusiasm for Ploetz's race hygiene included unusual anti-Semitic views. He believed Jews were a "mental race" and attributed their assimilation into German society as an example of evolutionary mimicry.[17] Unlike most of humanity, which sought a control over nature, he claimed Jews sought a control over other people. Had it not been official Nazi policy to condemn the Jews, Lenz would have tolerated them. He thought Jews played a constructive role in history and that they fostered good family values. "Next to the Teuton, in fact, the Jewish spirit is the chief motive force of modern Western history."[18]

Lenz's views on race were more sophisticated than the Nazis'. He did not believe there were any pure races, but he believed in "the absolute value of race" as a force in society. He admired race hygiene and saw it as a path to socialism. He was the first professor of racial hygiene in Germany (1923) and he founded the Archive fur Rassen-und Gesellschaftesbiologie

(Archive for Racial and Social Biology). Lenz was an ardent supporter of the Nazis and their basic racial policies of developing a Nordic ideal. He tolerated Hitler's anti-Semitism and praised him as "the first politician of truly great import who has taken racial hygiene as a serious element of state policy."[19] By 1931, Lenz was fully committed to Nazism and equated National Socialism with applied biology.[20]

Eugen Fischer was raised in a Catholic, conservative home and reflected those views in his early career. His interest in anthropology led him to embrace racial theory and race hygiene, although he did not believe that racial mixture was harmful. He accepted the views of many geneticists that interracial hybrids could show hybrid vigor, including human racial miscegenation.[21] Fischer studied human genetics, including twin studies, and published widely in that field. He was an expert witness and later a jurist in the eugenic courts initiated by the Weimar Republic and brought to fruition in the Nazi eugenics program.

When the Baur-Fischer-Lenz text was translated into English (1931), it received favorable reviews from many biologists and eugenicists around the world, but it was severely reprimanded for its racism and sexism by the American geneticist, H.J. Muller. After praising Baur's coverage of basic genetics, he denounced Fischer and Lenz for their social views on genetics, claiming these were "...less and less scientific, and soon we find them acting as mouthpieces for the crassest kind of popular prejudice." He criticized their reliance on "...intelligence quotients, which we know to be strongly influenced by training....."[22] It was widely known by then that the first edition of the Baur-Fischer-Lenz text was read with approval by Hitler while he was in prison after his failed Munich "putsch."

ORIGINS OF GERMAN EUTHANASIA

Euthanasia was not originally conceived as a eugenic measure. The "right to die" was an issue that developed as life expectancy increased in the last half of the 19th century, and incurable and chronic illness became part of the aging process. The right-to-die movement considered the decision a personal one chosen by the ill patient and often requiring the collaboration of a sympathetic physician or relative. In 1895 Adolf Jost extended the concept in his book *The Right to Death* by arguing that the state should play an active role in culling its most enfeebled citizens.[23] Jost argued that the state calls on its healthiest men to sacrifice their lives during war and that the incurably ill were a burden on the state and its people and hence a threat as real as a marching army ready to descend and inflict harm on the public's health.[24] Ploetz embraced Jost's views as consistent with the winnowing process that eugenics required for purification of the people perceived as an organic whole.

In 1920 a small book appeared with the title *The Permission to End Life Unworthy of Life*.[25] The authors were a legal scholar and a physician. Karl Binding was a Doctor of Jurisprudence and Philosophy at Leipzig and Alfred Hoche was a physician in Freiburg. A generation later both the prosecution and the defense referred back to this book as a basis for their position in the Nuremberg war crimes trials. The prosecution claimed the euthanasia movement was part of a systematic program of killing politically undesirable and innocent people. The defense claimed that euthanasia predated Hitler's rise to power and was part of acceptable medical ethics. Binding cited legal precedence dating to 1835 in Wurtemberg that exempted from murder the "killing in response to a strong desire of a terminally ill person or a mortally injured person."[26] Binding posed a question for the German people to consider: "Is there any human life that has to such a degree lost its legal rights, that its continuation is of no value for itself or for society?"[27] Binding reassured the reader that "one very important conclusion becomes necessary: the full respect of everyone's desire to live, no matter how sick, how tortured, and how useless."[28] But not all those with devalued life are capable of making a decision to stay alive, and Binding assigned to those individuals the relatives who wished to see them "die with dignity" or "the managers of these institutions."[29] Binding hoped that "compassion" for the incurably ill person would be the major motivation and that the person who carried out the killing would be a physician who would otherwise have treated the patient in a non-hopeless situation.

Dr. Hoche saw the problem from a medical view and identified two classes of "life unworthy of life." The terminally ill, if conscious and in chronic pain, usually requested euthanasia. The "mentally dead," a broad category that included the mentally retarded and nonfunctional psychotics,[30] were not in pain and could not request a release from life. Dr. Hoche justified the euthanasia of both classes on humanitarian grounds, the former out of compassion for the suffering individual and the latter out of compassion for society. "But one of these days maybe we will come to the conclusion that elimination of the mentally dead is no crime, nor an immoral act, and no unfeeling cruelty, but a permissible and necessary act."[31]

NAZI EUTHANASIA OF INFANTS WITH BIRTH DEFECTS

No official laws were passed by the Nazi government to justify or legally mandate euthanasia for either of Dr. Hoche's categories of the unfit. Hitler's personal letter, dated September 1, 1939, coinciding with the onset of the Second World War, allowed such killings. "Reichsleiter [Philip] Bouhler and Dr. [Karl] Brandt, M.D., are charged with the responsibility of enlarging the authority of certain physicians to be designated by name in such manner that persons who, according to human judgment, are incurable can, upon a most careful diagnosis of their condition of sickness, be accorded a mercy death."[32] Hitler was careful not to alienate Catholic and Protestant clergy opposed to euthanasia nor the German population itself. The date was chosen partly because the war would demand resources and personnel that were being devoted to care for those who represented society's least capable citizens. The war also provided a cover to permit the escalating mass murder of the unfit that the Nazi's top leadership had considered, at least in theory, as desirable for purging Germany of its degenerate stock.[33] Hitler expressed such sentiments as early as 1935 to Gerhard Wagner, the Nazi party's chief physician in the Reich.[34] Wagner died in 1939, and his replacement, Leonardo Conti, with Brandt and Bouhler, helped plan the child euthanasia and adult euthanasia projects in Germany.

Although the euthanasia program did not get started until 1939, the compulsory sterilization program had a much earlier start. Fritz Lenz in 1923 was warm in his praise for the US state programs of compulsory sterilization and he hoped that a similar, more embracing program would be initiated in Germany to preserve the integrity of the Nordic race.[35] Nazi ideology also championed the Nordic ideal but did not particularly stress how the differential breeding would be achieved to increase its numbers.

Both positive eugenics (larger Nordic family size) and negative eugenics (sterilization of the unfit or discouragement of marriage of Nordics to non-Nordics) would lead to the same goal within Germany. Hitler's early views, in *Mein Kampf,* speak of preservation of favored stocks and the slow, gradual emergence of them until they became the dominant racial type among the German people.[36]

Both eugenic approaches were eventually used and the projected rate of replacement was greatly accelerated as the Nazis gained control of the state. Shortly after Hitler took office as Chancellor in 1932, he asked Wilhelm Frick, his Minister of Interior, to draft a national sterilization law. The program was aimed at nearly a half million physically and mentally impaired patients in German institutions, including the feebleminded, psychotic, epileptic, blind, deaf, malformed, and chronic alcoholic.[37] The decision to sterilize these patients rested with the Hereditary Health Courts. These laws applied to Aryan Germans deemed unfit. In 1935 the government passed legislation, the Nuremberg laws, which also sought to prevent Germans from marrying Jews and to make it unlawful for Jews and non-Jews to engage in sexual intercourse. About 250,000 Germans were sterilized under these laws, mostly in the 1930s. Most of these people would have never married or had children because of their impairments, and it is not clear how much of a genetic difference their sterilization meant to the German population. After the concentration camps were organized for systematic killing, the sterilization of the mentally and physically infirm was less frequently used.

The killings began with infants and young children who were seriously malformed or retarded. An elaborate deception was arranged through which children were sent to special centers for their alleged treatment and false reports of diagnosis and treatment were relayed back to the parents and their physicians. When the children were starved to death or overdosed with drugs, the death certificates would report infections or major organ failures as the basis for the child's sudden downward turn in health. It soon became common knowledge that no child sent to these centers (30 of them were in operation) ever returned. The clamor by physicians and parents about the treatment of their children and the suspicion that they were deliberately killed was causing a serious wartime morale problem, and Hitler ordered the project to stop.[38]

EUTHANASIA OF GERMAN ADULTS

The adult killing project began in 1939 under the code name "T4." These adults were usually killed within 24 hours of their arrival at special killing centers, usually by carbon monoxide gassing. Karl Brandt headed the adult euthanasia project. With the help of the SS (Hitler's elite personal army), the prototype of later concentration camp gas chambers emerged. A fake shower room with benches permitted the medical staff to sit the patients in rows, and after the rooms were sealed, a vehicle would have its exhaust hose directed into an inlet in the room. The bodies were cremated and false death certificates were issued. As the wartime control over every aspect of life took place, virtually all of the patients in occupied territories were shot to empty out the hospitals for use by wounded German soldiers.[39]

HIMMLER'S EUGENIC PLANS FOR GERMANY

Hitler depended on Heinrich Himmler for the details that would lead to the realization of his hopes for a Europe free of Jews and a pan-Germanic population, purged of its hereditary defects, occupying most of Europe. Himmler was Munich-born and raised to be a farmer. He was too young to serve on the front lines in the First World War, and after the war he received a degree in agronomy and hoped to be a chicken farmer. He became active in the Nazi party and proved to be reliable, efficient, obedient, and enthusiastic, attracting Hitler's admiration for his work in the newly formed SS. In 1929, Hitler asked Himmler to lead that elite army dedicated to his protection. The SS was carefully screened to guarantee Aryan ancestry dating back to 1750. In addition to its primary duty, protection of Hitler, the SS was charged with carrying out espionage on all other aspects of German society. Himmler looked upon the SS as a genetic elite whose offspring would become the growing core of an Aryan people embodying the Nordic ideal.

Himmler devised many plans to bring this about. He developed the *Lebensborn* movement, an opportunity for SS men to father, out of wedlock, children from unmarried Aryan girls.[40] These children would be born in privacy and given out for adoption to SS families. Women who chose to raise their children would be given the status of married women (and called Frau) and many benefits of child care to permit them to raise their children without financial worry. Lebensborn women were usually about 15 or 16 years old and recruited from the League for German Girls, the female equivalent of the Hitler Youth. Himmler hoped, once the war was over, that a population of almost half a million girls would provide children for the rapid expansion of the certified Aryan stock in Germany. At least 12,000 children were born through this recruitment policy. Malformed babies were destroyed through the euthanasia program and carefully autopsied to determine the defects in the Aryan stock. Himmler hoped to produce 120 million Aryans by 1980 through this breeding program. He also wanted most of the Poles and Russians removed from Eastern Europe by exile or execution to provide room for this new Aryan expansion of Europe. Those remaining he hoped to sterilize and use as slave labor to carry out the heavy labor in farms and factories set up in the new communities. To obtain as many of the Teuton types as he could get, Himmler arranged for out-of-wedlock Norwegian babies to be transferred to the lebensborn program for adoption. All non-Jewish children of Poles and other Eastern Europeans, if blond and otherwise Nordic in appearance, were kidnapped and sent to Germany where they were issued birth certificates and adopted if still infants.[41] Young children were reeducated in special schools that tried to make them Germans and obliterate their memories of their origins.

THE HOLOCAUST AS GENOCIDE

The shift from killing as a humane act for the "mentally dead" and the hopelessly ill, to killing as a means of destroying the political prisoners in concentration camps, took place in 1941 through "operation 14f13."[42] The code was one of many used to describe causes of death at the concentration camps during the war. Prisoners allegedly in weak physical health or showing signs of mental disturbance were selected under this prison code for transfer to "rest homes for special treatment."[43] The standards for determining physical or mental impairment were arbitrary and not carefully supervised. The rest homes were killing centers and the "special treatment" another euphemism used to hide its synonym—execution. The term special treatment had circulated earlier and was introduced by

Gestapo leader Reinhard Heydrich (1904–1942) in 1939 for counterrevolutionaries and others posing dangerous threats to the state. The 14f13 program was a medical program and, however poorly it was administered, it was not a deliberate mass murder project. About 20,000 persons were selected, mostly as mentally impaired, and sent to their death in hospitals serving as killing stations.

Mass murder or genocide of undesired classes such as Jews, gypsies, and homosexuals began in earnest in January 1942 when death camps were set up in Poland. Six camps were set up for this purpose, of which the most notorious and the largest was at Auschwitz. The death camps were the direct outcome of a decision made on January 20, 1942 at Wannsee (a suburb of Berlin) where the task for the "Final Solution" was assigned to Heydrich on order from Hermann Goering. Heydrich convened SS chiefs (many of them scholars with Ph.D.s) to plan the mass killing of Jews and other undesirable elements kept in concentration camps or scattered in the occupied territories. Eichmann was secretary for this meeting. Himmler did not hide his feelings on their intent: "Jews are the eternal enemies of the German people and must be exterminated. All Jews within our grasp are to be destroyed without exception, now, during the war. If we do not succeed in destroying the biological substance of the Jews, the Jews will someday destroy the German people."[44]

There was no pretense that this was a medical operation. Selection was by class without regard to how healthy and sane the prisoners might appear to an examiner. The executions were not acts of euthanasia but deliberate murders of enemies deemed genetically unfit for life. A major technology for killing large numbers was worked out in the spring of 1942. The agent of choice was cyanide—zyklon B—a delousing agent that, piped into fumigation showers, quickly killed its victims before their bodies could experience convulsions that would unsettle the SS. Earlier attempts at mass murder in Russia by special SS shooting squads (einsatzgruppen) were psychologically devastating and led to suicides, psychosis, and alcoholism among those whose terrible duty it was to kill the innocent men, women, and children lined up over open pits. Cremation on a massive scale solved a disposal problem and avoided the burial of hundreds of thousands of bodies. It also made it impossible to determine the exact number killed and, if the Nazis had won the war, there would be little or no direct forensic evidence for later generations to determine that the Holocaust had taken place.

Die Spinne

Manch Opfer blieb im Netze hangen / Von Schmeicheltönen eingefangen
Zerreißt das Netz der Heuchelei / Ihr macht die deutsche Jugend frei

A major obsession of Nazi ideology was miscegenation with Jews. Hitler's Nuremberg laws included sterilization programs aimed at Aryans who were feeblemind-ed, psychotic, and physically impaired. A second set of laws were prohibitions against marriage or fornication between Jews and non-Jews. As Nazism pervaded German society, the propaganda against Jews increased. Julius Streicher was a virulent anti-Semite whose magazine *Die Sturmer* frequently depicted Jews as cunning, evil, and corrupting. Here the Jew as spider is enticing a German maiden. The doggerel may be roughly translated as "Many victims remain hung in the web, caught by tones of flattery; rip apart the web of hypocrisy, liberate German youth." (Reprinted, with permission, from E.A. Carlson [1984] *Human Genetics,* D.C. Heath and Co., Lexington, Massachusetts.)

Fortunately, the Allies overran the death camps before they could be destroyed, and immense forensic evidence remained for the victorious Allies to prosecute many of the Nazis involved as war criminals.

[1] Alan Bullock, *Hitler—A Study In Tyranny*. Hitler's genealogy is discussed on pp. 1–3. A second source on Hitler's life is Konrad Heiden's *Der Fuehrer: Hitler's Rise to Power* (Houghton Mifflin Co., Boston, 1944). Chapter III in Heiden's account gives Hitler's ancestry.

[2] Bullock, *Hitler*, pp. 13–15.

[3] Hitler's special gift was mass propaganda. He merged symbols, hierarchy, ceremony, pomp, nationalism, and uncompromising self-assurance as effective tools to appeal to a troubled audience tired of compromise, insecurity, drab life experience, and low expectations.

Rohm was among those purged during the 1934 "long knives" massacre of elements Hitler considered threatening to his power. The Brown Shirts, known for their street brawls and vigilante attacks on Jews and political enemies, were absorbed into the Nazi party but made submissive to the SS, Hitler's private security force pledged to his defense.

[4] The anti-Semitic and pan-Germanic nationalism were major movements in the 1870s and led to the unification of Germany and an uneasy detente between assimilated Jews and their fellow Germans. The eugenics movement in Germany had both European (Galtonian positive eugenics) and American (degenerate families requiring isolation or sterilization) antecedents of the late 19th and early 20th century.

[5] Vamberto Morais, *A Short History of Anti-Semitism* (W.W. Norton & Co., New York, 1976), p. 175.

[6] Virchow's influence was profound. The public hygiene movement was an application of the germ theory of disease developed by Louis Pasteur and Robert Koch. If infectious diseases were caused by microscopic agents, then such measures as pasteurization of milk, chlorination of the water supply, proper disposal and isolation of human wastes, compulsory vaccination, quarantine of persons bearing microbes that were not treatable, seizure and destruction of contaminated goods, and other containment and isolation measures would quickly minimize risk of a disease becoming an epidemic. That these measures sometimes conflicted with the political rights of the individuals (their civil liberties) was acknowledged, but the health of the population was almost always used by the courts to justify these temporary inconveniences imposed by the state. Henrik Ibsen's play, *An Enemy of the People*, illustrates this conflict between the zealous hygienist and the reluctant society to change its ways even to protect its own health. Germs, of course, do not respect laws, and their spread is very democratic. Each new disease that appears (like AIDS) revives the conflicts between those seeking draconian measures to nip an epidemic and those fearful of disrupting the laws that make society function. The success of public hygiene in minimizing infectious diseases made its supporters seek extensions to contaminated heredity.

[7] Clarissa Henry and Marc Hillel, *Of Pure Blood*, translated by Eric Mossbacher from the French [Au Nom de la Race] (McGraw-Hill, New York, 1976), p. 24.

[8] Robert W. Proctor, *Racial Hygiene: Medicine Under the Nazis* (Harvard University Press, Cambridge, Massachusetts, 1988), pp. 15–17. Proctor's thesis is disturbing. The German physicians and biological scientists flourished under National Socialism; they embraced its philosophy, appreciated the hygienic approach of Nazi racial theory, and felt little intimidation as the basis for their support. Denial of enthusiasm for Hitler after the war and a myth of intimidation were widely accepted by historians as the explanation for the participation of so many physicians in the SS and other Nazi programs. The theme of denial and coverup by participants in the race hygiene movement is explored also by Benno Muller-Hill in *Murderous Science [Todlische Wissenschaft]* (Harvard University Press, Cambridge, 1988).

[9] Ibid. p. 25.

[10] Ibid. p. 14.

[11] Baur was an excellent geneticist, his chief work being in cytoplasmic inheritance in plants, and he was respected as a leading geneticist throughout the world. Lenz and Fischer were too racist in their writings to have a world standing in human genetics, but Weinberg's work was respectable and his name lives on as the second part of the Hardy-Weinberg law, a basic tenet of population genetics (the frequency of a gene in a population remains constant as the population increases if there is no selection, no new mutations enter the population, breeding is random, and the population is very large).

[12] Proctor, *Racial Hygiene*, p. 22.

[13] Ibid. p. 24.

[14] Erwin Baur, Eugen Fischer, and Fritz Lenz, *Human Genetics* ([translated from the German *Grundriss der Menschlichen Erblichkeitslehre und Rassenhygiene*] London, 1931).

[15] Wilhelm Schallmayer, *Inheritance and Selection in the Life of Nations*, 1900. Schallmayer's views reflect those of David Starr Jordan's *The Blood of a Nation*. Galton also believed that nations differed in their populations' heredities and that this led to the rise and fall of civilizations. Chamberlain, and before him, Gobineau, developed this theme from degeneracy theory. Schallmayer and Galton rely more on an evolutionary model than an anthropological model of race mixture as the cause of these shifts in national character.

[16] Robert Jay Lifton, *The Nazi Doctors: Medical Killing and the Psychology of Genocide* (Basic Books, Inc., New York, 1986), p. 24.

[17] In mimicry, an organism looks like another organism or a natural object. A caterpillar may thus look like a stem or twig or floral array. A butterfly species that might be eaten by birds may acquire (through selection) the coloring of another butterfly species that is poisonous. For Jews it was a no-win situation. If they assimilated, they were using mimicry to hide their presence and subvert the nation; if they remained in their dress and custom as noticeable Jews, then they were aliens who refused to behave like ordinary citizens.

[18] Proctor, *Racial Hygiene*, p. 57.

[19] Ibid. p. 47.

20 Ibid. p. 61.

21 Hybrid vigor (heterosis) arises when two strains, usually inbred, are mated and their progeny show a robust assortment of traits superior to those of either parent. In hybrid corn, this is deliberately designed by breeders to create a highly heterozygous strain from several sources whose seed will yield a more abundant crop, better suited for the climate and more resistant to disease. Although the mechanism of heterosis is controversial (a simple dominance compensating for prior homozygous recessive defects versus some "overdominance" or intrinsic superiority of the heterozygote over either of its homozygous allelic forms), the observation is universally acknowledged among geneticists. It is ironic that many of the most rabid negative eugenicists, who championed hereditarian views, looked upon human racial mixing as a source of degeneracy ("mongrelization") rather than improved fitness.

22 Elof Axel Carlson, *Genes, Radiation, and Society: The Life and Work of H.J. Muller* (Cornell University Press, Ithaca, New York, 1981), p. 87. Muller was a Communist sympathizer and a eugenicist in the Galtonian tradition of improving the quality of humanity by selecting for intelligence, health, and a caring personality. He left the US in 1932 and went to the USSR where he became embroiled in the Lysenko controversy. Although Muller was a staunch anti-Stalinist and foe of Soviet policy from the time he left the USSR in 1937, he remained idealistic in his belief that human values and human ingenuity could be applied humanely to redirecting our own evolution.

23 A major difference between the early 19th century (the time of Malthus's belief that Providence would kill off the poor, weak, and ineffective element of humanity) and the late 19th century was the way a similar pessimism sought its resolve not through God or nature, but through human intervention.

24 Lifton, *The Nazi Doctors*, p. 46.

25 Karl Binding and Alfred Hoche [translated, with comments, by Robert L. Sassone from the German edition of Felix Meiner, Leipzig, 1920] *The Release of the Destruction of Life Devoid of Value* (Life Quality paperback, Santa Ana, California, 1975).

26 Ibid. p. 16.

27 Ibid. p. 17.

28 Ibid. p. 18.

29 Ibid. p. 25.

30 Ibid. p. 36. The term "mentally dead" should not be equated with the term used since the 1970s, "brain dead." A person who is brain dead has no cerebral function. A person who was called "mentally dead" is anyone with a less than normal Intelligence Quotient score or someone with a diagnosed psychiatric disease. Those are either arbitrary or ambiguous classifications compared to the clinically defined flat brain wave pattern of a comatose, brain dead, person. The spurious use of the term dead when applied to mental function misleads the reader. At worst, the retarded and psychotic person has a disturbed mental function, but it is difficult to equate that with death.

31 Ibid. p. 37.

[32] Ibid. p. 92 (Sansome's inclusion provides the War Crimes court citation for this document: *Trials of War Criminals Before the Nuremberg Military Tribunal Under Control Council* 10, Volume 1, 1950, p. 95).

[33] Lifton, *The Nazi Doctors,* p. 63.

[34] Ibid. p. 35.

[35] Ibid. p. 23.

[36] Ibid. p. 24.

[37] Ibid. p. 25.

[38] Ibid. p. 56.

[39] Lifton points out that these successive transfers of evil from one lower level to a higher level required a psychological adjustment of the physicians who were able to separate themselves as healers of the sick while acting as angels of mercy killing their patients. The continued medical services they performed allowed this valida- tion of the evil that was tucked, mentally, into the values of duty to the state, a high- er good, and the cleansing of the body politic. A similar process took place when Joseph Mengele shifted his medical attention to human genetics and carried out experiments on concentration camp prisoners. See Gerald Posner and John Ware, *Mengele: The Complete Story* (Dell Publishing Co., New York, 1986).

[40] The lebensborn project is a curious example of positive eugenics (i.e., a Galtonian approach) based on a dubious racial theory. Himmler believed naively that the out- ward traits (blond hair, blue eyes, sharply defined facial features) associated with the Nordic ideal type would be associated or linked to the behavioral traits he favored—military valor, obedience, leadership, creativity, intelligence, and self- assured elitism. Even if these behavioral traits were genetic, their genes are unlike- ly to be on the same chromosomes or closely linked to the genes for the superficial physical traits. Unless selection were intense for these behavioral traits (assuming that they are indeed inherited, a very unsettled argument), the Teuton types in Himmler's dreams would have had no more nor less diversity of behavior than Nazis who are brunettes. The primary socializing experience would be the educa- tion of the youth and not their heredity.

[41] Many of these adopted children were never traced, especially those whose parents perished during the war. Some of the forged birth certificates were crudely coded and the identification of the kidnapped child could be reconstructed. Despite efforts to indoctrinate the older children, there was fierce resistance, and most retained their language and memories of their homes and parents.

[42] The exact lines of command for the "final solution," as it is called by Holocaust scholars, is still controversial. Some, like Gerald Fleming in *Hitler and the Final Solution* (University of California, Berkeley, 1984), argue that Hitler ordered the elimination of the Jews and the process was staged through escalating procedures that followed an early plan. Others claim that the plan evolved haphazardly as events and personalities coincided, with Hitler initially having no grand plan and no particular preference for killing (rather than deporting) Jews.

[43] Lifton, *The Nazi Doctors,* p. 135.

[44] Ibid. p. 157. The Wannsee conference is discussed in Chapter 9 of Fleming, *Hitler and the Final Solution*. See also Hannah Arendt's account in *Eichmann in Jerusalem: A Study of the Banality of Evil*. Goering, in July 1941, had received a request from Hitler to set up a conference to determine the "final solution" of the Jewish question. Goering wrote to Heydrich on July 31, 1941, and an initial date was scheduled for December 9, 1941. The war with the United States and delays in getting all the participants together led to the January meeting. Hitler instructed his staff and subordinates not to put in writing his wishes for the extermination of all Jews during the war nor to refer to the killings in the concentration camps except through coded phrases through an appropriate chain of command.

19

The Abandonment of Eugenics by Genetics

E UGENICS AND HEREDITARIAN MODELS OF THE UNFIT appealed to middle-class and professional people in the late 19th century because they rang true to their values. They had no reason to doubt the widespread belief that like breeds like; they accepted the findings that scientists had disproved the theory of the inheritance of acquired characteristics; and they were sympathetic to the social belief that paupers, chronic criminals, the insane, and the retarded were of bad biological stock. All of these genetic errors carried over into the first third of the 20th century, and it required an erosion of these beliefs before the eugenics movement collapsed after World War II.

Long before the rediscovery of Mendel's laws of heredity in 1900, perhaps predating biblical times, it was widely believed that heredity involved like-for-like transmission. Variations, or departures from this like-to-like transmission, were assumed to arise from a different, unknown mechanism. Darwin and other mid-19th-century scholars considered them separate phenomena. Heredity was usually considered stable and gave the species its characteristics; variation was unstable and gave the individual its remarkable uniqueness. Heredity and variation were brought together as cognate studies in 1906, under the name of genetics, but the unity of these two subjects was not clarified until 1921 when H.J. Muller redefined mutation to fit the new findings in genetics. In a paper he presented to the

Second International Congress of Eugenics that year, he asserted that "The term mutation originally included a number of distinct phenomena, which, from a genetic point of view, have nothing in common with one another. They were classed together merely because they all included the sudden appearance of a new genetic type. Some have been found to be special cases of Mendelian recombination, some to be abnormalities in the distribution of entire chromosomes, and others to consist in changes in the individual genes or hereditary units. ...The usage most serviceable for our modern purpose would be to limit the meaning of the term to the cases of the third type—that is, to real changes in the gene.... In accordance with these considerations, our new definition would be 'mutation is alteration of the gene.'"[1] Heredity under this new definition was no longer distinct from variation; it was the heredity *of* variations that became a central theme of genetics.

MENDELISM COMPLICATES NAIVE EUGENIC VIEWS

In the like-for-like belief, healthy parents produced healthy children. Those with alleged hereditary physical or behavioral illnesses (such as the feebleminded, paupers, and innate criminals) only produced defective children. This model was supplemented, after 1900, with the false belief that a single-gene mutation ("unit-character") was involved in the production of any particular defective trait. Although this would have reduced transmission to 50% in like-for-like heredity because of the heterozygous state of the alleged dominant mutation, this aspect was frequently overlooked by early eugenicists. For recessive forms of single-gene traits, this would have reduced the risk to 25% for the birth of an affected child from normal (carrier) parents. Although some of the early 20th-century eugenicists recognized the Mendelian contradiction to like-for-like heredity, many ignored the implications that the parents of such children would be normal. None of the compulsory sterilization laws sought to sterilize healthy individuals whose only misfortune was a hidden recessive gene expressed in their children.[2]

The study of variable traits in fruit flies, especially by Muller, seriously weakened the simple model of single-gene mutations as the basis for most human behavioral traits and for complex traits such as longevity and

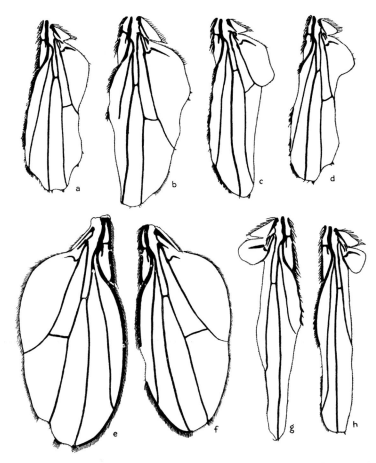

Muller's analysis of beaded wings revealed a complex relationship involving a chief gene, its genetic modifiers, and environmental modifiers resulting in a range of expression. Muller could predict from the combination of factors in the parents what would be the range and frequencies of different wing shapes among the progeny. Muller called this the gene-character problem and he believed it to be fundamental for explaining Darwinian natural selection at a genetic level. (Reprinted from T.H. Morgan, C.B. Bridges, and A.H. Sturtevant [1925] *The Genetics of Drosophila*, Martinus Nijhoff, The Hague, The Netherlands.)

overall health. Muller studied two unusual gene mutations found by Morgan, beaded wings and truncate wings.[3] Morgan could not make pure-breeding stocks of these mutations, and after spending some time with them, he turned them over to Muller, who had taken an interest in them. Beaded wings was the first mutation to yield decisive results on the mechanism of its peculiar heredity and variable expression. This dominant mutation showed a scalloped or ragged edge along the length and tips of the wings. Some flies showed a mild nicking of the outermost wing veins, some showed extensive scalloping. Muller demonstrated that the beaded wing mutation was homozygous lethal and was perpetually heterozygous over its normal allele. By constant selection, Muller obtained on a homologous chromosome an independent mutation, a recessive lethal, that was close to the site of the normal allele for beaded wings. When the chromosome bearing the beaded wings allele was associated with the chromosome bearing the newly arising, but nonallelic, lethal, the flies containing these two homologous chromosomes formed what Muller called "balanced lethals." In that state, almost all the flies had beaded wings. The rare exceptions were crossovers or recombinants occurring between these two nonallelic genes.[4]

Muller used the beaded wing analysis to demonstrate several genetic insights. The system of balanced lethals weakened the contending "mutation theory" proposed by Hugo de Vries to replace natural selection as the mechanism of evolution. The beaded wing case could be used to interpret de Vries's alleged mutations not as new species arising from a single event, but as recombinants arising among balanced lethal complexes of chromosomes. Second, the beaded case revealed that the intensity of expression of the beaded trait was based, not on fluctuations of the gene itself, but on the residual heredity or background genotype in the flies. Some of these background genes acted as modifiers to intensify or diminish the expression of the beaded trait.[5]

How Gene-Character Relations Weakened the Eugenics Movement

A more extensive study, carried out by Muller and Edgar Altenburg, demonstrated these modifier genes for the trait called truncate wings. Morgan's mutant gene for truncate wings, which makes the wings look obliquely sloped at the tips, did not express this trait unless one or more modifiers were present. Like beaded wings, the truncate wing gene was lethal when homozygous. Muller and Altenburg combined the truncate gene with intensifying modifiers they had mapped on the X, 2nd, or 3rd chromosomes. They could predict the percent of progeny that would be born with normal, mild, moderate, or extreme truncate wings by using Mendel's laws to account for the distribution of genes involved and converting the effects in each genotypic class into a corresponding phenotype for wing shape.[6]

Muller's recognition that the relation of genes to character traits was genetically complex was supplemented by his observation that the environment also affected character expression. For truncate wings, higher temperatures (about 28°C) produced more extreme truncation than lower temperatures (about 18°C). The chief gene, the truncate mutation, required higher temperature or a genetic modifier to express its trait. If a major gene for a variable trait like wing shape depended on genetic and environmental modifiers, it seemed plausible to Muller that highly complex and variable traits, such as intelligence, personality, or longevity would involve numerous genes and environmental factors for their expression.[7]

Muller's work on what he called "gene-character relations" was technical, involving a sophistication of genetic analysis few eugenicists of that time could follow without effort or the benefit of coursework in genetics. They preferred to interpret human character traits through simple pedigrees and single-gene effects. It was hard enough for them to follow Mendel's laws without straining under the added burdens of crossing over, balanced lethals, dominant traits (like beaded and truncate) that were simultaneously recessive lethals, or dominant traits (like truncate) that were only conditional and dependent on environmental or genetic modifiers that, by themselves, did not express the trait.

THE RISE OF POPULATION GENETICS

Muller's analysis of gene-character relations fitted a model of Darwinism by gradual, imperceptible increments, and this supported natural selection. It also helped clarify the way gradual selection for traits occurred through a bolstering of the chief gene's expression with homozygous modifier genes. Muller was not concerned that these two dominant traits, truncate and beaded wings, were also recessive lethal in their function. Morgan and his students had established that most spontaneously arising recessive genes, perhaps 90% of them, were lethal. If most functions were disrupted by the mutation process, genes associated with vital functions would be rendered lethal. Only a minority of genes would be associated with non-lethal functions or have nonlethal effects.[8]

THE ORIGINS OF POPULATION GENETICS

The relation of mutations in individuals to mutations in the popula-tion of their contemporaries or their descendants involved a more com-plex analysis leading to population genetics. The first attempts at popula-tion genetics were flawed by bad assumptions. The biometric school founded by Francis Galton, his student Karl Pearson, and W.F.R. Weldon assumed all heredity was blending or continuous in distribution.[9] Once they had committed themselves to this view in the late 1890s, they were unable to incorporate the discontinuities dictated and implied by Mendelism. A bitter fight erupted between William Bateson, who champi-oned Mendelism and the discontinuous origin and expression of traits, and the biometric school, especially Karl Pearson and W.F.R. Weldon, who identified their quantitative treatment of hereditary traits as Darwinian and consistent with natural selection. Both views, as it turned out several years later, were wrong. The entry of Mendelism into population genetics was provided independently by W. Weinberg and G.H. Hardy. Hardy con-sidered his contribution so trivial he did not mention it in his autobiogra-phy. One of Bateson's associates, Reginald Punnett, asked Hardy what the effect would be on gene frequency as a population increased. It took only a few minutes of Hardy's time to prepare an answer.[10]

The well-known Hardy-Weinberg law demonstrated that under ideal conditions the gene frequency for any given mutant or normal allele would not change. Thus, if a gene for the blood group rh is found in 40% (0.4) of the population and its allele Rh is found in 60% (0.6) of the population, the encounters of sperm (0.4 being rh and 0.6 being Rh) and eggs (0.4 being rh and 0.6 Rh) would yield 0.16 rh rh homozygotes, 0.48 Rh rh het-erozygotes, and 0.36 Rh Rh homozygotes. As long as the ideal conditions prevailed, the frequency would stay constant and would be calculated from the algebraic formula, $a^2 + 2 ab + b^2 = 1$, where a = frequency of the Rh

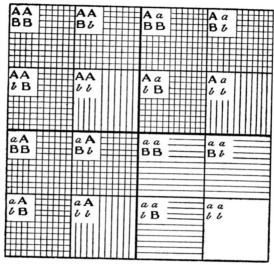

R.C. Punnett first used the Punnett square diagram to illustrate Mendelian ratios, in this case, the 9AB:3Ab:3aB:1ab expressed (phenotypic) ratio. (Reprinted from R.C. Punnett [1909] *Mendelism,* Bowes and Bowes, Cambridge, United Kingdom.)

allele and b = frequency of the rh allele. The ideal conditions are rarely met and require a large population breeding randomly with no selective advantage for either allele and no difference in mutation rate from Rh to rh or from rh to Rh. The development of population genetics involved precisely the consequences of varying these factors.[10]

The Hardy-Weinberg law helped to demolish the major assumptions of eugenicists. The like-for-like spreading among the Jukes or Tribe of Ishmael, if in excess of population growth as a whole, would require a selective advantage for socially undesired behavioral traits, a startlingly high mutation rate from normal to mutant genes, or differential breeding, the unfit having a huge family size compared to that of their more stable citizens. Since the first two assumptions were biologically unlikely, eugenicists attributed to the unfit an unrestrained sexual appetite and indifference to family planning.

Mendelism made it even more difficult for a rapid spreading of a newly arising mutant dominant or recessive gene. The elimination by sterilization of recessive traits (the alleged majority of all harmful disorders associated with the unfit) would be frustratingly long. To eliminate feeble-mindedness or to reduce it to a trivial incidence, if it were true that a single recessive gene caused the trait, dozens or hundreds of generations would be required. The dreams of a quick elimination of the unfit through sterilization would only work if the unfit behaved as if they obeyed a like-for-like inheritance or arose as Mendelian dominant mutations. Even more dismaying to those eugenicists who considered the applications of population genetics to social problems addressed by compulsory sterilization was the likelihood that for every homozygous feebleminded child born (e.g., due to phenylketonuria), there were about 100 normal individuals who carried the recessive gene in heterozygous form. The source of these unfit children (if most of them were due to single-gene defects) was not unfit parents but the vast majority of normal individuals, including the eugenicists themselves.[11]

THE HETEROGENEITY OF BEHAVIORAL TRAITS

Throughout the debate on the social policy to be adopted for dealing with the unfit, professional staff at mental hospitals became more familiar with the varieties of feebleminded individuals or psychotic patients. Some were dysmorphic with grotesque distortions of the facial features and skeletal organization. Many more were normal in their physical appearance. The clinical features of many were distinct enough to merit syndrome status. Down syndrome (then called mongolism) was one of the first to be so recognized, but Dr. John Langdon Down (1828–1896) attributed the features (small stature, epicanthal eye fold, slanting eyes) to an atavism or "throwback" to an ancestral, less robust form of human evolution originating in Asia and reflecting, in its name, Down's unexamined bias about Orientals.

By the early 1930s, the term feebleminded lost much of its meaning. Physicians skilled in genetics and diagnosis, although still rare, could distinguish many different disorders, not all hereditary, among the feebleminded. Lionel Penrose was one of the most skilled in this endeavor.[12] Just as geneticists using fruit flies, maize, or mice could distinguish many different genes affecting a similar trait (such as eye color, kernel color, or coat color), so too Penrose identified diverse kinds of mental retardation. This genetic heterogeneity arose from autosomal or X-linked genes, monogenic or polygenic inheritance, recessive or dominant modes of inheritance, and some familial patterns (e.g., occasional families with inherited Down syndrome) that defied genetic models available at that time. The more numerous and the more complex the genetic components were, the more difficult it was for eugenicists to make satisfactory predictions about the value of compulsory sterilization or the rate of reduction of the unfit in any one broad category of social disability.

SOME MUTANT GENES SHOW REDUCED PENETRANCE OR EXPRESSIVITY

In fruit flies some traits, whether homozygous recessive or heterozygous dominant, varied in the intensity of their expression or, like truncate, failed to express at all without the proper modifiers. N. Timofeef-Ressovsky called the appearance of the mutant trait, when the chief gene for its basis was present, the "penetrance," and its variable phenotype, the "expressivity" of a trait. Mutants could then be classified as having high or low penetrance with constant or variable expressivity. Those terms proved useful in describing human monogenic disorders.[13] Individuals receiving the gene for retinoblastoma, a cancer of the eye in children, had high penetrance when both eyes of the affected parent were involved. Other disorders, such as neurofibromatosis (often equated, erroneously, with the "elephant man syndrome"), have a lower penetrance and may skip a generation. Syndromes affecting several parts of the body such as the EEC syndrome (ectrodactyly or split hands and feet, skin and hair defects, and cleft lip or palate defect), may only show one or two of the traits in one sibling and all three in another, revealing the variable expressivity of this mutation in its background genotype. It is quite clear in small mammals such as rats, rabbits, and guinea pigs that the penetrance or expressivity of a mutant trait is usually a consequence of the chief-gene-with-modifiers model that Muller had worked out for beaded and truncate wings. A comparable genetic analysis through breeding cannot be done in humans, and the mechanism for reduced penetrance and expressivity of genetic disorders has only been worked out for a few disorders through the early 1990s.

THE SHALLOWNESS OF THE AMERICAN EUGENICS MOVEMENT

As geneticists and, to a lesser extent, eugenicists realized the complexity of predicting hereditary outcomes, they lost confidence in simple biological models used to interpret behavioral traits or to remedy them. By far, the most influential in shaping the American eugenics movement were Charles Davenport and the Eugenics Record Office at Cold Spring Harbor. Chief among his associates were Harry Laughlin and A.H. Estabrook. It was Laughlin (see Chapter 13) who framed the model sterilization law and served as an expert witness for the congressional hearings that led to the racially and ethnically biased immigration laws of 1921 and 1924. It was Estabrook who returned to the Jukes and Tribe of Ishmael and reinterpreted them exclusively in hereditarian terms, using Weismann's model of the germ plasm to negate any environmental benefits society might provide and an unproven but much asserted claim that the social failures of the unfit were germinal and not culturally or environmentally acquired. Of the three, Davenport had the best academic credentials and the managerial skills to support and promote the works of his associates. He was, according to one of his associates, MacDowell, too sensitive to criticism to tolerate or enjoy the style of critical debate that went on in Morgan's laboratory.[14] He preferred those who admired him and who extended his views. It was a self-defeating personality flaw, leading to professional indifference by his fellow geneticists and to occasional private disapproval. Davenport's uncritical acceptance of pedigree studies to identify genes for traits such as seafaring among New England families led to ridicule. His simplistic Mendelian model of feeblemindedness made the work of the Eugenics Record Office distrusted by British eugenicists, whose knowledge of social traits arose through the blending models of continuous variation that were later to be reinterpreted as polygenic.

DENUNCIATION OF THE AMERICAN EUGENICS MOVEMENT

The Third International Congress of Eugenics, held in New York City in August 1932, turned out to be the last.[15] (The three congresses are discussed in more detail in Chapter 15.) The first, held in London in 1912, was largely dominated by Galtonian idealism.[16] The second, in New York

City in 1921, was evenly balanced.[17] Geneticists were eager for an audience to discuss the findings of "classical genetics," as it later came to be called. At this youthful stage of their field's development, geneticists presented papers on mutation, non-disjunction, chromosome rearrangements, genetic mapping, and population genetics. Almost independent of these papers were those on social traits in humans, including pedigrees of the Pilgrim stock, the Jukes, and the Kallikaks. There were also abundant racial and ethnic studies. The Third International Congress of Eugenics limited itself to eugenic papers, because the Sixth International Congress of Genetics was to be held in Ithaca immediately following the Eugenic Congress. Muller submitted a paper with the provocative title, *The Dominance of Economics over Eugenics*.[18] Davenport, who chaired the program committee, was disturbed by the attack on eugenics and the Marxist values promoted by Muller. He sent a letter to Muller requesting the paper be cut from 1 hour to 15 minutes. Two weeks later, he cut the presentation time to 10 minutes and made its acceptance conditional, voicing the committee's objections that the paper was primarily sociological and not eugenic and this might not be acceptable. Muller replied in anger, insisting that he had "a right to be heard at the Congress, that the proposed talk is absolutely relevant to the purposes of the Congress, and that whether or not the views in it may be correct the members of the Congress should themselves have the privilege of considering."[19]

MULLER'S ATTACK ON THE AMERICAN EUGENICS MOVEMENT

Muller criticized the eugenics movement for stressing imbecility and a few exotic and rare defects for its major concern. Muller argued that the eugenic issues are more complex. "An individual's total genetic worth is a resultant of manifold characteristics, weighted according to their relative importance, positively or negatively, for society. It is a continuous function of all of these combined, so that there is no hard and fast line between the fit and the unfit, based on one or a few particular genes."[20] He criticized capitalist society for using economic criteria as the basis for genetic worth, such as success in business or acquiring personal wealth. This, he claimed, would lead to a reduced birth rate among those with more genetic talent. "The profit system leaves little place for children. In general, they are not profitable investments: their cost is excessive, but the dividends from them are uncertain, they are likely to depreciate in value, are non-transferable, and they do not mature soon enough. One child may be necessary for continuance of an estate, but each additional one weakens it. For the great masses who have no estates, each extra child commonly means more intensified slavery for the parents, and an additional unit of human unhappiness, in itself. And as the status of the middle class sinks, the parents hesitate to rear children with lesser privileges than they."[21]

Muller was not against birth control; he favored it. He wanted society to supplement family planning with a differential reproduction in which those with beneficial genetic endowments would have larger families than those without such endowments. He argued against the emphasis on negative eugenics and for a program that fostered positive eugenics: "Even more vital, from a biological standpoint, is an actual increase of those having more valuable genes, and it is the obtaining of this increase that is prevented by economic pressure, and by social pressure having an economic basis."[22] The first change in society, he argued, would be in our attitude toward women: "Do male eugenicists suffer from the illusion that most intelligent women love to be pregnant and to endure not only the physical disabilities but also the shame and humiliation, and the difficulties of maintaining a job, that pregnancy involves in our society. That they love the frightful ordeal of childbirth, so seldom relieved by competent medical treatment? That they love to spend forty or fifty thousand hours washing diapers, getting up in the night, tending colic, meeting in a city flat the little savages' just demands for safer outdoor play and companionship, stewing soups and milks, acting as household drudge and either abstaining from the life of the outer world entirely or else staggering under the double burden of a very inferior position outside and work in the home as well?"[23]

Muller pointed out the fallacy among eugenicists to overvalue the genetic component in human social traits. "The investigations of [Barbara] Burks on the resemblance between the intelligence of foster children and their guardians, checked by the calculations of [Sewall] Wright on this material, and [Horatio H.] Newman's converse findings concerning the considerable differences between the intelligence quotients of genetically identical twins who were reared apart, show clearly the important influence of environment as well as that of heredity upon intelligence as ordinarily measured."[24]

Muller used this gene-character relationship, one of his major research efforts over the previous 20 years, to attack the racism and spurious elitism of misguided eugenics. "The results, then, show us that there is no scientific basis for the conclusion that socially lower classes, or technically less advanced races, really have a genetically inferior intellectual equipment, since the differences found between their averages are to be accounted for fully by the known effects of environment."[25] The same line of reasoning he applied to temperament and moral qualities, including those that fall into the problem classes of poverty (paupers) and criminality. "Under these circumstances it is society, not the individual, which is the real criminal, and which stands to be judged."[26]

Muller's Marxism, then at its full tide, prevailed in his condemnation of his shocked audience. The class structure of capitalist society would change and the illusory problem of class and racial differences in birth rate, crime, poverty, illiteracy, and intelligence would be replaced, in a socialist society of long standing, by a new opportunity. "The new eugenics will then come into its own and our science will no longer stand as a mockery. For then men, working in the spirit of cooperation, will attain the social vision to desire great ends, and to judge of what is worthy. Then first, with opportunities extended as equally as possible to all, will men be able to recognize the best human material for what it is, and garner it from the neglected tundras of humanity. ...The possibilities of the future eugenics under these conditions are unlimited and inspiring. It is up to us, if we want eugenics that functions, to work for it in the only way now practicable, by first turning our hand to help throw over the incubus of the old, outworn society."[27]

Coming during the depths of the Great Depression, Muller's speech was received with enthusiasm by those sympathetic to Communism who read excerpts of it in the newspapers, and it was condemned by those who favored the view that able persons would surmount the economic hard times and that a harsh environment was as good a test of genetic worth as society could offer. Muller did not offer at this time his own proposal, for sperm banks to stock semen samples from the ablest and most eminent men in society and use that to generate "half-adopted" children among those women and their husbands willing to raise genetically well-endowed children. That proposal was offered a few years later in Muller's *Out of the Night,* a popularization of eugenics under socialism.[28] The book was naive

Out of the Night

A Biologist's View of the Future

BY H. J. MULLER

PROFESSOR OF ZOÖLOGY, UNIVERSITY OF TEXAS
MEMBER, NATIONAL ACADEMY OF SCIENCES OF THE U.S.A.,
FOREIGN MEMBER, ACADEMY OF SCIENCES OF THE U.S.S.R.

The Vanguard Press

NEW YORK, 1935

"We do not wish to imply that these men owed their greatness entirely to genetic causes, but certainly they must have stood exceptionally high genetically; and if, as now seems certain, we can in the future make the social and material environment favorable for the development of the latent powers of men in general, then, by securing for them the favorable genes at the same time, we should be able to raise virtually all mankind to or beyond levels heretofore attained only by the most remarkably gifted."

Excerpt and title page from H.J. Muller's *Out of the Night*. (Reprinted, with permission, from H.J. Muller [1935] *Out of the Night: A Biologist's View of the Future*, Vanguard Press, New York.)

in its appeal to a socialist pantheon of heroes: "It is easy to show that in the course of a paltry century or two (paltry, considering the advance in question) it would be possible for the majority of the population to become of the innate quality of such men as Lenin, Newton, Leonardo, Pasteur, Beethoven, Omar Khayyam, Pushkin, Sun Yat Sen, Marx (I purposely mention men of different fields and races), or even to possess their varied faculties combined."[29] Muller recognized that abuse was inherent in his scheme and feared his own society's choices if sperm banks were available. "It would be disastrous, therefore, to attempt such reproductive methods on a grand scale so long as the present ideology of individualism, careerism, charlatanism, unscrupulous aggression, and shallow hypocrisy prevails."[30] Unfortunately, as Muller was to learn by experience, precisely those latter qualities prevailed during his sojourn in the USSR. Two of his postdoctoral students, Solomon Levit and Isador Agol, were arrested and executed during the Stalin purge of 1936, and Muller found himself in hostile debate with Trofim Lysenko, whose anti-genetic movement resurrected a Lamarckian model of "shattered heredity" that could be "retrained" in any direction that the "selectionist" desired.

After fleeing the Soviet Union, Muller eventually found himself back in the United States but kept his eugenic philosophy muted until the late 1950s, when he felt sufficient time had passed since the Holocaust and he could reintroduce his eugenic vision. Muller called his outlook "germinal choice" and he believed it should be a voluntary system with education and the persuasion of well-founded argument as the means to promote it.[31] Shortly before his death in 1967, he negotiated with some backers to establish a Foundation for Germinal Choice (more popularly known as a sperm bank for geniuses). He quickly learned that the values of some of his backers in California were similar to those he had repudiated in his 1932 address, and he withdrew from the project, asking his wife to fight against the use of his name for that program after his death. Muller believed that having no eugenics program was preferable to having a seriously flawed one.[32]

FALSE APPRAISALS OF THE EUGENICS MOVEMENT

A great deal of the literature on the eugenics movement is based on a myth. The myth is comforting because it pits the forces of evil and power (bigotry, racism, privileged classes) against the forces of innocence and vulnerability (the poor, the retarded, the insane, the downtrodden). Seen in this light, eugenics is a weapon of oppression used to maintain the wealthy and the influential and to minimize the harm done by those who do not contribute to society, who demand help from society, or who inflict guilt, crime, social burdens, and higher taxes on society's most successful benefactors.

The myth is wrong and dangerous. It is wrong because the history of the unfit presented in this book has shown a much more complex development of ideas leading to the eugenics movement and the Holocaust. It is dangerous because it overlooks the numerous people of good will, many with outstanding credentials as social reformers, whose contributions became distorted or applied in mischievous ways. More difficult for us to understand is how caring people who did so much for humanity were able to include ideas that we now look on as inhumane. It is not easy for me to acknowledge that liberals, socialists, outstanding physicians, social workers, philanthropists, and brilliant scholars (some of them Jewish) were as much a contributing force to the eugenics movement and what led to its perversion in the Holocaust as were the mean-spirited, psychopathic, selfish, ignorant, and bigoted enthusiasts of the movement. Eugenics is a movement neither of the political left nor of the political right. In its social guise, it is a philosophy of preserving humanity or directing human evolution through some form of differential breeding. In its biological guise, it is a description of the consequences of the application or the failure of application of knowledge of human genetics to our present and future generations.[33]

Because the history of eugenics is complex, distortion is likely even without a willful intent to deceive. A liberal may blame conservative and reactionary thinkers for enacting restrictive immigration laws, involuntary institutionalization, and compulsory sterilization laws, while, at the same time, those conservatives are withdrawing public and private charity, relief, or welfare to aid those deemed unfit. A conservative may, by a similar selection of errors, blame those faddists and liberal social reformers whose "dreams of reason" provided the technologies of birth control, sterilization, sperm banks, prenatal diagnosis, genetic screening, and other reproductive options while those liberals are using the public's treasury to support persons who should be supporting themselves or receiving help from their own families.

CAN A HOLOCAUST OCCUR AGAIN?

No minority group is fully free from harm through bigotry, revenge, or an appeal to a greater good that can lead to its destruction. All three of these reasons for exterminating undesirable peoples have occurred in the past. Although this is a pessimistic appraisal, it should not be considered inevitable. The assignment or withdrawal of the right to live or to reproduce is done frequently in all societies. Ancient societies sometimes exposed their sickly infants to the elements rather than burden their society with those who could not contribute to it. Most societies permit killing in self-defense. Most nations permit the killing of enemy soldiers or even civilians during a war. Many nations forfeit the right to life of an offender who commits treason, murders another person, or participates in crimes deemed suitable for capital punishment, such as rape and kidnapping. In times past, such crimes included counterfeiting, mutiny, blasphemy, adultery, and bestiality. If one can justify the forfeiture of life for criminal or military behavior, it is not difficult to see how people can extend that justification to other situations that seem unimaginable to outsiders. There are Americans who do not feel morally troubled by the bombing of innocent civilians and their children in Hiroshima and Nagasaki. There are British who feel little if any guilt about the civilians killed in the saturation bombing of Hamburg, Berlin, and Dresden. Many Turks deny the scope of the massacre of Armenians in 1914 or claim that the Armenians fled or left of their own accord. Despite the overwhelming evidence of the systematic destruction of Jews during the Holocaust, there are non-Jews in Germany and throughout the world who deny it occurred or justify its occurrence. Almost any country has had its terrorist movements in which innocent people were destroyed to achieve political ends.[34]

No one knows why one minority becomes more vulnerable than another, even in the same country, to the ferocity of prejudice. The American Indian was deprived of land, livelihood, and life as European-Americans moved westward. The Irish-Americans and Afro-Americans escaped such decimation. Chronic conflict, such as that between Catholic and Protestant Irish, can lead to terrorism and a feeling by each side that the victims of terrorism brought about their own death. During the emotional climate of a tense period in a nation's history, the normal restraints that protect political, religious, or racial opponents as fundamentally people like us can be shattered, and the justification to destroy a hated or dangerous population can be overwhelming. Atrocities against civilians, like wars themselves, after many years have elapsed, may seem historically archaic and in times of peace, good will, and prosperity, an impossibility. A changing troubled economy, new social upheavals, and international conflict can suddenly trigger inappropriate hostility leading to the death of innocent groups.[35]

EUGENICS ON A PERSONAL SCALE

More troubling for those who try to draw lessons from the failures of the eugenics movement are those personal tragedies of birth defects and genetic disorders affecting the children of the rich and the poor, the saint and the sinner, the scientist and the lay person, the devoutly religious and the atheist. The biology of the human condition is democratic. To all classes of society it metes out a sterility rate of 10%. In all classes of society, about 20% of all women will experience a spontaneous abortion. For all parents there is a common risk that in 5% of births the baby will be born with a defect requiring medical attention. Abnormal chromosome numbers are distributed fairly to all ethnic groups, religions, races, and social classes. Unlucky shuffles of genes that strike down the young adult are equally distributed to upper and lower classes. This underlying biology that gives each fertilized egg a unique genetic constitution also gives a generous share of human misery to all segments of humanity.[36] Unlike the idea of the unfit who are blamed for their own misfortunes, the parents who deal with a congenitally ill child cannot blame themselves or society for their misfortune. They must live and suffer with their Down syndrome child, their baby paralyzed by spina bifida, their child who lives in an autistic universe, their infant who dies of Tay-Sachs disease, or their child who wastes away with Duchenne muscular dystrophy.

Parents can, of course, try to prevent bringing into the world another child with such a defect, and they can, in rare cases, prevent the birth of a child with a catastrophic illness if they have been screened beforehand and know they are at risk. They may take charge of their lives by not having any more children, by adopting a child, by using prenatal diagnosis with an option for abortion, or by using donor sperm.[37] Not all options are available to a couple, and parents must cope with what is available and what they can afford. In many instances, all options are unhappy. Some parents will cope with whatever fate has dealt them and, for religious or philosophic reasons, avoid the options of medical technology because they are deemed immoral or unnatural. These same persons may apply, with vigor, expensive medical technology to prolong the life or death of an infant too malformed to survive on its own and too impaired to have its essential functions restored.

It is difficult to find consistency in human moral positions. What is declared unnatural for reproduction is not declared unnatural when applied to the heart, lungs, or other vital organs. We approve of shunts that drain fluid from the brain of a hydrocephalic infant, heart-lung machines, kidney transplants, prosthetic hips and knees, pacemakers, and the gift relationship of donated blood. Why then is it unnatural to use donated sperm or eggs for the infertile? Why is it unnatural for a sterile couple to have their *own* gametes extracted, fertilized in a dish, and re-implanted in an oviduct or uterus?[38]

We fear differential breeding because it might alter the variability of humanity. Yet we tolerate assortative mating by choice, by custom, by bias, and by social change. We do not even wish for a random mating that would guarantee a maximum of variation in the population. For the greater part of the time that nations have existed, it was the father who chose the spouse for his child. Social classes have differed in their family size, sometimes the highly educated having fewer children than average, as in the first half of the 20th century and, since the end of the Second World War, the reverse, as those with better educations and incomes moved to suburbs and their own houses to raise more children than the average person could afford. We also tolerate a practical wisdom that makes it difficult for the blind, the mentally retarded, the grossly malformed, and the chronically sick person to find a spouse and raise a family. All of these tolerated methods that favor differential breeding far outweigh any change in variation brought about by any voluntary or enforced breeding program introduced in the past century.

FOOTNOTES

[1] H.J. Muller, "Mutation," *Eugenics, Genetics, and the Family* 1(1923): 106–112. The Congress was held in 1921, but its publication came out in 1923.

[2] There is a sentiment among physicians and scientists who work with new technologies that argues that no laws are better than bad laws. Often, in their zeal to protect the public, legislatures pass bad laws because false implications are drawn from the new technologies. Thus, when x-rays were first publicized in 1895, some state legislatures passed laws forbidding the x-raying of dressed women. The legislators erroneously believed that x-rays worked like the eyes of the 1940s comic book character Superman, and that a lascivious male would aim an x-ray at a woman to observe her private parts. While bad laws are indeed introduced into society, it is just as bad to neglect abuses and not protect the public. There is no perfect wisdom that permits legislators, or their advisers, to distinguish what is an appropriate protection and what is a foolish law.

[3] Both are single-gene mutations that express when the individual has only one dose of that mutant gene. Such a mutant state is called dominant, in contrast to the majority of all mutations that require two doses (one from each parent) of the defective gene in order to be expressed. H.J. Muller's analysis of beaded wings is in

"Genetic variability, twin hybrids, and constant hybrids, in a case of balanced lethal factors," *Genetics* 3(1918): 422–499.

[4] A balanced lethal situation could arise if a chromosome with the gene A encounters a chromosome with the gene B and the offspring bearing these two chromosomes show the traits AB. If two such AB individuals have progeny in turn, they should produce AA, AB, and BB offspring. But if homozygous AA and BB are lethal, only AB (mutually heterozygous) balanced lethal offspring will survive development. Thus, AB individuals will only produce AB progeny unless the AB is crossed to some other individual with a different genetic constitution, in which case A or B offspring will occur.

[5] During this formative period of classical genetics, there was considerable debate about the constancy of the gene. Morgan's school (at Columbia) argued for its stability (except for rare mutations) and William Castle's school (at Harvard) argued that the gene was unstable and fluctuated every generation, throwing off minor variations of a Darwinian sort. Castle was wrong, but the fight led to a permanent dislike of Castle for Muller and Muller for Castle. Muller, against Morgan's wishes, carried the debate into print. See the discussion of the Muller-Castle debate in my book, *The Gene: A Critical History* (Saunders, Philadelphia, 1966), Chapter 5.

[6] E. Altenburg and H.J. Muller, "The genetic basis of truncate wing—an inconstant and modifiable character in Drosophila," *Genetics* 5(1920): 1–59.

[7] That complexity was overlooked by many eugenicists who sought instead single-gene mutations as the cause of complex behavioral traits such as mental retardation, criminality, and insanity, most of which are polygenic or not genetic at all.

[8] The prevalence of recessive gene mutations arises from the likelihood that most genes make enzymes or structural proteins of cell organelles. The loss of function of an enzyme from one gene is compensated by a normal allele that makes the functional enzyme. Most of the time one dose of a normal gene does the job about as effectively as two doses of the normal gene. Dominant mutations arise in genes called regulators. Regulatory genes turn other genes on or off, and defects in such signals can be expressed in single dose. Mutations for recessive genes are due to loss of functions of the normal genes.

[9] For an account of the Pearson-Weldon debates with Bateson, see Carlson, *The Gene*, Chapter 2. Also see William Provine, *The Origins of Theoretical Population Genetics* (University of Chicago Press, Chicago, 1971).

[10] Hardy's contribution in 1908 was independently conceived by Wilhelm Weinberg. Curt Stern noted an oversight among non-German geneticists in citing Hardy alone and recommended that the fundamental equation be called the Hardy-Weinberg law. The Hardy-Weinberg law also enables mathematical geneticists to test models of natural selection based on mutation frequency, adaptiveness of a trait, and other factors that are more likely to prevail in nature than the ideal conditions.

[11] If a disorder is rare in the population (1 in 25,000 births), the incidence of the gene in the population is approximately twice the square root of that incidence, or about 1 in 80 people would carry that defective gene. Even if the parents of such a child

were sterilized, there would be 78 other carriers among every 25,000 people whose defective gene would be unseen in society. The rate of diminishing that incidence for any given recessive rare mutation would be excruciatingly slow by even the most diligent sterilization process of child and parents! Few eugenicists in the American eugenics movement wanted to bother about the consequences of population genetics because they wanted to believe the simpler fantasy that like breeds like and the defective only come from defective stock.

[12] An excellent account of Penrose's contributions to the scientific development of human genetics is given in Daniel J. Kevles's *In the Name of Eugenics: The Human Uses of Human Genetics* (Alfred Knopf Inc., New York, 1985). See Chapter 10, "Lionel Penrose and the Colchester Survey."

[13] They do not have to apply exclusively to dominant disorders. Sometimes a homozygous recessive disorder fails to express or is expressed in a milder or more severe form. Some siblings with cystic fibrosis, an autosomal recessive condition, will show dramatic differences, one manifesting severe symptoms at birth (a distended colon) and dying young, the other not being diagnosed until later childhood and living 30 or more years with few hospitalizations. Although the siblings are both homozygous for the same alleles that give them cystic fibrosis, their modifier genes differ. In some disorders, such as Huntington disease and Duchenne muscular dystrophy, the variation in expression is caused by reiteration of a small sequence of nucleotides within the gene.

[14] E. Carleton MacDowell, "Charles Benedict Davenport," *Bios* 17(1946): 1–35.

[15] *A Decade of Progress In Eugenics* (Williams and Wilkins Co., Baltimore, 1934). The editing was probably done by Davenport and Laughlin. The Third International Congress of Eugenics was held at the American Museum of Natural History, August 21–23, 1932. Davenport was president of the congress.

[16] *Problems in Eugenics*, two volumes, 1912. The First International Congress was held in London with Major Leonard Darwin as president.

[17] *Eugenics, Genetics, and the Family*, 1923. The Second International Congress of Eugenics was also held in New York City at the American Museum of Natural History. Henry Fairfield Osborn was president of the congress.

[18] H.J. Muller, "The dominance of economics over eugenics," *A Decade of Progress in Eugenics* (Williams and Wilkins, Baltimore, 1934), pp. 138–144.

[19] Elof Carlson, *Genes, Radiation, and Society: The Life and work of H. J. Muller* (Cornell University Press, Ithaca, New York, 1981), p. 179.

[20] Muller, *Dominance*, p. 138.

[21] Ibid. p. 139.

[22] Ibid. p. 140.

[23] Ibid.

[24] Ibid. p. 141.

[25] Ibid. pp. 141–142.

26 Ibid. p. 142.

27 Ibid. p. 144.

28 H.J. Muller, *Out of the Night: A Biologist's View of the Future* (Vanguard Press, New York, 1935).

29 Ibid. p. 113.

30 Ibid. p. 114.

31 H.J. Muller, "Germinal choice, a new dimension in genetic therapy," *Excerpta Medica*, 1961, Abstract No. 294, p. E135. See also H.J. Muller, "Evolution by voluntary choice of germ plasm," *Science* 134(1961): 643–649.

32 His widow, Thea, did go to court to have his name removed from Robert Graham's "sperm bank for geniuses" as the popular press called it. That sperm bank, in Escondido, California, which has used IQ and GRE or SAT college placement scores as a basis for accepting donors, has been used for the insemination of about 250 children, through the late 1990s. No article has appeared, however, to indicate how they are faring in school and socially. One of the great difficulties with a reliance on IQ and other achievement tests is their weak prediction of such socially desired traits as creativity, artistic talent, ambition, and compassion about whose genetic basis (if any) virtually nothing is known. A study of the now-defunct sperm bank is in progress. See Constance Holden, "Tracking genius sperm," *Science* 291(2001): 1893.

33 Galton intended both the scientific and the social aspects to be included in the term eugenics. This has created confusion when scientific fields like human genetics describe the frequencies of gene mutations or project them into the future. To hostile critics of eugenics, even the phrase "the eugenic implications of" smacks of a plot to rob humanity of its diversity, and the field of human genetics then is looked on with great suspicion. Geneticists, no matter how liberal their political philosophy may be, cannot deny the applicability of genetics to humans—and human genetics, like its medical specialty, medical genetics, is confronted with genetic problems that require genetic information. Down syndrome today would make little sense without a knowledge of cytogenetics. Ethnic disorders would be a racial puzzle without the rational explanation of a founder effect and a knowledge of the statistics of small numbers. The issue of treatment or prevention is a basic one that is not limited to eugenics. Many physicians and lay commentators derided the use of a Jarvik artificial heart for a man who had smoked heavily most of his life. Whether that was a causal association or not in this patient's condition is not at issue. What is at issue is the huge number of people with bad hearts, emphysema, and lung cancer who are smokers compared to the relatively few with these disorders among nonsmokers. The issue is complex: a physician does not like to blame the victim; a physician wants to treat.

34 Critics of elective abortion, in vitro fertilization, artificial insemination, and similar "genetic or social engineering" efforts that they deem murderous may simultaneously favor the death penalty for rape, kidnapping, murder, and espionage. The issue is not whether there should be "quality of life" values, but *whose* quality of life values should be used.

[35] I once spoke to Maurice Walsh, a Los Angeles psychiatrist who had treated the dictator Trujillo in the late 1930s (for a sexually transmitted disease) and had interviewed, at the Allies' request, Rudolph Hess for the Nuremberg trials. Walsh was puzzled how ordinary citizens could accept the political views of people who were psychotic when, if they saw these figures not as national leaders, but as private citizens, they would walk away from them as crazy. Walsh tried for years to interest scholars in looking at war, terrorism, and racist movements as a social pathology deserving scientific or psychiatric research, but he was usually rebuffed and told that his views were simplistic and that these issues are primarily economic or political and not psychological in nature. Those who lived through and reflected on the propaganda of a "total war" may have their doubts.

[36] See Chapter 1, "The human condition" in my text *Human Genetics* (D.C. Heath & Co., Springfield, Massachusetts, 1984.)

[37] Not all these options are available to any one couple. This will depend on cost (in US society where there is no national health insurance, the couple has to have a private insurance plan that covers those costs). In countries with socialized medicine, expensive and exotic therapies may not be available because the nation cannot afford such expenses. In countries like the US, adoptable children are difficult to obtain, and some childless couples have chosen children from third-world nations (especially interracial children rejected by the prejudices of the citizens of these nations for such hybrids) or they have chosen children with handicaps who are difficult to place. Critics of the medical technologies rarely appreciate the size of the problem. Ten per cent of marriages are infertile; that is a very large number of people, far more than the births of unwanted out-of-wedlock babies.

[38] The success rate, in the late 1990s, for in vitro fertilization (IVF) is about 40% at its best. Many hospitals have much less success. Some constraints include the observation that in normal males (and probably females) about 30% of all sperm and eggs are aneuploid, that is, they contain extra or missing chromosomes that abort the early embryo. For the woman who has tried 10 years to get pregnant and then succeeds through IVF, it is her life's most fulfilling experience when she gives birth to a healthy infant. Making choices is one of life's obligations, and society wrestles with what it permits the individual to do.

20

The Future of Eugenics

THE HISTORY OF "THE UNFIT" REVEALS how often that concept has recurred in human history since antiquity. Evolutionary psychologists may argue that this is something inherent in human nature, as a normal or expected xenophobic reaction to members of the human species who are neither kin nor ethnic members in a common culture. They would argue it was adaptive to the kinship or clan of related members who are assumed to have been the basic unit of reproduction in prehistoric times. Genes for a behavior of the fear of others (especially non-kin) and genes reinforcing a group loyalty to members of one's kinship would be the alleged components for our present tendencies to prejudice, civil war, and unease with our adjoining neighbors. There is no direct evidence for such "xenophobic genes." Those who reject this claim argue that our fears of "the other" (as philosophers like to describe the unfamiliar) are based on prejudices we learn. There is little likelihood of genes for anti-Semitism, fear of whites and blacks for each other, Irish fratricide based on religious upbringing, or other genetically determined inharmonious relations among people based on religion, culture, or race. Those who favor an environmental interpretation would argue, instead, that small children of different races, cultures, and religions regularly play together, and they can often identify the sources of their later prejudices.[1] Evolutionary psychologists would counterargue that infants and children are an exception because of their need for nurturing, and thus they are equipped with a genetically driven happy face for those who do the nurturing.

361

ORGANIZED EUGENICS IN THE UNITED STATES IN THE YEAR 2000

Eugenics (especially negative eugenics) was the latest and most notorious of a long history of attempts to justify the isolation or destruction of an unwanted, allegedly unfit group of people. We know that there are outbreaks of this tradition of hatred in our own time. We witness the Palestinian-Israeli conflict of a half century of failed attempts to live together with only partial successes over these years at achieving settlements among once warring nations. We witness the carnage on all sides in the Croatian-Bosnian-Serbian-Albanian conflicts of the Balkans with amazement as the catch phrases of the waning years of the 20th century (ethnic cleansing) bring back memories of a half-century earlier (a Jew-free Europe). In the half-century since the revelations of the horrors of the Holocaust, eugenics became a taboo topic, and its formal existence was dead or moribund. Only isolated attempts at a revival of eugenics exist. There is in the United States a small organization (assets of 5 million dollars), the Pioneer Fund, whose board members are sympathetic to the ideals of eugenics that led to its founding in 1937. Harry Laughlin was one of its founders. It supports a modest amount of research, mostly in twin studies and intelligence testing studies, with its major eugenic sympathies for the Galtonian ideals of promoting higher intelligence. It still supports (indirectly through contributions to other organizations) restrictive immigration, and it played a minor role in California's attempt to limit the influx of Mexicans into that state. Some of its support goes to members who contribute to a web site calling itself *Future Generations* (the web site is http://www.eugenics.net).[2]

There are hate groups in the United States based on white supremacy (including anti-Semitism and anti-miscegenation mandates). Hate groups also exist among victimized minorities, some who see whites as their enemies and attribute genetic flaws in their essential nature to such factors as a lack of genes for sufficient melanin, which they associate with the alleged defective character traits of whites. These groups receive public attention when their members flout the laws of the state or the nation, engage in criminal activities (e.g., robbing banks or using fraud to raise money), or identify themselves as an armed nation willing to defend their members with violence. None of these groups has sufficient following to match the eugenics movements of the 1920s in influence or resources.

A Temporary State Eugenics Program in Singapore

Worldwide, there are few eugenics movements active today as the 21st century begins. Only two have any sort of government backing. One was promoted by former Prime Minister Lee Kuan Yew, who ruled from 1959 to 1990 in the autonomous republic of Singapore.[3] Mr. Lee rejected democracy for an authoritarian administration to unite his ethnic minorities when Singapore was rejected from the Malaysian confederation. He imposed English as the national language to bring together the majority Chinese with the minority Tamil and Malay populations. He integrated the schools and the neighborhoods to break up, as effectively as he could, the ethnic isolation that predated his government. Mr. Lee received a bad press in reaction to his proposals of 1983 that, in some ways, many would consider not eugenic at all but its very opposite. Mr. Lee was concerned that women in Singapore who attended college became unmarriageable because of a prevailing male prejudice against educated women, who were seen as troublesome, lacking obedience, and challenging the value men favored of their making all the financial and other decisions about their family's welfare. Prime Minister Lee argued that this rejection of college-educated women deprived the nation of half its genetic worth. He proposed laws to change these values by guaranteeing subsidized good housing for families in which the woman held a college degree and free tuition for college for their children. What alarmed his critics was Mr. Lee's failure to disguise his motives by using a term other than eugenics as the basis for his proposals. If he had described his concern as only one of fighting sexism, he would have found much favorable reaction to this proposal. By arguing his case in genetic terms, that women who went to college had genes that were valuable for his nation, Mr. Lee became a Galtonian positive eugenicist. He compounded his difficulties with a second proposal, offering a financial incentive (a substantial down payment on a house) to women who had no more than a junior high school education and who had two or fewer children, if they would submit themselves to sterilization. That is classical negative eugenics, and it reflected Prime Minister Lee's confidence in the work of IQ theorists who held intelligence to be some 80% heritable. Environmentalists, of course, would have argued that the genes were not at issue because it is a given that there is no substantial genetic difference between men and women or between those of a small elite who attend college and almost all of those in Singapore who lack resources and opportunities to do so. Unfortunately for Prime Minister Lee, his political party lost ground in a subsequent election and most of his eugenics program was dropped.

CHINA'S EUGENIC EFFORTS SUPPLEMENTING FAMILY LIMITATION

The second government-supported eugenics program is in China. China has had a one-child-family policy for more than 30 years. It was based on the need of that very populous nation to provide sufficient resources, especially food, for its one billion citizens. China is not immune to the reality that 5% of children born will have a birth defect serious enough to require medical attention. The options for such parents are to raise such a child with limited prospects or to have an opportunity to have another child and to enjoy the benefits of an essentially normal child. In most such repeat pregnancies, the odds are closer to 19 out of 20 that the second child will be normal, because most of these children do not suffer from single-gene defects but from polygenic or chromosomal defects. Many Chinese physicians had argued that if the one-child family is a state law for the indefinite future, parents of a child with a birth defect should have the right to have a second child, and the state should provide funds for adequate prenatal diagnosis. China has many eugenic laws (or genetically based laws if one interprets their intent as not based on ideological grounds of superiority or inferiority). These include exemption from the one-family rule for the 5% of China that constitutes ethnic minorities (mostly in the western provinces). Among these are groups that are ethnically related to the Cambodian and Vietnamese, groups that share a genetic and cultural heritage with Indians and use a Sanskrit alphabet, and many relic populations that have the traditions and languages of peoples who lived in that area before the Han invasions conquered them. China also gives an exemption to farmers who require male laborers, and such farmers may request permission to have a son to help out.

The issue becomes eugenic when prenatal diagnosis is used to provide a male son, when female infanticide is practiced to abide by the one-child-family rule, and when prenatal diagnosis is mandated if a family has a child with a birth defect and wishes to try again. United States values, like those of most industrial nations, favor the autonomy of the individual to make such decisions. However benevolent a state may be, its choices of universal genetic obligation become a eugenic policy. Many nations are not democracies, and they believe that the state has the right to govern its people's welfare. They believe autonomy resides with the state and not the individual citizen. This view governed China's 1994 Maternal and Infant Health Care Law.[4] In it, China issued guidelines (as an ideal, rather than a law with specified penalties for noncompliance) offering genetic counseling before marriage to couples at risk, mandating sterilization or long-term contraception as a condition for marriage between two individuals at risk for having a child with a birth defect, and mandating prenatal diagnosis and acceptance of the recommendation of their attending physician.

There was considerable chagrin among American and European professionals involved in human genetics when this law was reported. In the tradition of industrial democracies, the decisions of what to do are left to the client or patient and not to the professionals. In all likelihood, the future of genetic services will be with these democratic traditions as more of the world observes the benefits of those values and traditions and demonstrates that informed people acting in their own self-interests tend to end up with the same outcomes as those who try to bring these about by coercion.

SPERM BANKS FOR GENIUSES

A minor, but nevertheless well-publicized, venture to use sperm from geniuses has met with very limited success. Originally inspired by Muller's papers calling for a voluntary program of germinal choice, the Foundation for Germinal Choice was endowed by Robert Klark Graham, a physician who made a fortune in the development of contact lenses.[5] He had met with Muller, but Muller became disenchanted with the values of people associated with Graham. It was too much like the eugenic values of the American Eugenics movement he had condemned in 1932. After Muller withdrew, Graham kept the foundation going through his own efforts and the support of some friends. He succeeded in getting a few Nobel laureates to donate sperm, and he switched to sperm of college students scoring 800 (the top 1%) on the verbal or mathematical sections of the GRE (a national examination testing the overall academic abilities of students hoping to enter graduate programs). In the 20 years of its operation, the foundation has sent sperm (it is free) to qualified requesters (women of high intelligence who have the permission of their husbands). About 250 births have resulted, but Graham's sperm bank has a policy of not following up on how the children are doing to assure privacy to the recipients. They did volunteer that their most frequent users are the younger wives of physicians. The physicians had vasectomies for their first marriage and when they divorced and married a younger woman desiring children, they elected to use a sperm bank for geniuses rather than take less potentially well-endowed sperm from other sperm banks. Most sperm banks in the United States do not identify the worth of sperm by ability. They check for disease and selected hereditary disorders.

THE ASCENDANCY OF ENVIRONMENTALIST VIEWS SINCE THE 1970s

Although formal eugenics movements are virtually nonexistent as the 21st century begins, there are numerous anti-eugenic critics whose influence has been much more effective.[6] Their concerns have made it difficult for those wedded to intelligence studies and twin studies to obtain funding for research from federal agencies and most large private philanthropies. The majority of these critics speak as individuals from universities, they write newspaper and magazine articles, or they publish books warning about the potential rebirth of eugenics. The arguments vary in their assumptions and values. Some oppose eugenics (real or alleged) because of fears of a "Brave New World." They see new technologies, including those used for prenatal diagnosis, assisted reproduction, genetic screening, gene therapy, or stem cell research as leading to genetic control of human reproduction or the modification of a natural genome by unnatural means. In their minds, the image of Dr. Frankenstein and his monster is not too far removed from the potentials genetic technology provides for eugenics.

A second group fears eugenics and its technologies more for their unnatural and unpredictable consequences than for some future dictatorial state stratified by eugenic classes. This includes some who prefer a laissez-faire policy to human reproduction (but not to the natural selection of bad outcomes, which are regarded as deserving of medical benefits). It also includes advocates of natural law who interpret almost all technologies associated with reproduction and human genetics as immoral because of their violation of natural law (whose central tenet is that sexual intercourse is intended for procreation and must remain open to that possibility even if the loving relationship of a couple is its primary motivation). Those who shy away from natural law arguments may support this fear because of the assumed loss of genetic variation in humanity that would ensue from the selection of favored genotypes and the simultaneous disappearance of those traits (and their genes) not favored by a eugenically minded humanity.

A third group sees medical genetics as "eugenics entering through a back door." Although genetic counselors by training are taught not to tell what clients should do and that they must be supportive of the autonomy of the client in making a reproductive decision, they or their colleagues on a genetic team are believed to subtly influence the clients into eugenic choices of abortion of birth defects. Some of these opponents of medical genetics (other than for diagnosis and treatment) argue that the allegedly abnormal children are victims of society's prejudices and should be welcomed by both the parents and the society as full human beings to be raised lovingly even if they will experience an early and painful illness and death or live a life with severe limitations. Others argue that making any exceptions for elective abortion based on genetic considerations will mark the beginning of a "slippery slope" leading to new eugenic courts and Nazi-like destruction of "life unworthy of life."

All of these critics are dismissed by those engaged in medical genetic services as misguided, ideologically rigid, ignorant of the genetics involved, or indifferent to the suffering experienced by the affected child (if born) and to the psychological suffering of the parents. They argue that there is no eugenic image or ideal in their minds as professional caregivers. They are in the field to relieve suffering, both physical and mental, of their clients. They argue that embryos are not autonomous, do not have legal status as persons, and that the future parents (especially the pregnant women), not society, should make the decisions on when to reproduce and whether to carry a pregnancy to term.

LONG-TERM CONSEQUENCES OF UNIVERSAL GENETIC SERVICES

Most of these debates concern the questions of how reproduction is carried out and how medical genetics attempts to diagnose, abort, or treat hereditary disorders and birth defects. Only a few individuals are genetically well informed enough to raise questions about the effect of these procedures and values carried out over many future generations, if not millennia, to come. There are several issues, each with its own outcome, that need to be explored. These include the long-term effects of:

- elective abortion of autosomal recessive traits (such as cystic fibrosis, sickle cell anemia, Tay-Sachs syndrome, Hurler syndrome).
- elective abortion of autosomal dominant disorders (such as Huntington disease, achondroplastic dwarfism, Apert syndrome, Marfan syndrome, retinoblastoma).
- elective abortion of X-linked disorders (such as Duchenne muscular dystrophy, X-linked mental retardation, Lesch-Nyhan syndrome).
- elective abortion of polygenic traits (such as neural tube defects, congenital cardiac defects, tracheo-esophageal defects).

For each of these genetic possibilities, I will make the assumptions that (1) the human genome will be worked out for each of its gene's functions by 2025 and (2) this will provide information on all single-gene defects that go to term (a significant number, perhaps more than half of all genes in the human genome may involve changes expressed before conception or birth that lead to failure to fertilize, failure to implant, failure to maintain an implanted early pregnancy, failure to produce normal organogenesis, and failure to maintain normal cell functions). We should also know those percentages with reasonable accuracy by 2025. If this information, along with the sequences of the normal genes, is available for genetic counseling, it is likely that dozens or hundreds of potential genetic defects could be screened for every pregnancy of a client asking for such information. The use of gene chips embedded with these normal gene sequences will pick up anomalous gene sequences for any of the sequences on the chip used. I will make as a first assumption that the worst fears of those who fear human genetics will be realized: All, or almost all, women who intend to become pregnant will request and receive a screening. Their husbands will also receive the same screening. If they are at risk for one or more genetic disorders, their embryos will be screened for any of these disorders. What would be the long-term consequences of a universal usage (with near-universal abortion of the affected embryos) for dozens of generations?[7]

The Consequences of Autosomal Recessive Disorders

It will come as a surprise to most of humanity that the long-term use of screening and elective abortion for autosomal recessive disorders will lead to little or no diminution in such a gene's frequency in humanity as a whole. What will disappear are the babies born with the birth defects. Let me illustrate this with an on-going example that matches the conditions I specified. About 1950–1975, Tay-Sachs disease was well known to Jews of Ashkenazic descent (about 90% of the world's Jews). It is a disorder that involves a missing enzyme (hexoseaminidase A) in the lysosomes of these babies. This enzyme cleaves neural lipids and recycles them after digestion in the lysosomes of cells. When the enzyme is missing, the lipids accumulate in the lysosomes, which become engorged (hence the name lysosomal storage disease). The neurons in which this happens begin to fail, and the child at about the age of six months begins to lose its ability to sit up. It becomes blind, suffers seizures and loss of capacity to learn, and slowly shifts to a vegetative state over the next two to four years. The sickness ends in death, often because the child can no longer swallow or because it chokes when it does swallow.

When amniocentesis became available for measuring the presence of hexoseaminidase A in the amniotic fluid of pregnant women, most Jews (Conservative, Reform, and Reconstructionist) had no moral difficulties with elective abortion. Synagogues frequently sponsored sessions for the young unmarried adults to have a blood test to screen for the presence of the gene in heterozygous form. In that Jewish population, about 1 in 25 adults carried the gene in heterozygous form. The chances two carriers would become reproductive mates was about 1/25 × 1/25 before genetic screening. The chances that two known carriers, represented by the cross **Tt** × **Tt**, would give rise to a homozygous child who is **tt** was 1 in 4. Thus, the odds of a Tay-Sachs birth by chance in the Jewish population was 1/25 × 1/25 × 1/4, or 1 in 2500 births.

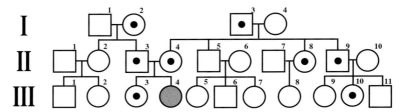

The occurrence and persistence of autosomal recessive traits. III-4 is the "bolt out of the blue" with Hurler syndrome. Neither side of the family had any recent occurrence of the disorder. Thus, I-2 and I-3 had no clue they were carriers and that granddaughter III-4 would bring the necessary genes together. The gene is in cousin carriers III-3 and III-10, who may pass that gene on to generation IV and farther. It may be a century or more before that mutant gene finds a matching partner to produce a child with Hurler syndrome.

If the **tt** baby dies before the age of five, it clearly does not pass on its genes. Thus, it does not matter if nature kills the **tt** baby or if a physician aborts the **tt** embryo. No change in gene frequency occurs. The odds for Jews today (who haven't seen such babies for a full generation) to have such a child is still 1 in 2500, and it will remain essentially that way no matter how many generations in the future abort the **tt** embryos.[8]

All autosomal recessive conditions that are lethal, or so severely incapacitating that the individuals with it fail to find a mate, will follow this example. Abortion of the homozygous embryo would only decrease (very slowly) the gene frequency of those autosomal traits that normally do reach reproductive maturity and have some chances of reproduction (such as albinos). Then the frequency (over several millennia) will level off to a balance between eliminations by abortion of the homozygote and additions to the gene pool of new mutations of the normal gene (normally about a 1 in 100,000 chance). By far the most common type of gene mutation of medical concern is the autosomal recessive (because we have 22 autosomes with some presently (2001) estimated 35,000 genes and better than a 90% chance that a mutation in any one of them will turn out to be autosomal recessive). Despite concerns about "loss of variation" from critics of eugenics, the use of prenatal diagnosis with elective abortion does not constitute a eugenic procedure because it does not change gene frequency for this very large category of inherited disorders.

CHANGES POSSIBLE FOR AUTOSOMAL
DOMINANT DISORDERS

Autosomal dominant disorders in humans, with very few exceptions, are of the parental genotypes **Aa** × **aa**, where **A** is the autosomal dominant trait (such as Apert syndrome, Huntington disease, retinoblastoma) and **a** is the normal allele that is recessive to the dominant allele. The outcome of this mating is a risk of 50% that a child will express the dominant disorder. If prenatal diagnosis could be applied to all known dominant disorders using appropriate DNA chip technologies, those at risk for familial cases (such as Huntington disease and many families with retinoblastoma or Marfan syndrome) will quickly elect to abort the embryos carrying the dominant mutant allele. In principle, after the Human Genome Project has worked out the sequences of all genes and their functions, all of the families at risk for dominant disorders could see these disorders disappear from humanity in one or two generations. It is important to recognize that this applies to familial cases. There will continue to be sporadic cases among two normal parents in which the mutation arises as a new variant in a once-normal gene. For many of these dominant disorders, this occurs about once in 100,000 gametes. These gametes are also more likely to be sperm than eggs because there is a higher likelihood of mutation in older sperm, which has many more rounds of replications of DNA than does the DNA of eggs. It is not likely that families who have no history of hundreds of dominant disorders would either know about them or worry about their occurrence by a rare mutational event.

The long-term consequence of this, if this analysis is correct, is that disorders like Huntington disease will nearly vanish and drop to a very low spontaneous mutation rate of occurrence (perhaps 1 in 100,000 births). Other disorders, like Apert syndrome, where almost all those with the condition do not reproduce (a few of the females with it are raped and become mothers with a risk of 50% of passing the disorder to the child), will not change because, for all practical purposes, the disorder is at its spontaneous mutation rate. In the United States about 60% of retinoblastoma cases are familial. These will drop to the spontaneous rate (which is what it is in many of the developing countries that lack medical facilities for early diagnosis).

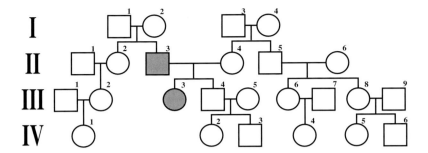

II-3 = Sporadic mutation to achondroplasia from a gamete of I-1 or I-2
III-3 = Daughter does not leave offspring
IV-1 through IV-6 = fourth generation of kindred is free of disorder

The natural elimination of autosomal dominant genes. A medically impaired individual with a dominant disorder is often traced for a few generations to a spontaneous mutation that starts the line of descent. If the condition is lethal or severely incapacitating, the line stops at generation II. If it impairs and the chances of finding a mate are low, it may stop at generation III or IV. The less there is an impediment to finding a mate, the greater the number of generations it will continue.

X-LINKED DISORDERS WILL DECLINE RAPIDLY

X-linked inheritance is a consequence of males having a single X and females having two X chromosomes. Less than 1% of the genes of the X are on the Y chromosome, and none of them is essential to life, because females do quite well without a Y chromosome. Males thus express whatever genes are on their X. This is called hemizygous inheritance. The classic case is red–green color deficiency. If a normal color-visioned male is **RY** and his mate is **Rr**, the father's Y-bearing sperm will produce sons who are **RY** (normal color vision) or **rY** (color deficient). The father's X-bearing sperm can only produce normal color-visioned daughters, half with the genotype **RR** and half **Rr**. It is the heterozygous mother who contributes the **r** allele. Because of hemizygosity, males whose X carries the **r** express color deficiency.

Color deficiency is not a life-threatening condition in most circumstances. For reasons not known, it is quite common in the human popula-

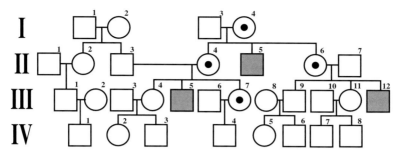

The natural elimination of X-linked traits. If muscular dystrophy arose as a mutation from a sperm of the father of I-4, it may get passed to her two daughters II-4
and II-6, who, like their mother, do not express it. However, II-5 inherits the
mutant gene and dies of the disorder. If II-4 passes the gene to her daughter III-7,
she may pass it on again. III-5 and III-12 die from the disease and cannot pass it
on. III-7 has one child, a normal son, and thus the gene is not passed on by that
line of descent. III-11, the daughter of carrier II-6, did not receive the gene and
thus cannot pass it on. Hence, the mutation that first appeared in I-4 is extinct by
generation IV.

tion; about 6% of American men are red–green color-deficient. In contrast, a condition like Duchenne muscular dystrophy (which results in a
wasting of the muscles, whose bundles of fibers detach from the cell mem-
brane of the muscle cells) will either lead to the death of an adolescent
male before he reaches reproductive maturity, or so incapacitate him that
he would not be able to mate. Let us assume a female carries this as a new
mutation that just arose in one of her X chromosomes (from her father's
X-bearing sperm, let us say). Her genotype is **Dd**. If she marries, her husband will be **DY** (normal). The **d** allele is the mutant allele for Duchenne
muscular dystrophy. How long will this gene last in her line of descent? Let
us assume she has two children. She could have two daughters, both heterozygous like her or both homozygous **DD**, or she could have one **DD**
and one **Dd** daughter. If she only had two sons, she could have bad luck
and have two that are **dY**. She could as easily have very good luck and have
both that are **DY**. She is most likely to have one son **DY** and one son **dY**.
Very clearly, the sons get rid of a **d** allele if it appears in the hemizygous
boy, or if good luck has given the boy a **DY** genotype, that mutant **d** allele
is gone from his line of descent.

The female line perpetuates the **d** allele. We can thus determine how many generations it will take if a line of descent each with two daughters finally gives us two daughters each of whom is **DD**. The alternative is **Dd**; the **dd** is not possible if, as is almost always the case, the father is normal or **DY**. The chances of extinction are thus 50% for each generation she reproduces. It is similar to a problem of tossing heads or tails. We can have a rare streak of 10 or more heads in a row, but we know this is unlikely. In fact, in seven generations the odds fall to 1% that the gene is still in that line of descent. In China, with a one-child family, the elimination is faster because it stops whenever a son is born or whenever a daughter is **DD**. Most serious hemizygous disorders will be culled out of the Chinese population (without elective abortion) in three or four generations. They are being culled out at a slower rate, on average, about seven or eight generations in any one line of descent that arises in an industrial democracy where the two-child family is the ideal. This is why most of these rare, very harmful hemizygous disorders are likely to have arisen only one or two centuries ago at most. A good example is the sudden appearance of factor VIII deficiency (hemophilia) in Queen Victoria's family (either her father or her maternal grandfather initiated that new mutation). Her large family allowed it to be spread to several royal families through marriages to the ruling houses of Europe. If factor VIII transfusions had not been developed in the mid 20th century, in all likelihood the gene would have disappeared from Victoria's descendants somewhere in the 22nd or 23rd century. Because those hemizygous sons with hemophilia will live to reproduce, the gene will have an indefinite perpetuation into the future.

If there is no effective treatment for a hemizygous condition and it does kill a male or limit his prospects of reproducing, most of those disappear on their own over a two-century period. Elective abortion would greatly shorten this to one generation, or at most to two generations, in families at risk. Despite the relative disappearance of such ill children, the spontaneous mutation rate on the X chromosome would not change and there would be substantial de novo cases that were not screened.

We can conclude from this analysis that X-linked disorders are similar to dominant disorders in their vulnerability to extinction. The world one or two millennia from now will still have its sporadic cases arising, but at-risk female carriers will be able to prevent a recurrence in their own families.

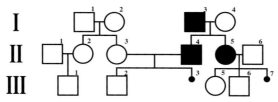

How dominant disorders are eliminated by prenatal diagnosis. If I-3 will die of Huntington disease and passes the gene on to II-4 and II-5, they could be screened. Let us assume both choose this option and they know that they will express this disorder in 10 to 15 years. At-risk pregnancies for II-4 included III-2 (prenatally diagnosed as normal) and III-3 (prenatally diagnosed and aborted). Similarly, II-5, who knows she carries the mutant gene, uses prenatal diagnosis for III-5 (normal), III-6 (normal), and III-7 (carries the mutant gene and aborted). Thus, for generation III, survivors III-2, III-5, and III-6 are assured the gene is no longer in the family.

POLYGENIC DISORDERS ARE MOST LIKELY TO DIMINISH

A polygenic disorder is one that has several genes involved in the expression of a trait. They are usually compiled into empiric risk tables. A genetic counselor might be able to tell a parent of a child born with a con-

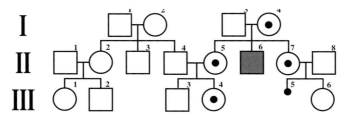

How X-linked traits are eliminated by technology or by natural selection. If muscular dystrophy appears with II-6, daughters II-5 and II-7 can be screened. Both are carriers. II-5 has her two pregnancies checked by prenatal diagnosis. III-3 is healthy and III-4, although a carrier, is born and she, in turn, will choose prenatal diagnosis for her pregnancies. The pregnancies of II-7 reveal by prenatal diagnosis that III-5 will have the disease and the parents abort the pregnancy. Prenatal diagnosis reveals III-6 is not a carrier. Because each carrier has the option of monitoring each pregnancy, the mutant gene will eventually cease to appear because it will only be carried in half the females. The same result will occur if no prenatal diagnosis is chosen. Only if carriers are aborted will the elimination occur more rapidly.

genital heart defect that the chances of a subsequent child having a cardiac defect is 3% (instead of the 1 in 3000 chance that the general population has). This is true for a variety of known defects, such as cleft lip and palate or blocked guts (e.g., pyloric stenosis) or neural tube defects (such as spina bifida and anencephaly). Of these, only the neural tubes can be successfully screened by a biochemical test; in the communities using voluntary screening (Great Britain and more affluent counties and cities in the United States), the incidence of babies born with these conditions has fallen dramatically. That is because almost all the pregnant women are given a blood test to look for a biochemical signal (called alpha-fetoprotein), and almost all the women who learn they have an embryo with a serious neural tube defect elect to abort it. Very little is known about the number of genes in such traits or what they do. There is a great deal of hope that the Human Genome Project will allow the first insights into such complex inheritance. As far as is known, there are far fewer types of polygenic disorders requiring major medical attention than birth defects associated with single-gene defects. We just know too little to be certain. But if the alpha-fetoprotein model is the polygenic model of the future, there will be a rapid disappearance, because only a few key genes on a chip could be used for routine screening of all who wish to use that assay.

Even if there is an elimination by elective abortion of such infants, this does not mean that their incidence (accompanied by a constant screening out by elective abortion) will dramatically disappear among conceptions. It may take millennia before such combinations are substantially reduced.[9]

WHAT DOES THIS ALL MEAN FOR THE HUMANITY OF THE FUTURE?

I chose a worst-case scenario for those who fear eugenics, who favor environmentalist interpretations of almost all human behavior, who urge a morality based on natural law, who believe scientists should not play God, who support right-to-life ideology, or who believe that nature should be left alone at the reproductive level. I did so to show that many of the claims made against genetic screening and prenatal diagnosis are false. Even with a full use of the Human Genome Project and the introduction of cheap, accurate, and easy-to-use techniques for screening large numbers of genetic disorders, and with almost all of humanity choosing to ignore the positions of those opposed to using this knowledge, there will be no major diminution of genetic variation in the reproductive cells of future generations. What will diminish is the number of very sick babies born or conditions that have debilitating effects on children or young adults. Each generation will add more conditions they would prefer to see prevented rather than treated. This will not lead to a super race. It will not lead to a uniformity of personalities, physical appearances, or a new species of humanity.

In none of these scenarios did I invoke positive eugenics, the selection or addition of allegedly beneficial traits to one's own gametes, or other measures that some call "designer children." I am not convinced that a model of positive eugenics is desired by most of humanity (it certainly is not desired as the 21st century begins).[10] No one can predict the values of humanity several generations from now. There may indeed be a future sentiment for humanity taking evolution into its own hands (by voluntary means), and this could lead to changes. We must remind ourselves, however, that we have no way, with our limited knowledge today, of predicting outcomes, good or bad, from such conscious efforts to redesign our bodies and brains.

MUCH ADO ABOUT CLONING

Advances in medicine have led to contradictory predictions of their long-range consequences. One school, we saw, believes that applications of new medical technologies will lead to a deterioration of the human gene pool. Another school believes that the applications will lead to a loss of variation should people choose options that prevent the births of babies with genetic disorders. As is often the case when complex social problems are studied, the answers are suspect. There is too much uncertainty and too little understanding of the components of human reproductive practices to provide reliable conclusions. Those who believe that deterioration will result base their claim on the restoration to normal function of individuals who would have died or remained unmarried because of their inherited medical problems. By living on to reproduce, persons with these less adaptive genes may transmit them to their offspring instead of having them selected out as they would have been prior to the introduction of new medical technology.

Those who believe medical technology fosters an undesirable eugenic agenda claim that parents who try to spare themselves of malformed and sickly children deprive the world of a bounty of variation that may be useful in changed world environments. Also adding to their fears is the belief that deliberate eugenic practices such as using sperm banks supplied by donations of geniuses, or cloning classes of humanity, would dramatically alter the world into genetic castes similar to those described by Aldous Huxley in his satire, *Brave New World*.[11]

Huxley's vision of hordes of epsilon laborers exploited by more privileged classes in a controlled society has made eugenics a mischievous doctrine in the view of most of humanity. Such a world is not the world that some advocates of positive eugenics sought. It may be no more representative than Kafka's *Amerika* is of American culture or Gilbert and Sullivan's *Mikado* is of Japan. Huxley's anti-Utopia is less ordered by inherent behavioral traits than by psychological conditioning, mind-altering drugs, and skillful propaganda. One does not need a genetic clone to produce a goose-stepping horde of obedient and enthusiastic troops. Huxley's argument that an island of alpha-plus geniuses would degenerate into warring factions resentful of doing common labor is shallow and falsely assumes that bright people are lazy, spurn physical work as undignified, or lack the imagination to replace drudgery with technically more efficient ways to minimize such work.

Cloning came back with a vengeance as the 20th century ended, with the cloning of Dolly, Ian Wilmut's famous sheep, in 1997. Since then, the work has been confirmed by the cloning of more sheep, cattle, monkeys, mice, and pigs. The initial belief, since the work of Briggs and King and John Gurdon on frogs and toads in the 1960s, seemed pessimistic. The smaller the cells and the later the stages of the embryos used for nuclear transfer, the less likely it was that they would form a clonal embryo going to term. Mammalian cloning was thought impossible on the supposition that too many genes had been switched on or off and no differentiated cell would be able to have its nucleus work inside an enucleated egg. Wilmut proved this was false.

Although human cloning has not been done (so far) for technical reasons and fears of impaired health, there are many who believe the moral reasons for its prohibition are likely to be ignored. These moral reasons include a prohibition on "playing God," interfering with natural processes (a sort of Frankensteinian outcome that might be an untoward consequence), violations of natural law in the creation of such clones, creating psychological difficulties for the participants involved, and sliding down a "slippery slope" to a world largely dominated by cloned populations.

Each of these prohibitions has been challenged. We all play God when we act as parents and decide everything from our child's religion to eating habits. Some play God by sending armies into battle (presidents, generals). Some play God by sentencing people to terms in prison or death (the judicial system). Some play God by creating and abandoning the means of work hundreds or thousands may depend on (corporation officers, speculators in stock markets). Life is filled with people making decisions over which most citizens have little or limited control.

We live in a very unnatural world and wouldn't want to be in a state of nature, say several millennia ago. We want our public health, germ theory, antibiotics, green revolution, transportation by cars, trains, and planes. Thousands owe their lives and ability to survive to antiseptic surgery and the gifts of organ transplants, artificial heart valves, and a variety of prosthetic devices from hearing aids to pacemakers. We reject natural law for our reproduction, and most of the world uses artificial means to regulate family size and compensate for infertility by assisted reproductive technologies. We consider (if we are free to exercise that option) how many children we bring into the world our business and not the act of fate nor the dictate of political or religious leaders.

Slippery-slope arguments invoked by philosophers are speculative. We have had prenatal diagnosis in the United States since 1968, and virtually none have used it to produce boys rather than girls (or the reverse). Only in nations with strong prejudices about gender are such sex selection methods used. Sperm banks for geniuses have largely been ignored despite widespread publicity given to them when they were first introduced. It is as easy to detect a carrier of cystic fibrosis, Hurler syndrome, or any other disastrous single-gene mutation as it is to detect an individual homozygous for it. Yet in all these years, hardly anyone has asked for a eugenic abortion to protect the future by having only homozygous normal embryos going to term. Just because new technologies are introduced does not mean they are automatically abused in a slippery-slope slide to foolishness or disaster.

The strongest argument for a moratorium on human cloning is the psychological argument. What does it do to a partnership (married or otherwise) if only one partner's genotype is used to generate a child? What does it do to the parent–child relationship when a clonal twin turns out to be different in behavior from the donor parent's expectations? What does it do to the clonal child to see the behavioral outcomes and physical disabilities and appearances of the donor parent raising that child? We know very little about such reactions.

No doubt there will be an avalanche of new knowledge about our human biology as the human genome is carefully analyzed and compared to those of other organisms. The new cloning techniques will be widely used in those agribusinesses seeking better quality control of commercial products (including human pharmaceuticals from human gene insertion) from animals. It is very difficult to make predictions from a very limited knowledge of the future. Nothing in these new technologies suggests an abandonment of sexual reproduction between two individuals to produce their own children. Somatic changes by gene therapy will have no effect on the genes present in the ovaries or testes of those individuals. We have had the knowledge of differential breeding in humans for over a century, and eugenic movements have come and gone with virtually no effect when those programs were based on individual choice. Our fears of a future eugenics should not be based on the introduction of new technologies that have been ongoing since human culture began, but on the very real threats of abuse of power by totalitarian systems, imposed mischievous laws in democracies, and foolish fads more likely to harm a few than to convince informed people to abandon their reason.

FOOTNOTES

[1] Evolutionary psychology is an offshoot of sociobiology. It uses arguments based on inferred adaptations to interpret innate behavior. E.O. Wilson is the acknowledged leader of this school of evolutionary thought. See E.O. Wilson, *Sociobiology: The New Synthesis* (Harvard University Press, Cambridge, Massachusetts, 1975), for the scientific evidence for this view.

[2] There are about 8000 web sites (in the year 2000) for eugenics. *Future Generations* has received a considerable web traffic; it uses its site both to promote its views and to rebut the work of its critics. The quality of these sites varies from polemics to

presentation of data, but since there is no peer review on web site dissemination of views, the reader is likely to feel bewildered by the mutual accusations of distortion and intellectual dishonesty.

[3] Mr. Lee's work is discussed in Chapter 9 of *Ethics, Reproduction, and Genetic Control*, edited by Ruth F. Chadwick (Routledge, London, 1987). See Chee Choon Chan, *Eugenics on the Rise: A Report from Singapore*, pp. 164–171.

[4] For a discussion of the 1994 eugenic law, see "American Society of Human Genetics Statement: Eugenics and the misuse of genetic information to restrict reproductive freedom," *American Journal of Human Genetics* 64(1999): 335–338. The major discussant and author for that statement is Philip Reilly. Of interest is the observation that a very similar eugenic law was enacted in Taiwan several years earlier with little or no outcry from the world press.

[5] *The Foundation for Germinal Choice* was in Escondido, California. They provided a 15-minute videotape interview with the late Dr. Graham. The tape is part biographical and part procedural information on the obtaining of sperm donors and the procedures for obtaining the sperm. The foundation also provided a copy of its lengthy questionnaire used for its prospective donors or recipients. After Dr. Graham's death, the foundation closed in 1999.

[6] Among the major critics of genetic determinism in human behavioral traits are Stephen J. Gould and Leon Kamin. Gould's *The Mismeasure of Man* (Norton, New York, 1981) reveals a history of unconscious bias in data recording and interpretation among many 19th-century advocates of human inequality based on race and social class. Kamin's *The Science and Politics of IQ* (Lawrence Erlbaum Associates, Hillsdale, New Jersey, 1974), examines the alleged genetic differences in IQ scores among populations based on race and social class and finds those data and their interpretation to be spurious. Kamin was the first to question the integrity of the core of separated monozygotic twin studies compiled by Cyril Burt. Burt's coworkers on his papers are either nonexistent or they (or their relatives) have never come forward in his defense. Kamin also shows that many of the data were faked.

[7] It is well known, through the Hardy-Weinberg law, that gene frequencies remain stable in large populations. Factors that alter these frequencies, such as mutation of the normal gene, selection for any of the genotypes of the genes involved, or size of the population, have been studied, and their consequences are mathematically predictable. At the level of the family, a simple Mendelian model of probability prevails in determining the likelihood of a gene being transmitted in a pedigree. The extinction of genes over several generations was studied by H.J. Muller, who focused on both elimination by homozygosity and elimination by partial dominance of the recessive genes. See H.J. Muller, "Our Load of Mutations," *American Journal of Human Genetics* 2(1950): 111–176. Muller referred to this heterozygous route for extinction of a deleterious mutation as genetic load.

[8] For some Orthodox Jews (for whom abortion is not permitted), a system of arranged marriages still prevails. The marriage broker has a file on the carrier status of all eligible young men and women in the Orthodox community. When parents seek a spouse for their child, the broker supplies names of eligible candidates, making sure that the matching partner is not heterozygous if the client's child is

heterozygous. This system thus permits **Tt** × **TT** and **TT** × **TT** but not **Tt** × **Tt** matings. The result over a long time (dozens of generations) would be a slight increase in gene mutation frequency (from mutations of the normal **T** to the mutant **t** allele) because the prevention of homozygosity would eliminate one route for genetic extinction. This method would not alter eliminations through genetic load, the effect of the partial dominance of the **t** allele in the population.

[9] The polygenic traits are similar in their genetic analysis to the traits studied by Johannsen for continuous quantitative traits such as bean size. In Johannsen's classic studies, intense selection and inbreeding took approximately 10 or more generations to render a selected size (smallest or largest) homozygous. No such intense inbreeding exists in human populations, and in all likelihood with selection alone there would be little loss of variation in the population for a millennium or more. The overwhelming bulk of these individual genic components would still be in the population. Only their rare combinations (like those of winning lottery numbers) would decrease slowly each generation through elective abortion of traits that allow some opportunity for surgical repair and reproduction. Wilhelm Johannsen's work was mostly in Danish during the active years of his best-known research (1903–1909). He summarized his views in book length in 1909 in German: *Elemente der Exakten Erblichkeitslehre* (G. Fischer, Jena).

[10] I base my judgment on the poor success of Graham's Foundation of Germinal Choice. Graham said he had great difficulty getting Nobelists to donate their sperm. Despite the national publicity given to the foundation over the years, there have been few requests for its sperm. One could argue that potential users are wary of its old-line eugenic values. There has been no strong demand by sperm banks (or the clients of these banks) for the sperm of famous artists, poets, novelists, musicians, sports champions, actors, scientists, or corporation executives (assuming, of course, the unproved belief that there are genetic components for their eminence). I do not deny that values and preferences might change. The technology for obtaining such sperm has been around for almost a half-century, but there has been no effective social movement motivating people to use such semen sources.

[11] Aldous Huxley's *Brave New World* appeared in 1932. Huxley reacted to the scientific predictions of his brother Julian and J.B.S. Haldane. The satire describes the political development of totalitarian societies that govern all aspects of its citizens' behavior with little room for individuality or dissent. To achieve this harmony of obedience to society, conditioning plays a major role. Huxley also assumed that what eugenicists were telling him is true—much of personality, talent, and intelligence is inherited. Curiously, those who criticize eugenics the most often do so not because its assumptions about heritability are wrong or unproven, but because they believe the central thesis of eugenics is correct and the major differences in human behavior are genetic!

CHAPTER

21

Dealing with Life's Imperfections

ALL STAGES OF THE HUMAN LIFE CYCLE are vulnerable to suffering, disease, and death. It is our good fortune to live in an age in which these personal tragedies are less frequent and less apparent than in all previous generations of humanity. We are beneficiaries of public health, the near-universal acceptance of the germ theory, a year-round wholesome nutrition, the pasteurization of milk, compulsory immunization programs, and antibiotic therapy. Virtually none of this was available a century ago. Just 150 years ago, when Thomas R. Malthus and William Godwin were debating the merits of "vice and misery" as checks to population growth rather than the "perfectibility of man," both agreed that half of all children born would die before reproductive maturity, most of them as infants.

What so profoundly changed the human condition in the late 19th and early 20th centuries was the intervention by science in human life. From every discipline, ideas fed into applications, and instead of half our children dying, almost all of them survived to become adults; instead of uncertain yields of food, hybrid corn and rust-resistant wheat created an era of plenty; instead of seven or eight children, the family size of industrialized nations has shrunk to two or three. Gone are the scourges of typhoid, diphtheria, smallpox, pneumonia, tuberculosis, and dysentery. Accompanying the greatly enhanced survival of the newborn was the introduction of mass universal public education; the shift from the farm to industry and service occupations; the rapidity of travel and communication among nations; and the multiplication of the universities based on a new model of scholarly research.

It is difficult for us to imagine a world without these benefits, and we would reject those who would return us to a more natural world or a less artificial world, if we could not graft onto this the longevity, freedom from want and disease, and educational opportunities we cherish. In achieving these much desired changes in the human condition, we have actively or passively changed our values. We rejected the view that the natural is better than the artificial; we rejected the view that fate (the will of God) should predominate over human intervention; and we rejected the view that there are parts of the natural world, especially life processes, that are inappropriate for inquiry. We have taken charge of our health, of our reproduction, and of our psyches. We have sacrificed the hand-holding intimacy of the family physician for the detached professional competence of the medical specialist.

SYMPATHY AND HOPE AS MOTIVATION FOR WELCOME CHANGES

Most of the misery and vice a century or more ago was caused by malnutrition, infectious diseases, and inadequate housing. In an age when these were part of the human condition and taken for granted, the only recourse was prayer. The natural human sympathy for suffering individuals was supplemented by the Golden Rule. The scientist applying this rule eased pain and suffering by preventing disease and, to a much lesser extent, by treating it. Effective treatment of the major causes of morbidity and mortality did not occur until the 1930s, when the sulfa drugs proved to be the first antibiotics.[1]

The decision to intervene in what had been nature's monopoly on human misery was an assessment that a natural wrong should be righted. If it was unfair for an innocent child to get sick and die (assuming the cause was an agent of nature's or society's inefficiency but not God's judgment to punish the parents or somehow educate the public), then a person having the knowledge to treat the child should do so. The development of effective public health and medicine as a means to prevent natural agents (especially germs) from destroying otherwise normal individuals can be interpreted as an application of the Golden Rule. By extension, this would also include special education, programs for the physically handicapped, and welfare programs to keep dependent children in good health.

The consequences of the application of the Golden Rule bring more than a righting of a natural wrong. By preventing early death, parents are left with larger families and more mouths to feed. Here the Golden Rule no longer applies, and a second, compensatory ethic must be applied. A larger family is not itself a disease or a natural wrong. It can be regulated by family planning, but that requires an acceptance of human control over reproduction. That, in turn, requires a knowledge of the reproductive cycle and the potential to intervene by natural or artificial means. For most of humanity today, at least in the industrial nations, that choice has been through artificial means. Vasectomies; salpingectomies; chemical, hormonal, and mechanical means; as well as abortions, are favored over the less reliable means of abstinence during an uncertain and often irregular pattern of ovulation.[2] The value associated with a smaller family size is not health but the quality of life. We would rather put more effort in a few children so that they can be well educated and receive our attention than find ourselves spread thin with too few resources, too little time, and a self-sacrificial image of parenthood. We see ourselves as having the right to a full life; one that gives father and mother opportunities for careers, for intellectual and aesthetic enjoyments, for social lives that extend beyond the immediacy of child care. We prefer "the perfectibility of man" rather than "vice and misery" as our image of the human condition.

If we have fewer children in order to provide them and ourselves with better lives, it is not sympathy that guides our values but expectation or hope. We expect these well-nourished, well-clothed, well-housed, and well-educated children to be a blessing to ourselves and society. We nurture their academic successes, their creativity, their enrollment in colleges, and their choices of professions. We want a world that is more mechanized, with less drudgery, with more jobs requiring thinking and aesthetic skills, and with more leisure time to pursue our interests.

HOPELESSNESS AND RESIGNATION AS FORCED MOTIVATIONS FOR MAINTAINING THE QUALITY OF LIFE

Infectious diseases and malnutrition are external agents or forces that deprive individuals of their potential for a good life. These have been, for the most part, successfully removed in the industrial nations. Remaining as part of the human condition are internal agents or processes that mete out a modest share of human misery, including sterility to about 10% of marriages, spontaneous abortion to about 20% of all women who experience pregnancy; birth defects to some 5% of newborn babies, and premature illness and death to adults from cancer, degenerative diseases, strokes, and heart disease. These too call for sympathy as a response. Where the technology is available to provide pacemakers, plastic valves, dialysis machines, organ transplants, hearing aids, or artificial joints, we accept the lesser of two evils and function with our artificial prostheses or donated organs. We know that the defective organ system cannot be restored miraculously to normal function, and we must get by with the substitute. There is no hope for the defective organ, but there is still hope for regained function. We resign ourselves to the loss but retain hope through technology for a replacement.

At all stages of the life cycle, this treatment by replacement has become an accepted alternative to the much more desirable cure that seems so easy to achieve for most infectious diseases. The novelty of each substitute method is at first shocking and meets some resistance because of the disturbed self-image it creates, but after living with the artificial hip, accepting the jiggling valve in the heart, or appreciating the now-functioning donated kidney, the beneficiaries convince the rest of us of the legitimacy of these artificial devices.[3]

The prevention and cure of infectious disease and the treatment of lost, malformed, or hopelessly ruined organs by replacement are procedures that are suffused with sympathy and hope. Not all disorders, however, fall into these two categories of cure and replacement. The terminally ill cancer patient, the aged patient with congestive heart failure, the profoundly retarded infant, the dying child with a lysosomal storage disease, or the newborn with an extra chromosome 18 in all of its cells are examples where no medical technology exists to right the wrong.[4] Neither science nor prayer restores these individuals to normalcy. For the relatives of those involved, acceptance replaces grief; hope is abandoned. It is this residue of misery that spawns the most controversy. Should one prevent recurrence of the Tay-Sachs child or risk another? Should one elect to abort a fetus without hope? Should one withhold heroic measures on those who will not benefit from the treatment? Should one institutionalize the retarded child or raise it at home? Should one correct defects that prolong life but do not restore health? Should one withhold nutrition from the terminally ill if they no longer can feed themselves? Here, in all its misery, we wrestle with the image of the person that cannot be, and all our choices are unpleasant.

It is also here that a wealth of values confront each other. Neither the Golden Rule nor the hope for a better life prevails. Instead, we pit relief from pain and suffering against our guilt at being active or acquiescent mercy killers. We rationalize that a child may get better care at an institution than we can provide in our solitude and torment ourselves that what we are really asking to do is to abandon our child. We feel inwardly that taking on parenthood should be an opportunity, not an inborn disaster over which we have no control. We withdraw personhood from an embryo or fetus when we learn it is abnormal and justify its abortion by convincing ourselves that it is to spare the potential child (and less so ourselves) from misery that we act now. No matter how bad these examined choices may be, they seem preferable to the futility of prolonging dying in the most artificial and undignified setting, bringing into the world those who cannot function effectively in it, remaining sterile when odds favor a technology that can bring us a normal child, or devoting our lives to a profoundly sick or retarded child. When all hope is abandoned, values collide.

THE RISKS OF ABANDONED HOPE

Whether people have hope or have abandoned it, we have assumed that the victims of the misfortunes of the human condition have acted in good faith. When belief is transformed into factual certainty or when faith becomes fixed as ideology, however, moral mischief is a possibility. Several times in the late 19th and early 20th centuries these transformations have occurred. It was believed that there were dangerous classes who represented the paupers, criminals, and feebleminded. In the mid-19th century, where hope was pervasive, these classes were thought to be morally diseased and curable through reforms of poor laws, conversion of prisons into "moral hospitals," and construction of wholesome asylums with trained professionals. By the end of the century there were still the poor, the criminal, and the retarded. All were abandoned. The paupers were declared innately lazy and parasites of society; criminals were said to be born with associated stigmata of their facial features; the feebleminded were seen as a corrupting menace, transmitting their defective minds in uncontrolled hordes as they outbred the more rational middle class.

These fixed degenerate classes, by the 20th century, had become the unfit. In the United States, legal and social remedies were sought. Some, especially the feebleminded, were isolated by containment in asylums, not for treatment, but to protect the public from their presence. From 1907 on, some 30 states enacted laws permitting the sterilization of different classes of unfit. The false underlying assumptions were that the unfit were not curable, and their origin was from their own kind. Sterilization was seen as a preventive medicine. In the Third Reich, Nazi ideology was fixated on purification of Aryan blood. There was extensive sterilization of the unfit supplemented by a secret program of euthanasia. The insane, the retarded, and the malformed disappeared. Much more horrifying was the mass murder of those deemed racially unfit—the Jews especially—with long-range plans for the murder of Poles and Slavs and the repopulation of Europe with a Teutonic race.

It would be wrong to suppose that the moral evil of Nazi ideology could be seen in milder form among the earlier efforts at classifying the unfit or seeking ways to remedy their problems. The eugenics movement had wide and popular support among the middle class. It included liberals and conservatives, the hopeful and the pessimistic, idealists and reactionaries, environmentalists and hereditarians; it included physicians, clergy, social workers, professors, publishers, journalists, and philanthropists. The belief in human betterment was as reasonable a belief to many of the educated professionals of the early 20th century as their belief in the necessity and intrinsic good of public health and medical science.

Judged by the Holocaust, all efforts at human betterment are tarnished by a eugenic brush. This can lead to an outright condemnation of new technologies applied to health. What distinguished Nazism was its massive application by the state of medical and scientific technology to an ideology of bigotry. The destruction of Nazism did not destroy bigotry, technology, nor the power of the state over the individual. The potential for abuse is real, and the fearful have good reason to remain on guard.

The Ambiguity of Moral Guidance for Life Crises

The most important choices of our lives are made with incomplete knowledge and insufficient time to sift all the information to guide us to a confident decision. The parents who are confronted with a baby with a profound birth defect must make decisions to operate or not to operate, to call in specialists or stay with the recommendations of their attending physician, to extend hopeless life or to let nature take its course. Whether informed consent to proceed is granted or denied, it can hardly be called informed. The relatives of the terminally ill face similar crises of choice, and their own values may be negated by a legal system fearful of offending contending values.

To what extent does abandonment of hope for the terminally ill adult, the trisomic fetus, or the malformed newborn resemble the assignment of status as an unfit person? Note how, by our choice of terms, we foreclose options. To call abortion murder is to forbid it. To apply the term euthanasia to the withholding of medical procedures that are not themselves treatments for the person's disorder heaps guilt and immorality upon the next of kin. To equate prenatal diagnosis with eugenics makes it a search and destroy mission. To see artificial insemination as a sophisticated adultery or eugenic abuse condemns it. Much of our moral decision making is guided by the overladen vocabulary we, our physicians, or our culture uses to describe our crises and choices. We can reverse the connotations of our vocabulary and force the moral decisions in a different direction by referring to abortion as a medical option; by describing the crisis of the terminally ill patient as a time to die; by calling artificial insemination a procedure of half-adoption; and by looking upon genetic counseling as providing the options for informed decision making. We mask our values by the way we phrase our sentences.[5]

If we sometimes become confused by the description of our choices in a time of crisis, we are more likely to be so when we confront new technologies and their applications. Here we do not have the advantages of multiple presentations in the media with their varied criticisms. Nor do we have the luxury of time that permits examination in novels, poetry, drama, and essays. How do we respond to the compilation of a human gene library? Is it an intrinsic good to know all our potential genetic disorders?

Is it an evil or a good to have an arsenal of probes to identify potential carriers and potentially affected fetuses? How much can we read into our destiny from a knowledge of all our genes? Will this lead to a spurious genetic palmistry? We cannot answer these questions because we do not know enough about gene action in the formation of organ systems and their functions. Even if we did know the biological consequences of our genes, we cannot answer these questions because the uses of all knowledge are not fully predictable, nor are the applications of knowledge inevitably destined to follow a path our imagination discerns.[6]

Although the term eugenics has suffered many definitions and its use today is usually pejorative, the underlying theme of differential reproduction and its consequences is a valid biological concern. All human reproduction has eugenic consequences in this basic sense of the term. Consider the changes in our own century. It is more common today than a century ago for a person to marry someone who lives hundreds or thousands of miles away. It is more likely that a college-bound youth will meet and marry someone at that college rather than from his or her own hometown. Persons who would have succumbed to infectious diseases a century ago are now enabled to marry and reproduce. Although we acknowledge that differential reproduction is taking place through assortative mating and the survival of those who would have been carried away in the not-too-distant past, we do not know the exact consequences of these processes. We just do not know enough about the genetic differences involved to make meaningful predictions. We do not know how much of a genetic involvement there is in complex human behavioral traits such as intelligence, talents, and personality. It is risky and unwise to build personal and social policy on indirect and uncertain estimates of such heritability. Although this interpretation may be true and may serve as sufficient reason for restraint, it is very different from the two most prevalent claims of opponents of eugenics—outright denial of any significant hereditary influence on behavioral traits or the proffering of an unproven anti-eugenic hypothesis that sperm banks, differential breeding, germinal choice, or other voluntary means of eugenic breeding will lead to a harmful diminution of variability in the population.

There is an embarrassment of riches for the ways in which sterile couples might become parents. The husband's (or someone else's) sperm may fertilize the wife's (or someone else's) egg in the wife's (or someone else's) oviduct (or in a glass dish) and be gestated in the wife's (or someone else's) uterus with all third parties either freely giving their gametes (or uterine time) or being paid for this service. Despite all these possibilities, the success rates for those involving surgical procedures for their infertility hover at about 10% to 40%. It is also an expensive and time-consuming process for sterile couples to become fertile. The easy solution of the past, adoption, is no longer easy because there are so many more sterile couples than adoptable children. As the techniques in this field improve and costs go down and success rates go up, the demand for reproductive options will increase. Moral opposition to those who elect to circumvent their sterility by these biomedical techniques is largely academic to those who ignore this opposition. That it is an indignity to the family; that it is against natural law; that it involves masturbation; that it may involve morulacide or blastocystocide may be transiently acknowledged by the sterile couple, but it does not convey the same gravity of decision making as the withholding of medical procedures to the dying or the malformed. On a spectrum of increasing guilt, they fall just to the right of birth control by artificial means. Nothing could be more potent a justification in the minds of the parents than the adoring hugs and smiles of the child they brought into being by unconventional means. We are not a generation that likes to bear crosses.

Equally ambivalent is our attitude toward molecular medicine. The initial reaction was one of fear. Somehow isolating genes, cloning them, or splicing them into bacterial cells was looked upon as risky because we did not know the changes that might take place in the recipient cells. This was a wisely debated controversy, and it led to a much-needed temporary caution that provided more careful laboratory habits and a step-by-step escalation of risk as each stage of success in testing for safety reassured the scientists that their research plans would not be put on permanent hold.

Little objection is now made for the applications to health of synthetic human hormones, blood clotting factors, interferons, clot dissolvers, and other natural biological products that are copied from isolated human genes or their appropriate DNA sequences. More concern is focused on the transfer of the normal gene into cells containing a mutant allele that results in a genetic disorder. Certainly caution is required for technical reasons; at present the normal gene cannot replace the defective gene by switching places with it. Multiple copies of the normal gene may be inserted in different chromosomes, and their effects on the functioning of these host sites are not presently predictable. A go-slow process is both scientifically and socially required. This, however, is not the reason critics of gene therapy are worried. They often fear that the molecular physician may substitute the normal gene for the mutant gene in the germinal tissue and thus practice eugenics.[7] Even if this were the intent of the physician, it would be a costly and labor-intensive means to alter every potential sperm or egg to achieve that end. Elective abortion of a prenatally diagnosed affected fetus already accomplishes this goal. Gene replacement would be the equivalent of trying to construct a towering cathedral using one-centimeter bricks and a pair of tweezers to align them.

Far more important in reflecting on molecular and prosthetic medical practice today and in the years to come is how far we are willing to be substituted, repaired, or artificed for our various ailments before we lose an identity with ourselves. It might be acceptable, although costly, to find one's declining years assisted by false teeth, hearing aids, eye glasses, a pacemaker, a metal hip, a metal knee, reconstructed fingers, a donor's kidney, a smidgeon of fetal neural tissue to ward off Parkinson's disease, an inserted insulin pump, and a policeman's belt loaded with the daily medications to carry out what our spent cells have long failed to do. Those who cling to their lives will no doubt say, "A man's a man for 'a that." Those who develop an identity crisis may entertain doubts about life for life's sake.

What then are we to do with the deluge of options and techniques that modern medicine provides? We cannot reject quality-of-life values because that is what we invoke when we treat to cure. We want to restore a quality of life compatible with what existed before. We cannot assume that prolongation of life is an absolute good, and we must temper our hopes for recovery with the bitter possibility that this is not possible. We must weigh our personal happiness or purpose in life against the limited potentials of a disordered fetus or newborn. We must weigh our capacity to function and find some meager satisfaction out of life against the burden of pain, indignity, and loss of identity associated with too much medical attention. In sum, we must learn to live in an uncertain tension between the curative values of sympathy and hope and the stoic values of hopelessness and acceptance.

FOOTNOTES

[1] A good example of this is Lewis Thomas's description of his father's medical practice in *The Youngest Science*. His father told him that the medicines he gave his patients did not do any good but were necessary for the patients' sense of well-being. He told his son that a physician's contribution was a good diagnosis and prognosis. Later, in the 1930s, Thomas describes the excitement among young physicians when the first cures of pneumonia occurred after the administration of sulfa drugs.

[2] There is considerable individual and religious ambivalence about the significance of biological parenting. An environmentalist should not really care whose sperm or egg gives rise to his or her child, because it is the environment and not innate factors that determine the child's intelligence, personality, talents, and success. Yet environmentalists, like most people, want their own children if at all possible. It is not clear what forms the religious basis for revulsions and condemnations of "unnatural" or "illegal" or "sinful" means of procreating. The terrible exclusion of the *mamzerim* (Chapter 1) and the status of bastards through history are examples of this preference for the children contracted for in marriage. Children conceived by their own parents' gametes, as in in-vitro fertilization, cannot fall into the category of bastards, but their "unnatural" state of conception marks them or their parents as transgressors at least to the most orthodox of religious belief. This is particularly puzzling when one considers the number of married couples who have children the natural way but whose motivation may be suspect, including insobriety, jealousy, "wife rape," and fantasizing intercourse with someone else! No similar sin, stigma, or admonition seems to have accompanied that motivational basis for procreation. In contrast, one may claim that many infertile couples have the most caring and positive attitudes and love to share with their IVF child. Consistency is apparently not a hallmark of the basis for religious or moral belief.

[3] The strangeness of new technologies is not limited to the patients who use it. When Helmholz was a young physician, he was berated by older physicians for using mechanical instruments, such as the thermometer, the speculum, a pocketwatch (for counting the pulse), or a rubber hammer (to tap the chest or knees), because these degraded the patient and demeaned the human into a mere machine. See John Theodore Merz's *European Thought in the Nineteenth Century* volume 2, pp. 388–399 (William Blackwood & Sons, London & Edinburgh, 1912).

[4] It is difficult to predict which individual genetic disorders or developmental birth defects will eventually find "cures" or treatments that address the primary disturbances of the disorders. It is difficult to conceive of replacing parts that are missing, as in infants born without kidneys or lungs. It is difficult to imagine a combination of surgery and medical treatment that would address the 80 or more defects found on autopsy in the typical trisomy 18 or trisomy 13 child who lives a median life expectancy of 3 days. And it is difficult, in many lysosomal storage diseases (like Tay-Sachs) to conceive of a way to deliver the enzyme hexoseaminidase A so that it will enter all the appropriate neurons of the fetal or newborn baby's brain (or, if extreme opponents to genetic therapy were to permit it, to introduce the normal

gene to manufacture that enzyme so that a copy of each normal gene resides in the nuclei of those neurons).

5   One of my colleagues, Frank Myers, who is far from being a conservative, once told me that "we have too many rights." He felt that rights, as in the Bill of Rights to the U.S. Constitution, are nonnegotiable (e.g., freedom of speech). When movements like "Right to life" or "Right to clean air" arise, he believes they trivialize our basic guarantees, because by claiming these moral or personal philosophies as "rights," they become demands and are removed from the normal political process of negotiation and compromise. The phrasing of moral positions on reproductive and health issues is often presented as a conflict of "rights."

6   My personal view on the significance of the total nucleotide sequence of the human genome may be summed up by the following thought. We cannot read Shakespeare's sonnets in the sequence of his nucleotides. What is essentially human, to us, is what we do as persons. Much of what we do is learned and shaped by the particular lives we lead. At best, there are constraints and potentials that our genotypes provide, but it seems doubtful to me that a full knowledge of those genotypes would guide us on which of many dozens of careers and outcomes any individual would chose for his or her life.

7   I am not sure what the worry is. Gene frequencies change if we let a patient die or rescue a patient so that the person lives to reproduce. Gene frequencies are not likely to change much from very expensive genetic therapies for rare genetic disorders. It is also difficult to imagine either parents or physicians wanting to change both somatic cells and germinal cells of a patient with a severe genetic disorder. It is so much cheaper to use artificial insemination or elective abortion for preventing that genetic disorder from being passed on. Since the people who are likely to choose an "unnatural" method for their reproduction are not disturbed by those who consider them sinners, I doubt that they would seek a less effective and more uncertain process than one, like prenatal diagnosis, that has proven to be so reliable. The fear that somatic genetic therapy will lead to widespread germinal genetic therapy I would classify as a red herring, like widespread human cloning, that favorite theme of science fiction writers.

Appendices

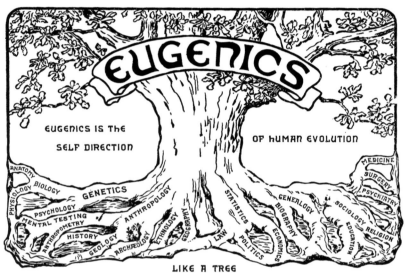

EUGENICS IS THE SELF DIRECTION OF HUMAN EVOLUTION

LIKE A TREE
EUGENICS DRAWS ITS MATERIALS FROM MANY SOURCES AND ORGANIZES
THEM INTO AN HARMONIOUS ENTITY.

Flow Diagrams and the History of Ideas

THERE ARE SEVERAL APPROACHES to the history of science. I am an anomaly in this field, one of the few working scientists who have also attempted to be historians of science. Scientists have a suspicion of historians of science, who lack, many scientists believe, an appreciation of the way science works in a scientist's career. At the same time, there is a suspicion by historians of science (those who earned their doctoral degrees in this field and practice it as a full-time occupation) that scientists are too wedded to the review article approach to science history to appreciate fully the influences of the social, cultural, and historical circumstances on science. Both of these suspicions have some validity, and both scientists as historians and professional historians of science making inquiry into the development of a branch of science or a major concept in science must be careful not to fall into these errors.

In addition to these hazards, there are differences in approach that are controversial among historians of science, whatever their background. There was an older tradition that flourished until the 1980s which explored the history of ideas. The underlying assumption of this tradition is that the personal lives of the scientists are often quite varied and none too different from the lives of poets, painters, and politicians, not to mention ordinary folk who are not public figures. What matters to practitioners of this form of inquiry is the way ideas were transformed and what circumstances led to these shifts in ideas. Very often these ideas (and the shifts) are tied to specific persons, and less interest is paid to what might be called social psychology or minor works and contributors. Scientists recognize, of course, that in the history of ideas (e.g., of the gene, cell division, genetic recombination), huge numbers of scientific papers on these topics (and conferences and correspondence) are written, and all of these minor bits become a major part of the support that often leads to new ideas. The working scientist is immersed in such activities, and even minor papers or conversations can suggest major ideas. Not all of these tangential influences are acknowledged in print by the scientist who comes up with new ideas. In the waning decades of the 20th century, this history of ideas approach has frequently been criticized as naive or biased in bringing out a few names and claiming an exaggerated influence for them on a field. This may be true in the sense I just discussed, but it may also be true that scientists (as in their review articles) do believe that these intellectual pathways are real, and that is how they learned how their field developed. That apparent contradiction between the perceived reality of a scientist through coursework and review articles and the alleged reality of the historian of science who tries to reconstruct the more complex relations of an older science is not easy to resolve.

This appendix presents a series of flow diagrams to assist the reader in tracing some of the ideas (or persons) whose work was influential in my discussion of ideas regarding the unfit in this book. I believe they are useful because they quickly summarize the way things came together if read in a historical direction. But more accurately, they should be read in an *archeological direction*, from the present back, because the *historical direction* implies causality, a view that I do not endorse. I believe, but cannot prove, that Hitler would have had a final solution for Jews whether or not germ theory, public health, Mendelian genetics, or the term eugenics had been introduced before 1940. The idea of unfit people long predates any of these more recent ideas. Ideologues and bigots use what they can to justify their fixed obsessions. In the 18th and 19th centuries, it was degeneracy theory that was frequently invoked to explain the unfit. Prior to that it was usually some moral failing or transgression or long-standing curse that served as the pretext for regarding other peoples as unfit. In using these flow diagrams, the reader is reminded to heed the cautions historians of science offer—they are an abbreviated version of history based on a few of the major ideas and players in a much more complex story told in these chapters and in the works of those referred to for each chapter.

THE RISE OF NEGATIVE EUGENICS

Negative eugenics seeks to cull humanity of its alleged defective members by restricting them from breeding. It has its earliest origins in degeneracy theory. Malthus and Spencer both accepted this model as an accurate depiction of the human condition; both thought of keeping the species healthy. Darwin was influenced by Malthus, but his views were limited primarily to one aspect of a theory of natural selection leading to evo-

THE RISE OF NEGATIVE EUGENICS

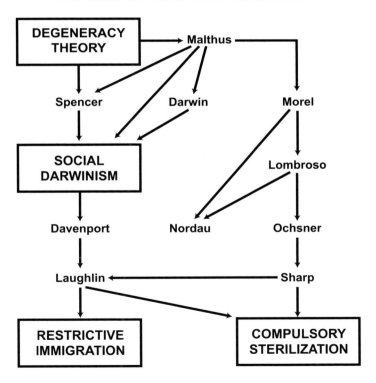

lutionary change. Darwin played little direct role in the movement called Social Darwinism, which is really a nonevolutionary theory of degeneracy championed by Spencer just before the appearance of the Darwin–Wallace papers. Malthus's views on degeneracy were extended by Morel, whose book on degeneracy influenced Lombroso's views, especially on criminals and degenerates. Morel and Lombroso influenced Nordau, who did not have a major role in the application of degeneracy theory to social policy. Nordau's influence was on classifying degenerate types, especially in the arts and popular culture. This may have influenced Nazi ideology, such as the infamous "Degenerate Art" exhibit in Munich. Lombroso's views were influential on Ochsner, who in turn influenced Sharp to begin the vasectomies of alleged degenerates as a eugenic measure. Social Darwinism influenced Davenport and Laughlin, who played major roles in the two accomplishments of the American eugenics movement—compulsory sterilization laws and restrictive immigration. If this assessment is correct, the roots of eugenics arose about 1700 (the first degeneracy theories based on onanism), gathered momentum about 1800 (when Malthus promoted his theory on the causes of misery and vice), and culminated in the first third of the 20th century as an international movement modeled on the efforts of the American eugenic programs of compulsory sterilization and restrictive immigration policy based on alleged eugenic deficiencies.

The Rise of Positive Eugenics

Darwin's theory of natural selection leading to new species had a powerful influence on Galton. Galton was also influenced by the idea of progress, which stemmed from the Enlightenment philosophy of Condorcet. Galton coined the term "eugenics" and stressed its importance for improving the human species by selection for genius and eminence. His views influenced the psychological theories of Terman. Muller was influenced by Galton's idealism and proposed a voluntary program he called germinal choice. Graham sought to apply Muller's ideas by establishing a bank for genius sperm in the Repository for Germinal Choice. No widespread use of positive eugenics has occurred through any of these conscious efforts, but Osborn has argued that prosperity and opportunities to

THE RISE OF POSITIVE EUGENICS

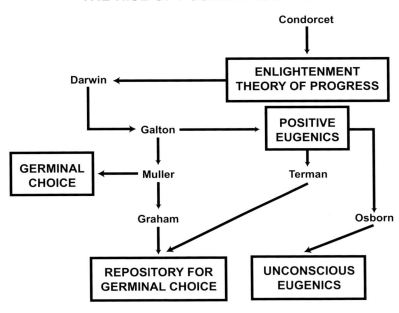

men and women lead to an unconscious differential reproduction with positive eugenic outcomes.

THE AMERICAN EUGENICS MOVEMENT

What can be called the American eugenics movement (or old-line eugenics) had its origins about the 1870s with the introduction of Social Darwinism and degeneracy theory into American social thought. Dugdale's study of the Jukes influenced Jordan and McCulloch, who were also influenced by the parasitism theory developed by Lankester. Weismann's theory of the germ plasm led to the idea of fixed behavioral traits, and Davenport and Laughlin used these in promoting restrictive immigration laws and compulsory sterilization. Jordan influenced both father and son in the Holmes family, and the son supported compulsory sterilization on constitutional grounds. The movement lasted through the 1930s and then went into eclipse.

THE AMERICAN EUGENICS MOVEMENT

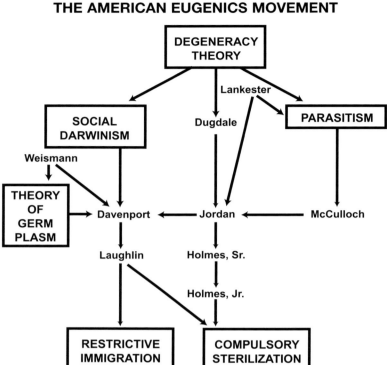

The racial theories of Gobineau and Chamberlain included anti-Semitism, a movement strongly influenced by Marr in Germany. Grant much later in the United States promoted racist ideology, embracing white (primarily Aryan) supremacy. This lent international support for Nazi race theory. Negative eugenics and anti-Semitism joined racism and the Hoche-Binding euthanasia movement to feed into Nazi ideology, which introduced the Holocaust as a final solution to what was called the Jewish question.

Nazi eugenics was part of Nazi ideology. It had its origins in the public hygiene movement begun by Virchow, which strongly influenced Schallmayer, Fischer, and Lenz. Schallmayer's views were influential on early Nazi

ORIGINS OF THE HOLOCAUST

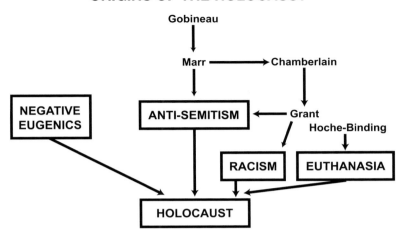

racial ideologists Wagner and Conti. Fischer's school included Verschuer and Mengele. Ploetz was influenced also by the public hygiene movement and introduced the idea of race hygiene. This was embraced by top Nazi officials, with Himmler taking a leading role in its implementation.

THE RISE OF NAZI EUGENICS

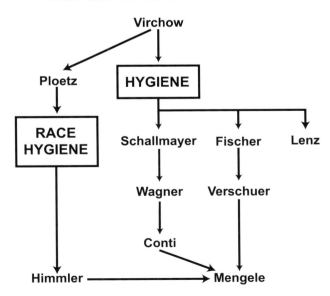

HUMAN BIOLOGY AND NAZI IDEOLOGY

Several components were brought together in shaping Nazi ideology and human biology. Hegel contributed the idea of a national spirit tied to its people and manifested through its will. His views influenced both Nietzsche and Wagner, and they led to a peculiar form of Teutonic nationalism. Supplemented by Gobineau's and Chamberlain's racial views, they led to a myth of Aryan superiority. Quite separate was a movement leading to the germ theory through the work of Semmelweis, Pasteur, Koch, and Lister. Their contributions were absorbed by Virchow's public health movement. Ploetz brought these two movements together in his concept of race hygiene. Schallmayer, Fischer, and Lenz contributed to the Nazi race hygiene view, bringing in their ideas of eugenic courts and a social policy to promote Aryan eugenics. The means to implement this (eventually through the Final Solution) were provided by the Hoche-Binding theories of euthanasia.

HUMAN BIOLOGY AND NAZI IDEOLOGY

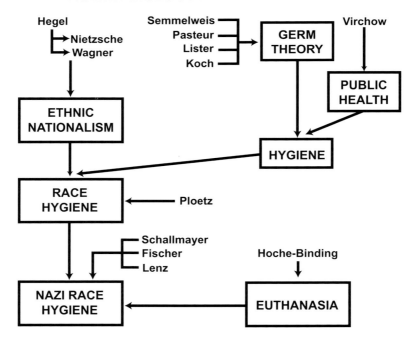

Useful Books on the History of Eugenics

T HE LITERATURE ON EUGENICS is voluminous. This is a sampling of books I found particularly helpful in shaping my own views on the history of eugenics. It should be noted that eugenics arose out of degeneracy theory, and few works are devoted to the history of that older idea.

Adams, Mark B., Ed. 1990. *The Wellborn Science.* Oxford University Press, New York. This collection of essays on eugenics around the world is very helpful for those who wish to assess the worldwide impact of eugenic thought.

Bajema, Carl J., Ed. 1976. *Eugenics: Then and Now.* Halsted Press, New York. Bajema's collection of source articles on eugenics makes it an excellent reference on hard-to-find documents and chapters in out-of-print books. It is a must for historians of eugenics.

Blacker, C.P. 1952. *Eugenics: Galton and After.* Duckworth, London. This is chiefly an account of the eugenics movement in the United Kingdom and the role of British intellectuals in establishing eugenical societies and publicizing the aims of the eugenics movement.

Burley, Justine, Ed. 1999. *The Genetic Revolution and Human Rights* (The Oxford Amnesty Lectures). Oxford University Press, New York. This is a superb collection of diverse views on eugenics and human cloning concerns for the 21st century.

Gallagher, Nancy. 1999. *Breeding Better Vermonters: The Eugenics Project in the Green Mountain State*. University Press of New England. Gallagher's study of the ambitions and limitations of one bit player in the eugenics movement illustrates clearly how well-intentioned professionals lose their values when enticed by the ideas of higher social goals.

Gould, Stephen Jay. 1981. *The Mismeasure of Man*. Norton, New York. This well-known work shows how self-deception often leads to results that are consistent with one's biases, especially in the human sciences where the prevailing cultural biases are not easy to identify as such.

Haller, Mark H. 1963. *Eugenics: Hereditarian Attitudes in American Thought*. Rutgers University Press, New Brunswick, New Jersey. This is one of the earliest full histories of eugenics. It is scholarly, avoids polemics, and focuses mostly on American eugenics.

Kevles, Daniel. 1985. *In the Name of Eugenics: Genetics and the Uses of Human Heredity*. Knopf, New York. Both British and American eugenics are the focus of this work which deals with 20th-century eugenics, mostly negative eugenics and its failures. Kevles uses eugenics as a cautionary tale and raises concerns about the intentions of modern human genetics which might lead to bad eugenic outcomes.

Ludmerer, Kenneth. 1972. *Genetics and American Society*. Johns Hopkins University Press, Baltimore, Maryland. This is an excellent work on the role of professionals, especially in medicine and hygiene, in the history of eugenics in America.

Müller-Hill, Benno. 1988. *Murderous Science: Elimination by Scientific Selection of Jews, Gypsies and Others, Germany, 1933–1945*. Oxford University Press, New York. Müller-Hill uses primary sources in Germany to reveal the role of German physicians and intellectuals in their version of eugenics (race hygiene) and how the Nuremberg laws and later secret arms of the Nazi party worked out the tools, procedures, code names, and cover stories to disguise the mass killings of both Aryan defectives and those, especially Jews, deemed unworthy of life.

Reilly, Philip R. 1991. *The Surgical Solution: A History of Involuntary Sterilization in the United States.* Johns Hopkins University Press, Baltimore, Maryland. This is a fine analysis of the role of physicians in the eugenics movement (or, more precisely, the American negative eugenics movement) and the enthusiasm they had for using their skills to solve a social problem through compulsory sterilization of the allegedly unfit.

Schneider, William. 1990. *Quality and Quantity: The Quest for Biological Regeneration in Twentieth-Century France.* Cambridge University Press, New York. This is an excellent account of French eugenics that tells how diverse were the contradictory groups that assembled under a common movement. The role of Lamarckian heredity is a major theme in much of the eugenics movement in France before World War II.

Selden, Steven. 1999. *Inheriting Shame: The Story of Eugenics and Racism in America.* Teachers College Press, Columbia University, New York. Eugenics propaganda in the classrooms was part of the effort by the American eugenics movement. I remember well those illustrations and cautionary tales from my own childhood, reading my assigned texts in New York City classrooms.

Smith, J. David. 1985. *Minds Made Feeble: The Myth and Legacy of the Kallikaks.* Aspen Systems Corporation, Rockville, Maryland. This is a useful account of the Kallikaks and how Goddard used them as an alleged controlled scientific experiment to show the persistence of heredity across many generations. Later Goddard repudiated his early work, but the damage had already been done.

Bibliography

Anonymous. 1889. A word upon the regulation of prostitution and sexual hygiene (translated abstract of August Forel in *Correspondenzblatt fur Schweitzer Aertze* [1889]). *Medical Record* **36**: 320–321.

Anonymous. 1890. Procreation of the criminal and degenerate classes. *Medical Record* **37**: 562.

Anonymous. 1892. Castration for neuroses and psychoses in the male. *Medical Record* **41**: 43.

Anonymous. 1892. Castration for melancholia. *Medical Record* **42**: 736.

Anonymous. 1893. An experiment in castration. *Medical Record* **43**: 433–434.

Anonymous. 1894. Castration of sexual perverts. *Medical Record* **45**: 479–480.

Anonymous. 1957. Albert John Ochsner 1858–1925. *Dictionary of American Biography* **13**: 616. C. Scribner's Sons, New York.

Anonymous. 1886. Castration in nervous and mental disease (editorial). *Journal of the American Medical Association* **7**: 547–549.

Anonymous. 1892. Castration recommended as a substitute in capital punishment. *Journal of the American Medical Association* **18**: 499–500.

Anonymous. 1925. Albert John Ochsner (obituary). *Journal of the American Medical Association* **85**: 374.

Anonymous. 1959. Amalekite. *Encyclopedia Britannica* **1**: 724. William Benton, Chicago.

Anonymous. 1959. Ashdod. *Encyclopedia Britannica* **2**: 509. William Benton, Chicago.

Anonymous. 1959. Karl Ernst von Baer. *Encyclopedia Britannica* **2**: 920. William Benton, Chicago.

Anonymous. 1959. William Godwin. *Encyclopedia Britannica* **10**: 465–466. William Benton, Chicago.

Anonymous. 1959. Physiognomy. *Encyclopedia Britannica* **17**: 886–887. William Benton, Chicago.

Anonymous. 1959. Poor Law. *Encyclopedia Britannica* **18**: 215–224. William Benton, Chicago.

Anonymous. 1959. August Weismann (1834–1914). *Encyclopedia Britannica* **23**: 491–492. William Benton, Chicago.

Anonymous. 1894. Richard L. Dugdale. *Appleton's Cyclopedia of American Biography* **2**: 250.

Anonymous. 1894. Elisha Harris. *Appleton's Cyclopedia of American Biography* **3**: 9.

Anonymous. 1898. *First Biennial Report of the Board of Managers of the Indiana Reformatory, Jeffersonville* (from November 1, 1896 to October 31, 1898, inclusive). Reformatory Printing Trade School, Jeffersonville, Indiana.

Anonymous. 1900. *Second Biennial Report of the Board of Managers of the Indiana Reformatory, Jeffersonville* (from November 1, 1898 to October 31, 1900, inclusive). Reformatory Printing Trade School, Jeffersonville, Indiana.

Anonymous. 1902. *Third Biennial Report of the Board of Managers of the Indiana Reformatory, Jeffersonville* (from November 1, 1900 to October 31, 1902, inclusive). Reformatory Printing Trade School, Jeffersonville, Indiana.

Anonymous. 1904. *Fourth Biennial Report of the Board of Managers of the Indiana Reformatory, Jeffersonville* (from November 1, 1902 to October 31, 1904, inclusive). Reformatory Printing Trade School, Jeffersonville, Indiana.

Anonymous. 1906. *Fifth Biennial Report of the Board of Managers of the Indiana Reformatory, Jeffersonville* (from November 1, 1904 to October 31, 1906, inclusive). Reformatory Printing Trade School, Jeffersonville, Indiana.

Anonymous. 1984. Gemara: Sanhedrin, 8. In *The Talmud*, vol. 31, Sanhedrin and Makhot, pp. 262–266 (translated by Jacob Neusner). University of Chicago Press, Chicago, Illinois.

Anonymous. 1984. Gemara: Bruchot, page 3:4. In *The Talmud*, vol. 1, Yerushalmi Berakhot, pp. 130–136 (translated by Tsvee Zahavy). University of Chicago Press, Chicago, Illinois.

Anonymous. 1904. Harry C. Sharp (obituary). *The New York Times*, Friday, November 1, 1940 (p. 25, col. 4).

Anonymous. 1756. *ONANIA, or the Heinous Sin of Self Pollution and All Its Frightful Consequences in Both Sexes Considered*, 18th edition. Charles Corbett, London.

Anonymous. 1719. *Onanism Display'd, Being, I. An Enquiry into the True Nature of Onan's Sin. II. Of the Modern Onanists. III. Of Self-Pollution, its Causes, and Consequences. IV. Of Nocturnal Pollutions. V. The Great Sin of Self-Pollution. VI. A Dissertation concerning Generation. With a Curious Description of the Parts* (English translation from the Paris edition), 2nd edition. E. Curll, London.

Anonymous. 1885. Private Aid to Public Charities. *Andover Review* **4:** 220–230.

Anonymous. 1907. Surgeons to deal with criminals. *The Indianapolis Star,* Thursday, March 7, 1907 (p. 10).

Anonymous. 1877. Tramps and pedestrians. *Blackwood's Magazine* **122:** 325–345. W. Blackwood, Edinburgh.

Anonymous. 1891. The problem of the slums. *Blackwood's Magazine* **149:** 123–136. W. Blackwood, Edinburgh.

Anonymous. 1910. The sterilization of criminals and other degenerates. *Indianapolis Medical Journal* **13:** 163–165.

Anonymous. 1909. Sterilization of human beings: The Indiana plan. *New York Medical Journal* **90:** 1241–1242.

Anonymous. 1950. *Trials of War Criminals Before the Nuremberg Military Tribunal Under Control Council* 10, Nuremberg, October 1946–April 1949, vol. 1.

Allen, Margaret Andrews. 1892. Jennie Collins and her Boffin's Bower. *The Charities Review* **2:** 105–115.

Appleman, Philip, Ed. 1975. Malthus's *Essay on the Principles of Population.* Norton Critical Edition, W.W. Norton & Co., New York.

Arendt, Hannah. 1976. *Eichmann in Jerusalem: A Study of the Banality of Evil.* Penguin Books, New York.

Bajema, Carl J., Ed. 1976. *Eugenics: Then and Now.* Halsted Press, New York.

Baker-Benfield, Ben. 1972. The spermatic economy: A nineteenth century view of sexuality. *Feminist Studies* 1: 336–372.

Barbour, Levi L. ca. 1881. Vagrancy. In *Proceedings of the Eighth Annual Conference of Charities and Corrections,* Boston, July 25–30, 1881, pp. 131–138.

Baron, Walter. 1974. Johann Friedrich Blumenbach (1752–1840). In *Dictionary of Scientific Biography* (ed. Charles Coulson Gillispie), vol. 2, pp. 203–205. C. Scribner's Sons, New York.

Barr, Martin W. 1904. *Mental Defectives: Their History, Treatment, and Training.* Blakiston, Philadelphia.

Baur, Erwin, Eugen Fischer, and Fritz Lenz. 1931. *Human Genetics* (translated from the German [1923] *Grundriss de Menschliche Erblichkeitslehre und Rassenhygiene,* Lehmann, München). G. Allen & Unwin, London.

Binding, Karl and Alfred Hoche. 1975. *The Release of the Destruction of Life Devoid of Value* (translated, with comments, by Robert L. Sassone from the German edition of Felix Meiner, Leipzig, 1920). Life Quality Paperback, Santa Ana, California.

Bloch, A.J. ca. 1894. Clitoridectomy in a two and a half year old child. In *Transactions of the Louisiana Medical Society, New Orleans* 1894, p.333.

Bloch, Iwan. 1928. *The Sexual Life of Our Time* (English translation of the 1908 edition). Allied Book Co., New York.

Boal, Robert. 1894. Emasculation and ovariectomy as a penalty for crime and the reformation of criminals. *Journal of the American Medical Association* 23: 429–432.

Boies, Henry M. 1893. *Prisoners and Paupers.* G.P. Putnam's Sons, New York.

Bolles, Mrs. E.C. 1893. Would direct personal influence diminish pauperism? *The Charities Review* 2: 410–419.

Bonar, James. 1885. *Malthus and His Work.* MacMillan & Co., London.

Bowditch, Henry I. 1877. *Public Hygiene in America.* Little, Brown, & Co., Boston.

Brace, Charles L. 1875. Pauperism. *The North American Review* 120: 316–334.

Brent, Peter. 1981. *Charles Darwin: A Man of Enlarged Curiosity.* Harper & Row, New York.

Brower, Daniel R. 1899. Medical aspects of crime. *Journal of the American*

Medical Association **32**: 1282–1287.

Brown, Frederick. 1995. *Zola: A Life.* Ferrar, Straus, & Giroux, New York.

Browne, Janet. 1995. *Charles Darwin: Voyaging.* Alfred A. Knopf, Inc., New York.

Bullock, Alan. 1962. *Hitler—A Study in Tyranny.* Harper & Row, New York.

Burley, Justine, Ed. 1999. *The Genetic Revolution and Human Rights.* Oxford University Press, New York.

Burlingame, L.J. 1970. Jean Baptiste Lamarck. *Dictionary of Scientific Biography* **7**: 584–594. C. Scribner's Sons, New York.

Carlson, Elof A. 1966. *The Gene: A Critical History.* W.B. Saunders, Philadelphia.

Carlson, Elof A. 1980. R.L. Dugdale and the Jukes family: A historical injustice corrected. *Bioscience* **30**: 535–539.

Carlson, Elof A. 1981. *Genes, Radiation, and Society: The Life and Work of H.J. Muller.* Cornell University Press, Ithaca, New York.

Carlson, Elof A. 1984. *Human Genetics.* D.C. Heath & Co., Lexington, Massachusetts.

Carmony, Donald F. 1942. Genesis and early history of the Indianapolis Fund. *Indiana Magazine of History* **38**: 17–36.

Chamberlain, Houston Stewart. 1911. *Foundations of the Nineteenth Century* (reprinted in 1968 by H. Fertig Co., New York). John Lane, The Bodley Head, Ltd., London.

Chapin, Henry Dight. 1892. The survival of the unfit. *The Popular Science Monthly* **41**: 182–187.

Chase, Alan. 1976. *The Legacy of Malthus.* Alfred A. Knopf, Inc., New York.

Chevalier, Louis. 1973. *Laboring Classes and Dangerous Classes in Paris during the First Half of the Nineteenth Century* (translated by Frank Jellink). H. Fertig Co., New York.

Clark, Martha Louise. 1894. The relation of imbecility to Pauperism and crime. *Arena* **10**: 788–794.

Cook, Joseph. 1879. *Boston Monday Lectures.* Houghton, Osgood & Co., Boston.

Cottman, George. 1911. Old time slums of Indianapolis. *Indiana Magazine of History* **VII**: 170–173.

Cushing, E.W. 1887. Melancholia; masturbation; cured by removal of both ovaries. *Journal of the American Medical Association* **8**: 441–442.

Daniel, F.E. 1893. Should insane criminals, or sexual perverts, be allowed to procreate? *New York Medico–Legal Journal* (1893): 275–292.

Darwin, Charles. 1859. *The Origin of Species.* John Murray, London.

Darwin, Charles. 1868. *Variation of Plants and Animals under Domestication* (popular edition [1905] two volumes). John Murray, London.

Darwin, Charles. 1871. *The Descent of Man.* John Murray, London.

Darwin, Charles. 1871. Letters to the editor. *Nature* III: pp. 502–503.

Darwin, Leonard, Ed. 1912. *Problems in Eugenics* (two volumes): Papers communicated to The First International Eugenics Congress, University of London, July 24–30, 1912. Eugenics Education Society, London.

Davenport, Charles, Secretary. 1910. Report of the Eugenics Committee of the American Breeder's Association. *American Breeder's Magazine* 1: 128–129.

Davenport, Charles, Secretary. 1911. Report of the Eugenics Committee of the American Breeder's Association. *American Breeder's Magazine* 2: 62.

Davenport, Charles, Ed. 1934. *A Decade of Progress In Eugenics: Proceedings of the Third International Congress of Eugenics, New York, 1932.* Williams & Wilkins Co., Baltimore.

Deslandes, Leopold. 1839. *On Onanism and other Sexual Abuses Considered in Relation to Public Health* (translated from the 1835 French edition). Otis, Broaders & Co., Boston.

Deslandes, Leopold. 1841. *A Treatise on the Disease Produced by Onanism, Masturbation, Self-Pollution, and other Excesses* (translated from the French, second edition). Otis, Broaders & Co., Boston.

Duffy, John. 1963. Masturbation and Clitoridectomy. *Journal of the American Medical Association* 186: 246–248.

Dugdale, Richard L. 1875. *Report of special visits to county jails for the year 1874.* Thirtieth Annual Report of the Prison Association of New York, Albany.

Dugdale, Richard L. 1877. *The Jukes: A Study in Crime, Pauperism, Disease, and Heredity.* G.P. Putnam's Sons, New York.

Dugdale, Richard L. 1881. Origin of crime in society. *The Atlantic Monthly* 48: 452–462; 735–746.

Ehrlich, Paul. 1968. *The Population Bomb.* Ballantyne Books, New York.

Estabrook, Arthur. 1916. *The Jukes in 1915.* The Carnegie Institution of Washington, Washington, D.C.

Estabrook, Arthur. 1923. The tribe of Ishmael. In *Eugenics Genetics and the*

Family: Proceedings of the Second International Congress of Eugenics, September 22–28, 1921, vol 1, pp. 398–404. Williams & Wilkins Co., Baltimore, Maryland.

Fawcett, Henry. 1871. *Pauperism: Its Causes and Remedies*. MacMillan, London.

Fleming, Gerald. 1984. *Hitler and the Final Solution*. University of California, Berkeley.

Fromm, Erich. 1941. *Escape from Freedom*. Farrar and Rhinehart, New York.

Galton, Francis. 1865. Hereditary talent and character. *MacMillan's Magazine* **12**: 157–166; 318–327.

Galton, Francis. 1869. *Hereditary Genius* (second edition 1892). MacMillan, London.

Galton, Francis. 1871. Experiments in pangenesis, by breeding from rabbits of a pure variety, into whose circulation blood taken from other varieties had previously been largely transfused. *Proceedings of the Royal Society* (Biology) **19**: 393–404.

Galton, Francis. 1871. Letters to the editor. *Nature* **IV**: pp. 5–6.

Galton, Francis. 1883. *Inquiry into Human Faculty and Its Development*. MacMillan, London.

Galton, Francis. 1889. *Natural Inheritance*. MacMillan, London.

Galton, Francis. 1901. The possible improvement of the human breed under the existing conditions of law and sentiment. *Nature* **64**: 659–665.

Galton, Francis. 1905. Eugenics: Its definition, scope and aims. In *Sociological Papers*, pp. 45–50. MacMillan & Co., London.

Galton, Francis. 1908. *Memories of My Life*. Methuen & Co., London.

Gobineau, Arthur de. 1967. *The Inequality of Human Races* (translated from the 1853 French by Adrian Collins). H. Fertig Co., New York.

Godwin, Richard. 1820. *Of Population: An Enquiry Concerning the Power of Increase in the Numbers of Mankind, Being an Answer to Mr. Malthus's Essay on That Subject* (reprinted in 1964 by A.M. Kelly, New York). Longman, Hurst, Rees, Orme, and Brown, London.

Goerke, Heinz. 1973. *Linnaeus: A Modern Portrait of the Great Swedish Scientist*. C. Scribner's Sons, New York.

Goertzel, Victor and Muriel Goertzel. 1962. *Cradles of Eminence*. Little, Brown, & Co., Boston.

Goldstein, Marc and Michael Feldberg. 1982. *The Vasectomy Book.* J.P. Tarcher, Inc., Los Angeles.

Goodman, Richard M. 1979. *Genetic Disorders among the Jewish People.* Johns Hopkins University Press, Baltimore.

Gould, Stephen Jay. 1981. *The Mismeasure of Man.* W.W. Norton & Co., New York.

Gould, Stephen Jay. 1987. An Universal Freckle. *Natural History* **96:** 14–20.

Granger, Gille. 1970. Condorcet, Marie Jean Antoine Nicolas Caritat, Marquis de (1743–1794). *Dictionary of Scientific Biography* **3:** 383–388. C. Scribner's Sons, New York.

Grant, Madison. 1916. *The Passing of the Great Race.* C. Scribner's Sons, New York.

Greenwood, James. 1870. *The Seven Curses of London.* Stanley Rivers & Co., London.

Gridgeman, N.T. 1970. Francis Galton (1822–1911). *Dictionary of Scientific Biography* **5:** 265–266. C. Scribner's Sons, New York.

Haller, John S., Jr. 1989. The role of physicians in America's sterilization movement, 1894–1925. *New York State Journal of Medicine* **89:** 169–179.

Haller, Mark H. 1963. *Eugenics: Hereditarian Attitudes in American Thought.* Rutgers University Press, New Brunswick, New Jersey.

Hamilton, A.E. 1914. Pioneers in eugenics. *Journal of Heredity* **5:** 370–372.

Hare, E.H. 1962. Masturbatory insanity: The history of an idea. *The Journal of Mental Science* **108:** 1–25.

Harris, Elisha. 1868. *Report of the Council of Hygiene and Public Health of the Citizen's Association of New York upon the Sanitary Condition of the City.* C. Westcott & Co., New York.

Harris, Marvin. 1968. Race. *International Encyclopedia of the Social Sciences* **13:** 263–268.

Harrison, Reginald. 1899. *Selected Papers on Stone, Prostate, and Other Urinary Disorders.* Churchill, London.

Hart, Hastings Hornell. 1913. *Sterilization as a Practical Measure* (pamphlet no. 11). Department of Child Helping of the Russell Sage Foundation, Inc., New York.

Hartsfield, Larry K. 1985. *The American Response to Professional Crime: 1870–1917.* Greenwood Press, Westport, Connecticut.

Hassencahl, Frances Janet. 1971. *Harry H. Laughlin, "Expert Eugenics*

Agent" for the House Committee on Immigration and Naturalization, 1921 to 1931 (Ph. D. dissertation). Case Western Reserve University, Cleveland, Ohio (available from University Microfilms International, Ann Arbor, Michigan).

Hazard, Rowland. 1884. *What Social Classes Owe to Each Other* (book review). *Andover Review* 1: 159–174.

Heiden, Konrad. 1944. *Der Fuehrer: Hitler's Rise to Power.* Houghton Mifflin Co., Boston.

Heilbrun, Bernhard. 1883. Untitled and translated note from *Centralblatt fur Gynakologie,* Sept. 22, 1883. *Journal of the American Medical Association* 1: 591.

Henry, Clarissa and Marc Hillel. 1976. *Of Pure Blood* (translated by Eric Mossbacher from the French [Au Nom de la Race]). McGraw-Hill, New York.

Henry, Matthew. 1925. *Commentary on the Whole Bible: Genesis to Deuteronomy* (reprint of 1704 edition). Fleming H. Revell Co., Old Tappan, New Jersey.

Holland, J.G. 1876. The dead-beat nuisance. *Scribner's Monthly* 12: 592–593.

Holland, J.G. 1877. The disease of mendicancy. *Scribner's Monthly* 13: 416–417.

Holland, J.G. 1877. The pauper poison. *Scribner's Monthly* 14: 399–400.

Holland, J.G. 1877. The public charities. *Scribner's Monthly* 15: 272–273.

Holland, J.G. 1878. Once more the tramp. *Scribner's Monthly* 15: 882–883.

Holland, J.G. 1880. The shadow of the negro. *Scribner's Monthly* 20: 304–305.

Holland, J.G. 1880. Industrial education again. *Scribner's Monthly* 20: 464–465.

Hollond, E.W. 1870. The vagrancy laws and the treatment of the vagrant poor. *The Contemporary Review* 13: 161–176.

Holmes, Oliver Wendell. 1875. Crime and automatism: With a notice of M. Prosper Despine's *Psychologie Naturelle. Atlantic Monthly* 35: 466–481.

Homan, L. Beecher. 1875. The Suffolk County Alms House. In *Yaphank as It Is,* pp 209–213. John Polhemus (Printer), New York.

Howe, Joseph William. 1883. *Excessive Venery, Masturbation, and Continence.* Birmingham & Co., New York.

Hutchinson, Jonathan. 1890. A plea for circumcision. *British Medical Journal*, September 27, 1890: p. 769.

Hutchinson, Jonathan. 1890–1891. On circumcision as preventive of masturbation. *Archives of Surgery*, 1890–1891, part ii: p. 838.

Huxley, Aldous. 1932. *Brave New World*. Chatto & Windus, London.

Huxley, Julian. 1970. *Memories*, vol. I. G. Allen & Unwin, London.

Iltis, Hugo. 1922. *Life of Mendel* (translated by Eden and Cedar Paul). G. Allen & Unwin, London.

Johnson, Edgar. 1952. *Charles Dickens: His Tragedy and Triumph*. Simon and Schuster, New York.

Jordan, David Starr. 1888. *The Ethics of the Dust* (Commencement Address, 1888). Indiana University Library, Bloomington (call number: LD2524 J8).

Jordan, David Starr. 1898. *Footnotes to Evolution: A Series of Popular Addresses on the Evolution of Life*. D. Appleton & Co., New York.

Jordan, David Starr. 1901. The blood of a nation. *Popular Science Monthly* **59**: 90–100; 129–140.

Jordan, David Starr. 1907. *The Human Harvest*. American Unitarian Association, Boston.

Jordan, David Starr. 1912. *Unseen Empire: A Study of the Plight of Nations That Do Not Pay Their Debts*. American Unitarian Association, Boston.

Jordan, David Starr. 1922. *The Days of a Man: 1851–1899*, vol. 1. World Book Co., Yonkers-on-Hudson, New York.

Jordanova, L.J. 1984. *Lamarck*. Oxford University Press, Oxford, United Kingdom.

Kantor, William M. 1937. Beginnings of sterilization in America. An interview with Dr. Harry C. Sharp, who performed the first operation nearly forty years ago. *Journal of Heredity* **28**: 374–376.

Katz, Jacob. 1986. *The Darker Side of Genius: Richard Wagner's Anti-Semitism*. Brandeis/University of New England Press, Hanover, New Hampshire.

Kevles, Daniel J. 1985. *In the Name of Eugenics: Genetics and the Uses of Human Heredity*. Alfred A. Knopf, Inc., New York.

Keyes, E.L. 1895. *The Surgical Disease of the Genito-urinary Organs Including Syphilis*, p. 439. D. Appleton & Co., New York.

Koestler, Arthur. 1971. *The Case of the Midwife Toad*. Hutchinson, London.

Koestler, Arthur. 1976. *The Thirteenth Tribe: The Khazar Empire and Its*

Heritage. Hutchinson, London.

Lallemand, Claude-Francais. 1839. *Involuntary Seminal Discharges* (translated [1841] from the French by William Wood). A. Waldie, Philadelphia.

Lankester, Edwin Ray. 1880. *Degeneration, A Chapter in Darwinism.* MacMillan, London.

Lankester, Edwin Ray. 1910. The feeble-minded. In *Science from an Easy Chair,* pp. 271–282. Methuen & Co., London.

Laughlin, Harry Hamilton. ca. 1914. Calculations on the working out of a proposed program of sterilization. In *Proceedings of the National Conference on Race Betterment,* January 8–12, 1914, pp. 478–494. Race Betterment Foundation, Battle Creek, Michigan.

Laughlin, Harry Hamilton. 1914. Legal, legislative, and administrative aspects of sterilization. *Eugenics Record Office Bulletin No. 103* (February 1914), Cold Spring Harbor, New York.

Laughlin, Harry Hamilton. 1922. *Eugenical Sterilization in the United States.* Psychopathic Laboratory of the Municipal Court, Chicago.

Laughlin, Harry Hamilton. 1929. Legal status of eugenical sterilization. Supplement to *The Annual Report of the Municipal Court, Chicago.*

Laws of the State of Indiana Passed at the 65th Regular Session of the General Assembly. 1907. Chapter 215, pp. 377–378.

Leaming, Hugo P. 1977. The Ben Ishmael tribe: A fugitive nation of the old Northwest. In *The Ethnic Frontier* (ed. Melvin G. Holli and Peter d'A. Jones), pp 97–142. Wm. B. Eerdmans Publishing Co., Grand Rapids, Michigan.

Lee, Vernon. 1893. The moral teaching of Zola. *The Contemporary Review* **63**: 196–212.

Leonard, E.M. 1900. *The Early History of English Poor Relief.* Cambridge University Press, United Kingdom.

Lifton, Robert Jay. 1986. *The Nazi Doctors: Medical Killing and the Psychology of Genocide.* Basic Books, Inc., New York.

Lindemann, Albert S. 1988. Anti-Semitism: Banality or the darker side of genius? *Religion* **18**: 183–195.

Linnaeus, Carl. 1735. *Systema Naturae,* Leyden (reprinted in 1964 by Weinheim for J. Kraner, New York [Stechert-Hafner Service Agency]).

Lombroso, Cesare. 1918. *Crime, Its Causes and Remedies* (translated by Henry P. Horton). Little, Brown, & Co., Boston.

Lombroso-Ferrero, Gina. 1972. *Criminal Man, According to the Classification of Cesare Lombroso* (reprint of the 1911 edition with introduction by Leonard D. Savitz). Patterson Smith Publishing Co., Philadelphia, Pennsylvania.

Lucas, Prosper. 1847–1850. *Traité Philosophique et Physiologique de l'Hérédité Naturelle dans les États de Santé du System Nerveux* (2 volumes). Chez J.B. Ballière, Paris.

Ludmerer, Kenneth. 1972. *Genetics and American Society.* Johns Hopkins University Press, Baltimore. Maryland.

Lydston, G. Frank. 1905. *Disease of Society and Degeneracy.* J.P. Lippincott, Philadelphia.

Lydston, G. Frank and E.S. Talbot. 1891. Studies of criminals. *Journal of the American Medical Association* 17: 903–923.

MacDonald, Robert H. 1967. The frightful consequences of Onanism: Notes on the history of a delusion. *Journal of the History of Ideas* 28: 423–431.

MacDowell, E. Carleton. 1946. Charles Benedict Davenport 1866–1944: A study of conflicting influences. *Bios* 17: 1–50.

Malthus, Thomas Robert. 1798. *An Essay on the Principle of Population.* Reaves and Turner, London.

Malthus, Thomas Robert. 1826. *An Essay on the Principle of Population,* seventh edition. Reaves and Turner, London.

Marr, Wilhelm. 1879. *Der Sieg des Judenthums über das Germanenthum* (The Victory of Jewry over Germany). R. Costenoble, Bern.

McCassey, J.H. 1896. Adolescent insanity and masturbation with excision of certain nerves supplying the sexual organs as the remedy. *Cincinnati Lancet Clinic* 37: 341–343.

McCook, John J. 1893. Tramps. *The Charities Review* 3: 57–69.

McCulloch, Oscar C. 1891. The tribe of Ishmael: A study in social degradation. Reprinted from *Proceedings of the Fifteenth National Conference of Charities and Correction,* Buffalo, July, 1888. Charity Organization Society, Plymouth Church Building, Indianapolis, Indiana.

Meagher, John F.W. 1924. *A Study of Masturbation and Its Reputed Sequelae,* p. 28. William Wood & Co., New York.

Medawar, Peter. 1986. *Memoir of a Thinking Radish.* Oxford University Press, Oxford, United Kingdom.

Merz, John Theodore. 1912. *European Thought in the Nineteenth Century,* vol. 2, pp. 388–399. W. Blackwood & Sons, London and Edinburgh.

Millhauser, Milton. 1959. *Just before Darwin: Robert Chambers and Vestiges*. Wesleyan University Press, Middletown, Connecticut.

Money, John. 1985. *The Destroying Angel: Sex, Fitness, and Food in the Legacy of Degeneracy Theory; Graham Crackers, Kellogg's Corn Flakes, and American Health History*. Prometheus Press, Buffalo, New York.

Morais, Vamberto. 1976. *A Short History of Anti-Semitism*. W.W. Norton & Co., New York.

Morel, Benedict Auguste. 1857. Traité des Dégénérescences Physiques, Intellectuelles, et Morales de l'Espèce Humaine (reprinted in 1976 by Arno Press, New York). Chez H. Baillière, Paris.

Morner, Magnus. 1967. *Race Mixture in the History of Latin America*. Little, Brown, & Co., Boston.

Muller, H.J. 1918. Genetic variability, twin hybrids, and constant hybrids, in a case of balanced lethal factors. *Genetics* 3: 422–499.

Muller, H.J. 1923. Mutation. *Eugenics, Genetics, and the Family* 1: 106–112.

Muller, H.J. 1932. The dominance of economics over eugenics. In *A Decade of Progress in Eugenics*. Williams & Wilkins Co., Baltimore.

Muller, H.J. 1935. *Out of the Night: A Biologist's View of the Future*. Vanguard Press, New York.

Muller, H.J. 1961. Germinal choice, a new dimension in genetic therapy. *Excerpta Medica*, Abstract No. 294, p. E135.

Muller, H.J. 1961. Evolution by voluntary choice of germ plasm. *Science* 134: 643–649.

Muller, H.J. 1973. *Man's Future Birthright: Essays on Science and Humanity*. State University of New York Press, Albany.

Muller, H.J. 1973. *The Modern Concept of Nature: Essays on Theoretical Biology and Evolution*. State University of New York Press, Albany.

Muller, H.J. and E. Altenburg. 1920. The genetic basis of truncate wing— An inconstant and modifiable character in *Drosophila*. *Genetics* 5: 1–59.

Müller-Hill, Benno. 1988. *Murderous Science: Elimination by Scientific Selection of Jews, Gypsies, and Others, Germany, 1933–1945*. Oxford University Press, New York.

Murphy, Leonard J.T. 1972. *The History of Urology*. Charles Thomas, Springfield, Illinois.

Myers, Gustavus. 1943. *History of Bigotry in the United States*. Random House, New York.

Nicoll, Henry J. 1882. Prison reform: John Howard. In *Great Movements and Those Who Achieved Them*, pp. 1–47. Harper and Brothers, New York.

Nordau, Max. 1895. *Degeneration*. D. Appleton & Co., New York.

Nordau, Max and Gustav Gottheil. 1904. *Zionism and Anti-Semitism*. Scott-Thaw Co., New York.

Nordenskiold, Eric. 1935. *The History of Biology* (reprint of 1928 edition, Alfred A. Knopf, Inc.). Tudor Publishing Co., New York.

Nye, Robert A. 1976. Heredity or milieu: The foundations of modern European criminological theory. *Isis* **67**: 335–355.

Ochsner, Albert John. 1899. Surgical treatment of habitual criminals. *Journal of the American Medical Association* **32**: 867–868.

Oppenheim, Nathan. 1896. The stamping out of crime. *Popular Science Monthly* **48**: 527–533.

Orel, Vitezslav. 1996. *Gregor Mendel: The First Geneticist*. Oxford University Press, Oxford, United Kingdom.

Osborn, Henry Fairfield, Ed. 1923. *Eugenics, Genetics, and the Family: Proceedings of the Second International Congress of Eugenics*, September 22–28, 1921, vol. 1 (reprinted in 1985 by Garland, New York). Williams & Wilkins Co., Baltimore.

Paré, Ambroise. 1982. *On Monsters and Marvels* (translated from the 1573 edition by Janis Pallister). University of Chicago Press, Chicago, Illinois.

Pashley, Robert. 1852. *Pauperism and the Poor Laws*. Longman, Brown, Green, and Longman, London.

Peek, Francis. 1875. *Our Laws and Our Poor*. John B. Day, London.

Peyton, David C., General Superintendent. 1915. *Annual Report of the Indiana Reformatory at Jeffersonville* (for the year ending September 15, 1915). Printed at the Reformatory Press (57 pp.). Reprinted in *Indiana Magazine of History* (1917) **13**: 195–197.

Popenoe, Paul, 1934. The progress of eugenic sterilization. *Journal of Heredity* **25**: 19–26.

Posner, Gerald and John Ware. 1986. *Mengele: The Complete Story*. Dell Publishing Co., New York.

Preuss, Julius. 1911. *Biblical and Talmudic Medicine* (translated in 1978 by Fred Rosner). Sanhedrin Press, New York.

Proctor, Robert W. 1988. *Racial Hygiene: Medicine Under the Nazis*. Harvard University Press, Cambridge, Massachusetts.

Provine, William. 1971. *The Origins of Theoretical Population Genetics.* University of Chicago Press, Chicago, Illinois.

Putnam, G.H. 1915. *Memories of a Publisher,* pp. 171–172. G.P. Putnam's Sons, New York.

Reed, Myron. ca. 1891. Tributes to Oscar C. McCulloch. In *Proceedings of Charities and Correction at the Nineteenth Annual Session,* Denver, Colorado, pp. 247–250. George Ellis, Boston.

Reilly, Philip R. 1987. Involuntary sterilization in the United States: A surgical solution. *Quarterly Review of Biology* **62**: 153–170.

Reilly, Philip R. 1991. *The Surgical Solution: A History of Involuntary Sterilization in the United States.* Johns Hopkins University Press, Baltimore, Maryland.

Rentoul, Robert Reid. 1906. *Race Culture; or, Race Suicide (A Plea for the Unborn).* Walter Scott Publishing Co., London.

Rightmire, Robert W. 1980. *Suffolk County's Adoption of a County Alms House.* Ph. D. dissertation, State University of New York, Stony Brook.

Roberts, Blair O. 1971. History of cleft lip and palate treatment. In *Cleft Lip and Palate: Surgical, Dental, and Speech Aspects* (ed. W.C. Grabb, S.W. Rosenstein, and K. Bzoch), pp. 142–169. Little, Brown, & Co., Boston.

Roberts, Junius B. 1911. Plymouth Church, Indianapolis. *Indiana Magazine of History* **VII**: 52–60.

Robinson, Charles D. 1876. Insanity and its treatment. *Scribner's Monthly* **12**: 634–648.

Roper, A.G. 1913. *Ancient Eugenics.* Oxford University Press, Oxford, United Kingdom.

Rosen, George. 1958. *A History of Public Health.* MD Publications, Inc., New York.

Rosenthal, Michael. 1924. *The Character Factory: Baden-Powell and the Origins of the Boy Scout Movement,* p. 187. Pantheon Books, New York.

Salk, Jonas and Jonathan Salk. 1981. *World Population and Human Values: A New Reality.* Harper & Row, New York.

Schallmayer, Wilhelm. 1900. *Inheritance and Selection in the Life of Nations.* G. Fischer, Jena.

Schmiedeler, Rev. Edgar. 1943. *Sterilization in the United States* (pamphlet, 38 pp.). Family Life Bureau, National Catholic Welfare Conference, Washington, D.C.

Schneider, William. 1990. *Quality and Quantity: The Quest for Biological Regeneration in Twentieth-Century France.* Cambridge University Press, New York.

Seymour, Horatio. 1873. On the causes of crime. *Popular Science Monthly* **2**: 589–596.

Sharp, Harry C. ca. 1898. Neurasthenia and its treatment. In *Proceedings of the Mississippi Valley Medical Association*, Nashville, Tennessee, October 11–16, 1898.

Sharp, Harry C. 1902. The severing of the vasa deferentia and its relation to the neuropsychopathic constitution. *New York Medical Journal* **75**: 411–414.

Sharp, Harry C. 1908. *Vasectomy: A Means of Preventing Defective Procreation.* Indiana Reformatory Printing Trade School, Jeffersonville.

Sharp, Harry C. 1908. *The Sterilization of Degenerates.* National Christian League for Promotion of Purity (pamphlet).

Sharp, Harry C. 1909. Vasectomy as a means of preventing procreation in defectives. *Journal of the American Medical Association* **53**: 1897–1902.

Shepard, Edward Morse. 1884. *The Work of a Social Teacher Being a Memorial of Richard L. Dugdale* (Economic Tracts No. XII). The Society for Political Education, New York.

Simon, Walter M. 1960. Herbert Spencer and the "Social Organism." *Journal of the History of Ideas* **21**: 294–299.

Simpkins, Diana M. 1970. Malthus, [Thomas] Robert (1766–1834), p. 67. *Dictionary of Scientific Biography* **9**: 67–71. C. Scribner's Sons, New York.

Smith, Gerald L.K., Ed. 1957. *The International Jew: The World's Foremost Problem* (reprinted, with additions, from the 1921 edition of the *Dearborn Independent*). Christian Nationalist Crusade, Los Angeles.

Smith, J. David. 1985. *Minds Made Feeble: The Myth and Legacy of the Kallikaks.* Aspen Systems Corp., Rockville, Maryland.

Spencer, Herbert. 1875. *The Principles of Biology,* volume 1 (first edition appeared in 1864). D. Appleton & Co., New York.

Spencer, Herbert. 1954. *Social Statics* (reprint of the 1850 edition). Robert Schalkenbach Foundation, New York.

Spencer, Herbert. 1961. *The Study of Sociology.* Ann Arbor Paperback, Ann Arbor, Michigan.

Stimson, Henry A. 1885. The new charity. *The Andover Review* **3**: 107–120.

Stern, Curt. 1967. The problem of complete Y linkage in man. *American Journal of Human Genetics* **9**: 147–166.

Stevenson, D.R. 1985. Joseph Arthur Gobineau. *Historical Dictionary of the French Second Empire 1852–1870* (ed. W.E. Echard), pp. 265–266. Greenwood Press, Westport, Connecticut.

Suddith, W. Xavier. ca. 1880. *The Law of Heredity* (department leaflet, 4 pp., unnumbered, undated, ca. 1880). Woman's Temperance Publishing Association, Chicago.

Swingewood, Alan. 1984. *A Short History of Sociological Thought.* St. Martin's Press, New York.

Tauffer, Wilhelm. 1884. The castration of women (translated from *Zeitschrift fur Geburtshulefe und Gynakologie*). *Journal of the American Medical Association* **2**: 632.

Terman, Lewis. 1925. *Genetic Studies of Genius.* Stanford University Press, Stanford, California.

The Birth of Genetics: Mendel, de Vries, Correns, Tschermak. 1950. Supplement to *Genetics*, vol. 35(5), part 2, 47 pp. Brooklyn Botanic Garden, Brooklyn, New York.

Thenen, R. 1968. Joseph Arthur, Comte de Gobineau. *International Encyclopedia of the Social Sciences* **6**: 193–194.

Thomas, Lewis. 1983. *The Youngest Science: Notes of a Medicine-watcher.* Viking Press, New York.

Thompson, Elizabeth. 1882. *Heredity: Its Relation to Human Development.* Institute of Heredity, Boston.

Tissot, Samuel. 1985. *Onania, or a Treatise upon the Disorders Produced by Masturbation* (translated into English from the 1758 Latin edition [reprinted in English edition]). Garland Press, New York.

Townsend, Peter. 1964. *The Last Refuge: A Survey of Residential Institutions for the Aged in England and Wales.* Routledge and Kegan Paul, London.

Weber, John B. 1894. Pauperism and crime. *The Charities Revue* **3**: 117–126.

Weeks, Genevieve C. 1976. *Oscar Carleton McCulloch 1843–1891 Preacher and Practitioner of Applied Christianity.* Indianapolis Historical Society, Indianapolis, Indiana.

Weismann, August. 1883. On Heredity. In *Essays Upon Heredity and Kindred Biological Problems*, two volumes (ed. E.B. Poulton, S. Schönland, and A.E. Shipley), pp. 67–106 (translated into English in 1891–1892).

Clarendon Press, Oxford, United Kingdom.

Weismann, August. 1885. The Continuity of the Germ-plasm as the Foundation of a Theory of Heredity. In *Essays Upon Heredity and Kindred Biological Problems*, two volumes (ed. E.B. Poulton, S. Schönland, and A.E. Shipley), pp. 163–256 (translated into English in 1891–1892). Clarendon Press, Oxford, United Kingdom.

Weismann, August. 1888. The Supposed Transmission of Mutilations. In *Essays Upon Heredity and Kindred Biological Problems*, two volumes (ed. E.B. Poulton, S. Schönland, and A.E. Shipley), pp. 431–461 (translated into English in 1891–1892). Clarendon Press, Oxford, United Kingdom.

Whipple, Edwin D. 1876. "Oliver Twist" (Commentary). *Atlantic Monthly* **38**: 474–479.

Wilbur, C. Keith. 1980. *Revolutionary Medicine 1700–1800*. The Globe Pequot Press, Chester, Connecticut.

Winship, A.E. 1900. *Jukes–Edwards: A Study in Education and Heredity*. R.L. Myers & Co., Harrisburg, Pennsylvania.

Wolfers, David and Helen Wolfers. 1974. *Vasectomy and Vasectomania*. Mayflower Books, St. Albans, United Kingdom.

Wright, Adele Williams. 1902. The extermination of the criminal classes. *The Arena* **28**: 274–280.

Wylie, B.G. 1876. Relations of hospitals to pauperism. *Popular Science Monthly* **9**: 738–743.

Young, Robert M. 1970. Franz Joseph Gall (1758–1828). *Dictionary of Scientific Biography* **5**: 250–256. C. Scribner's Sons, New York.

Zola, Emile. 1888. *The Fortune of the Rougons: A Realistic Novel*. Vizetelly and Co., London.

Zola, Emile. 1898. *Dr. Pascal*. MacMillan, New York.

Index

n indicates a reference to a note.

427